Rare Genetic Disorders

Muhammad Umair • Misbahuddin Rafeeq
Qamre Alam
Editors

Rare Genetic Disorders

Advancements in Diagnosis and Treatment

Springer

Editors
Muhammad Umair
Medical Genomics Research Department, King Abdullah International Medical Research Center (KAIMRC)
King Saud bin Abdulaziz University for Health Sciences, Ministry of National Guard Health Affairs (MNGH)
Riyadh, Saudi Arabia

Misbahuddin Rafeeq
Department of Pharmacology, Faculty of Medicine, Rabigh
King Abdulaziz University
Jeddah, Saudi Arabia

Qamre Alam
Department of Molecular Genomics and Precision Medicine
ExpressMed Diagnostics and Research
Zinj, Kingdom of Bahrain

ISBN 978-981-99-9322-2 ISBN 978-981-99-9323-9 (eBook)
https://doi.org/10.1007/978-981-99-9323-9

This Springer imprint is published by the registered company Springer Nature Singapore Pte Ltd.
The registered company address is: 152 Beach Road, #21-01/04 Gateway East, Singapore 189721, Singapore

Preface

Rare genetic disorders present a unique and challenging landscape within the realm of medicine. These conditions, often arising from elusive mutations in an individual's genetic code, affect a relatively small number of people individually but collectively impact millions worldwide. The quest to understand, diagnose, and treat these disorders is a journey characterized by complexity, diversity, and a relentless pursuit of knowledge and solutions.

The chapters that follow delve into the intricate world of rare genetic disorders, offering a comprehensive exploration of these conditions, their diagnosis, treatment, and the invaluable role of animal models in research. This preface provides an overview of the abstracts from each chapter, giving readers a glimpse into the depth and breadth of the topics covered.

Chapter 1 focuses on the enigmatic nature of rare genetic disorders. With over 7000 identified and many more yet to be uncovered, these conditions manifest in various forms, impacting millions of lives. Despite the diagnostic challenges posed by their diverse and often nonspecific symptoms, genetic tests offer hope for identifying these conditions. The absence of one-size-fits-all treatments underscores the need for tailored care, and ongoing research fuels optimism for the development of new therapies.

Chapter 2 delves into the diagnostic landscape of rare genetic disorders. Leveraging modern molecular biology techniques, particularly next-generation sequencing, offers hope in identifying these conditions, even when traditional clinical hypotheses fall short. By shedding light on the impact of recent advancements in genetic testing, the chapter paves the way for more accurate and timely diagnoses.

Genetic counseling takes center stage in Chap. 3, offering critical support to individuals and families affected by rare genetic disorders. The role of genetic counselors in providing information, assisting with reproductive decisions, and collaborating with healthcare practitioners is pivotal. Despite the obstacles to genetic testing, this chapter underscores the importance of addressing the psychological and emotional aspects of rare genetic illnesses.

Chapter 4 highlights the crucial role of animal models in rare disease research. These models, ranging from rodents to primates, and dogs to zebrafish, provide invaluable insights into disease mechanisms and therapeutic interventions. Advanced genetic engineering, especially CRISPR-Cas9 technology, has revolutionized the creation of precise animal models. The chapter emphasizes that understanding disease mechanisms, testing potential therapies, and bridging the gap between basic research and clinical applications are central to improving the lives of patients with rare diseases.

Chapter 5 addresses the diagnostic challenges of undiagnosed rare genetic disorders, where diversity and limited genetic knowledge pose significant obstacles. The chapter provides a comprehensive overview of research and clinical strategies aimed at identifying and treating these elusive conditions.

Chapter 6 emphasizes the growing importance of research and development for rare disorder therapies. With an increased focus on drug development and innovative approaches, including small molecules and biologics, the potential for accelerating the development of rare disease treatments is on the rise.

Chapter 7 offers a comprehensive review of two distinct rare genetic disorders, Fabry disease and Marfan syndrome, both characterized by multi-organ involvement and life-threatening complications. The chapter sheds light on pathophysiological mechanisms, gene mutations, and therapeutic approaches, highlighting the importance of early diagnosis and multidisciplinary care.

Mitochondrial disorders take the spotlight in Chap. 8, focusing on the challenges and potential treatments for these conditions. As gene therapy emerges as a precision medicine approach, the chapter explores the future possibilities and emphasizes the need for further research in this area.

Chapter 9 addresses the overarching challenges and opportunities of rare genetic disorders. The diagnostic odyssey faced by patients, economic obstacles in therapeutic development, the role of next-generation sequencing, personalized medicine, drug repurposing, and power of patient advocacy are all explored.

As we embark on this journey through the world of rare genetic disorders, we hope that these chapters will serve as a valuable resource for researchers, clinicians, and individuals and families affected by these conditions. The exploration of diagnostic methods, therapeutic approaches, and promising role of animal models will contribute to a deeper understanding of rare genetic disorders and, ultimately, offer hope for a brighter future in the field of rare disease research.

Riyadh, Saudi Arabia — Muhammad Umair
Jeddah, Saudi Arabia — Misbahuddin Rafeeq
Zinj, Kingdom of Bahrain — Qamre Alam

Contents

About the Editors and Contributors

About the Editors

Muhammad Umair currently holds the position of Associate Research Scientist and serves as the Team Leader of Functional Studies at the Medical Genomics Research Department within the King Abdullah International Medical Research Center (KAIMRC), part of King Saud bin Abdulaziz University for Health Science, Ministry of National Guard Health Affairs (MNGHA) in Riyadh, Saudi Arabia. Additionally, he serves as a "Research Advisor" at the Department of Life Sciences within the School of Science at the University of Management and Technology (UMT) in Lahore, Pakistan. Dr. Umair is a molecular geneticist with a particular focus on rare genetic disorders (RGDs). His research involves multiple projects dedicated to rare genetic disorders and therapeutic genetics. He is currently engaged in research initiatives such as the Genetics and Rare Disease Program, Genetic and Functional Characterization of Rare Skeletal Disorders (GSDs), Genodermatosis, Preventative Genome Medicine for Inherited Genetic Disorders, Functional Characterization of Genes Associated with Neurodevelopmental Disorders (NDDs), Genetic and Rare Disease Registry, and Therapeutic Genomics.

Dr. Umair earned his M.Sc., M.Phil., and Ph.D. in Biochemistry/Molecular Biology, specializing in Human Molecular Genetics, from Quaid-i-Azam University in Islamabad, Pakistan. His extensive contributions to the field are reflected in his publication record, which comprises over 140 research articles in the domain of human genetics, accumulating an impressive impact factor exceeding 680. He also actively participates as an associate editor and reviewer for several international peer-reviewed journals. In recognition of his outstanding achievements in genetics, Dr. Umair was honored with the 2nd Dr. Sajjad Aslam Shami Gold Medal in Genetics by the Applied Zoological Society of Pakistan (AZSP) in March 2021.

Misbahuddin Rafeeq is a distinguished physician, accomplished scientist, and dedicated medical educator. He earned his MBBS and MD degrees in 2009 from Aligarh Muslim University, India. Currently, he holds the position of Associate Professor at the Faculty of Medicine (Rabigh) within King Abdulaziz University, Jeddah, Saudi Arabia. Dr. Rafeeq is an esteemed member of the editorial boards of several reputable journals. His wealth of expertise has been showcased through his active participation in numerous scientific events spanning across regions such as

India, Southeast Asia, Asia Pacific, Europe, the UK, the Middle East, Africa, the Balkans, and the USA. His outstanding contributions have earned him several prestigious awards, including the Excellent Leadership Award from the Islamic Development Bank in Saudi Arabia, the ESP fellowship from NIHES in the Netherlands, the Distinguished Achievement Award from KAU in Saudi Arabia, and the Bronze Prize from the Diabetic Congress in the USA. Dr. Rafeeq is affiliated with esteemed organizations and scientific societies, including RCPEdin and the Faculty of Pharmaceutical Medicine at RCP London, ACCP and ASPET in the USA, ISoP, and the Royal Society of Medicine in the UK. Additionally, he serves as an honorary consultant on various educational boards and holds the position of International Secretary for SoPI in India. Dr. Rafeeq boasts an impressive publication record with over 60 high-impact and indexed publications to his credit. Furthermore, he has secured funding for several large grants, where he has served as both a principal and co-investigator. His research primarily focuses on the identification of novel genetic mutations in hereditary syndromes, exploration of novel therapeutic targets and proteins through molecular docking, PK/PD and toxicity studies, study of the effects of natural compounds on animal models, and broader field of lipid pharmacogenomics.

Qamre Alam currently serves as a Medical Research Scientist and Laboratory Supervisor at the Molecular Genomics and Precision Medicine Department of ExpressMed Diagnostics and Research, located in Zinj, Kingdom of Bahrain. Previously, from 2019 to 2022, he held the position of Medical Research Scientist at the Medical Genomics Research Department at King Abdullah International Medical Research Center (KAIMRC), part of King Saud bin Abdulaziz University for Health Science, Ministry of National Guard Health Affairs (MNGHA), Riyadh, KSA. Prior to that, he worked as a lecturer from 2011 to 2019 at the King Fahd Medical Research Center, King Abdulaziz University, Jeddah, KSA. Dr. Qamre Alam earned his M.Sc. from Jamia Hamdard University, New Delhi, India, and his Ph.D. in Biotechnology from JJTU, Rajasthan, India. He has an impressive publication record, with over 75 research articles in the fields of human genetics and cancer biology, accumulating an impact factor of more than 150. Dr. Alam is a molecular geneticist with a strong research interest in rare genetic disorders. He has actively contributed to various projects related to rare genetic disorders and preventive genomic medicine, including noninvasive prenatal testing (NIPT) and preimplantation genetic screening (PGS). Additionally, he serves as an associate editor and reviewer for several international peer-reviewed journals of high repute.

Contributors

Tasneem Abaza Biotechnology and Biomolecular Chemistry Program, Faculty of Science, Cairo University, Giza, Egypt

Suruthi Abirami Faculty of Science and Technology, Department of Genetic Engineering, SRM Institute of Science and Technology, Kattankulathur, Chengalpattu, Chennai, Tamil Nadu, India

Suman Adhikari Department of Chemistry, Govt. Degree College, Dharmanagar, Tripura, India

Abna Ajeesh Faculty of Science and Technology, Department of Genetic Engineering, SRM Institute of Science and Technology, Kattankulathur, Chengalpattu, Chennai, Tamil Nadu, India

Qamre Alam Department of Molecular Genomics and Precision Medicine, ExpressMed Diagnostics and Research, Zinj, Kingdom of Bahrain

K. N. Aruljothi Faculty of Science and Technology, Department of Genetic Engineering, SRM Institute of Science and Technology, Kattankulathur, Chengalpattu, Chennai, Tamil Nadu, India

Goutam Biswas Department of Zoology, Acharya Prafulla Chandra College, New Barrackpore, West Bengal, India

Sid Dsa Department of Genetic Engineering, Faculty of Engineering and Technology, SRM Institute of Science and Technology, Kattankulathur, Chennai, Tamil Nadu, India

Hadi Erfani Department of Chemical Engineering, Central Tehran Branch, Islamic Azad University, Tehran, Iran

J. Hemarangan Faculty of Science and Technology, Department of Genetic Engineering, SRM Institute of Science and Technology, Kattankulathur, Chengalpattu, Chennai, Tamil Nadu, India

B. K. Iyshwarya Human Cytogenetics and Genomics Laboratory, Faculty of Allied Health Sciences, Chettinad Hospital and Research Institute, Chettinad Academy of Research and Education, Kelambakkam, Chennai, Tamil Nadu, India

Megala Jayaraman Department of Genetic Engineering, Faculty of Engineering and Technology, SRM Institute of Science and Technology, Kattankulathur, Chennai, Tamil Nadu, India

Patheri Kuniyil Kaleena Department of Zoology, Presidency College, Chennai, Tamil Nadu, India

Sivakumari Kanakarajan Department of Zoology, Competitive Examinations Coaching Centre, Chennai, Tamil Nadu, India

Langeswaran Kulanthaivel Department of Biomedical Sciences, Science Campus, Alagappa University, Karaikudi, Tamil Nadu, India

K. Kumaran Faculty of Science and Technology, Department of Genetic Engineering, SRM Institute of Science and Technology, Kattankulathur, Chengalpattu, Chennai, Tamil Nadu, India

Nithar Ranjan Madhu Department of Zoology, Acharya Prafulla Chandra College, New Barrackpore, West Bengal, India

Amutha Parthasarathy Department of Biochemistry, School of Life Sciences, Vels Institute of Science, Technology and Advanced Studies (VISTAS), Chennai, Tamil Nadu, India

Soumosish Paul Department of Zoology, Acharya Prafulla Chandra College, New Barrackpore, West Bengal, India

Misbahuddin M. Rafeeq Department of Pharmacology, Faculty of Medicine Rabigh, King Abdulaziz University, Jeddah, Saudi Arabia

Archana Rajavel Membrane Protein Interaction Laboratory, Department of Genetic Engineering, School of Bioengineering, SRM Institute of Science and Technology, Kattankulathur, Chennai, Tamil Nadu, India

Muthuvel Raju Department of Biochemistry, Karuna Medical College and Hospital, Chittur, Palakkad, Kerala, India

Bhanumati Sarkar Department of Botany, Acharya Prafulla Chandra College, New Barrackpore, West Bengal, India

Raja Natesan Sella Membrane Protein Interaction Laboratory, Department of Genetic Engineering, School of Bioengineering, SRM Institute of Science and Technology, Kattankulathur, Chennai, Tamil Nadu, India

Rajesh Selvaraj Department of Clinical and Translational Sciences, Joan C. Edwards School of Medicine, Marshall University, Huntington, WV, USA

P. Shriya Faculty of Science and Technology, Department of Genetic Engineering, SRM Institute of Science and Technology, Kattankulathur, Chengalpattu, Chennai, Tamil Nadu, India

Gowtham Kumar Subbaraj Faculty of Allied Health Sciences, Chettinad Hospital and Research Institute, Chettinad Academy of Research and Education, Kelambakkam, Chennai, Tamil Nadu, India

Sreyank Tirunagari Department of Genetic Engineering, Faculty of Engineering and Technology, SRM Institute of Science and Technology, Kattankulathur, Chennai, Tamil Nadu, India

Muhammad Umair Medical Genomics Research Department, King Abdullah International Medical Research Center (KAIMRC), King Saud bin Abdulaziz University for Health Sciences, Ministry of National Guard Health Affairs #(MNGH), Riyadh, Saudi Arabia

T. L. Vasanth Kanth Faculty of Science and Technology, Department of Genetic Engineering, SRM Institute of Science and Technology, Kattankulathur, Chengalpattu, Chennai, Tamil Nadu, India

Ramakrishnan Veerabathiran Human Cytogenetics and Genomics Laboratory, Faculty of Allied Health Sciences, Chettinad Hospital and Research Institute, Chettinad Academy of Research and Education, Kelambakkam, Chennai, Tamil Nadu, India

Mohamed Y. Zaky Molecular Physiology Division, Department of Zoology, Faculty of Science, Beni-Suef University, Beni-Suef, Egypt

Introduction to Rare Genetic Disorders 1

Amudha Parthasarathy, Muthuvel Raju,
Langeswaran Kulanthaivel, and Gowtham Kumar Subbaraj

Abstract

Rare genetic disorders are a group of diseases that are caused by changes in a person's genes. These changes can be inherited from parents or occur spontaneously. Rare genetic disorders can affect any part of the body, and they can range in severity from mild to life-threatening. There are over 7000 rare genetic disorders that have been identified, and many more are still being discovered. These disorders affect an estimated 30 million people in the United States alone. The diagnosis of a rare genetic disorder can be challenging, as the symptoms can be varied and often nonspecific. However, there are a number of genetic tests that can be used to diagnose these disorders. There is no one-size-fits-all treatment for rare genetic disorders. The treatment plan will vary depending on the specific disorder and the individual's symptoms. However, there are a number of treat-

Muthuvel Raju and Langeswaran Kulanthaivel contributed equally and considered as first author.

A. Parthasarathy
Department of Biochemistry, School of Life Sciences, Vels Institute of Science, Technology and Advanced Studies (VISTAS), Chennai, Tamil Nadu, India

M. Raju
Department of Biochemistry, Karuna Medical College and Hospital, Chittur, Palakkad, Kerala, India

L. Kulanthaivel
Department of Biomedical Science, Science Campus, Alagappa University, Karaikudi, Tamil Nadu, India

G. K. Subbaraj (✉)
Faculty of Allied Health Sciences, Chettinad Hospital and Research Institute, Chettinad Academy of Research and Education, Kelambakkam, Chennai, Tamil Nadu, India
e-mail: gowtham.genetics@care.edu.in

M. Umair et al. (eds.), *Rare Genetic Disorders*,
https://doi.org/10.1007/978-981-99-9323-9_1

ments that can help improve the quality of life for people with these disorders. Research into rare genetic disorders is ongoing, and there is hope that new treatments will be developed in the future. In the meantime, there are a number of resources available to help people with rare genetic disorders and their families. This book chapter provides a brief overview of rare genetic disorders, including their causes, symptoms, diagnosis, treatment, and research.

Keywords
Genetic syndrome · Mental disorders · Disability · Chromosomal disorder · Genetic disorders

1.1 Aarskog Syndrome

Aarskog syndrome is a hereditary condition that mostly affects men. It is brought on by a mutation in the FGD1 gene, which makes a protein vital to the growth of several human tissues and organs. The additional terms for Aarskog syndrome are Aarskog sickness, Aarskog-Scott syndrome, AAS, faciodigitogenital syndrome, faciogenital dysplasia, FGDY, and Scott-Aarskog syndrome (Gorski 2003; Jones 1997).

1.1.1 Signs and Symptoms

Aarskog syndrome is predominantly observed in males, and it causes a distinct range of abnormalities in the face, skeleton, and genitals of affected boys. The clinical presentation may differ from one individual to another, even among families, indicating clinical heterogeneity. Males with Aarskog syndrome typically have a round face with a broad forehead, widely spaced eyes (ocular hypertelorism), droopy eyelids (ptosis), and downwardly slanting eyelid creases (palpebral fissures). They also typically have a little nose with flared nostrils (anteverted nares), an underdeveloped upper jawbone (maxillary hypoplasia), and a widow's peak. Additionally, those who are affected could have a wide nasal bridge and a protracted groove on their top lip (philtrum) (Orrico et al. 2015; Al-Semari et al. 2013).

Some children with Aarskog syndrome may have ear and tooth problems. Both the ear canal and the earlobes are abnormally large and meaty. Undeveloped teeth at birth, teeth not erupting at the proper time, and undeveloped tooth enamel (enamel hypoplasia) are all examples of dental anomalies (Verhoeven et al. 2012; Pilozzi-Edmonds et al. 2011). While intellectual disabilities have been documented in some males with this condition, it is not a universal symptom. Mild learning disabilities and/or behavioural problems such as delays in development in infancy, hyperactivity, attention deficit, impulsivity, and resistance may be present in those who are affected (Orrico et al. 2010; Bottani et al. 2007).

1.1.2 Causes

However, there is genetic and clinical variation in Aarskog syndrome, and the most studied type of the ailment is passed down as an X-linked characteristic due to mutations in the FGD1 gene. Although Aarskog syndrome is more common in males, girls who possess a single copy of an FGD1 gene mutation may also show some of the disorder's symptoms. Mutations in the FGD1 gene have been found in roughly 22% of afflicted males, suggesting that other genes may possibly be involved in the development of this disorder but have yet to be uncovered (Orrico et al. 2007; Kaname et al. 2006). The mutation of the FGD1 gene has been confirmed in around 60 people with Aarskog syndrome throughout the world. It is challenging to establish the real prevalence of the medical condition in the general community since some minimally afflicted youngsters may go unnoticed. Aarskog syndrome affects an estimated 1 in 25,000 people worldwide (Satoh and Yokoya 2006; Shalev et al. 2006).

1.1.3 Diagnosis

Aarskog syndrome is caused by a mutation in the FGD1 gene, and this mutation has now been proven in around 60 people throughout the globe. Due to the fact that some mildly affected children may go unrecognised, it is difficult to ascertain the true incidence of the medical condition in the general population. The prevalence of Aarskog syndrome is estimated to be 1 in 25,000 persons (Orrico et al. 2005; Lebel et al. 2002).

1.1.4 Treatment

The symptoms of Aarskog syndrome vary from person to person; hence, treatment must be individualised. Some of the associated congenital or structural abnormalities, such as hypospadias, umbilical or inguinal hernias, cryptorchidism, and severe craniofacial characteristics, may need surgical correction. Complete eye and dental examinations are recommended for patients with Aarskog syndrome. Some children have benefited from growth hormone therapy, and although this treatment has been found to increase their height, further research is needed to determine the best course of action and the typical response. A neuropsychiatric evaluation and consultation may be useful when dealing with neurodevelopmental problems. Other types of therapy include relieving symptoms and providing emotional support (Orrico et al. 2004; Schwartz et al. 2000).

1.2 Acromesomelic Dysplasia

Acromesomelic dysplasia is an uncommon genetic condition that causes low height and limb asymmetry due to improper bone formation and development. It falls within the category of skeletal dysplasia, which describes a collection of disorders that negatively impact bone and cartilage development (Khan et al. 2016).

1.2.1 Signs and Symptoms

Acromesomelic dysplasia (AMD) is a disorder that causes short-limbed dwarfism and exceptionally small forearms and lower legs because it stunts the development of certain long bones, such as those in the forearms and lower legs. This condition is usually noticeable in early childhood. The metacarpals, phalanges, and metatarsals in the hands and feet are not immune to abnormal cartilage and bone growth (Olney et al. 2006). Acromesomelic dysplasia typically does not affect the birth weight of infants. However, affected infants often display distinct facial abnormalities at birth, such as a relatively prominent forehead, pronounced back portion of the head, large head, a small pug nose, and slightly flattened midface. Children with AMD have an increased risk of developing vertebral anomalies, which may cause spinal curvature, as they age. The middle of the spine may curve forward and backward as a consequence, a condition known as low thoracic kyphosis, and the lower back may curve excessively inward, a condition known as lumbar hyperlordosis (Potter et al. 2006).

1.2.2 Causes

Researchers have identified five distinct forms of acromesomelic dysplasia. All forms save the Osebold-Remondini form of age-related macular degeneration are exceedingly uncommon and are passed down through families as autosomal recessive traits. Researchers have located the gene for the Maroteaux phenotype on chromosome 9 at the 9p13–12 region. The genetic link between Grebe dysplasia and Du Pan syndrome has been traced to the same region of chromosome 20: 20q11. Having a chromosome 4q23–24 location is associated with acromegaly and genital abnormalities (Szczaluba et al. 2005). The genetic mapping of Osebold-Remondini type is still pending. According to the medical literature from 2005, Hunter-Thompson-type ADM had approximately 10 cases, while Maroteaux-type AMD had 40–50 patients. Although the incidence of Grebe-type ADM is unknown, it is thought to be more common among the Brazilian population (Bartels et al. 2004).

1.2.3 Diagnosis

During the first few years of life, people with acromegaly are often diagnosed by a thorough clinical assessment, an extensive patient history, cutting-edge imaging methods, and recognition of diagnostic traits. Progressive abnormalities, such as forearm and lower leg bone abnormalities, short stature, additional broadening and shortening of bones in the hands and feet, limited elbow and arm extension, and progressive vertebral abnormalities, typically appear in late infancy or early childhood, despite the hands and feet displaying unusual broadness and shortness at birth (Al-Yahyaee et al. 2003).

1.2.4 Treatment

Acromesomelic dysplasia therapy is tailored to each patient in an effort to improve their unique set of symptoms and physical manifestations. Treatment focuses mostly on providing comfort and alleviating symptoms. Physical therapy, exercises, braces, casts, and, in extreme circumstances, corrective surgery may help with abnormal spine curvature such as low thoracic kyphosis and/or lumbar hyperlordosis. Physical therapy, supportive approaches, and orthopaedic surgery may all help with the disorder's symptoms. Individuals and their families who are impacted should seriously consider seeking genetic counselling. Supportive care and alleviation of symptoms are the focus of other therapies (Savarirayan et al. 2003).

1.3 Acute Eosinophilic Pneumonia

Acute eosinophilic pneumonia (AEP) is an extremely rare kind of pneumonia in which eosinophils (a type of white blood cell) accumulate in the lungs. Symptoms include sudden onset of fever, cough, breathing difficulties, and wheezing; if addressed, it may quickly progress to respiratory failure, idiopathic acute eosinophilic pneumonia (sometimes abbreviated as AEP or IAEP) (Rhee et al. 2013).

1.3.1 Signs and Symptoms

The onset of AEP often occurs within a week. However, in rare cases, symptoms might take up to a month to appear. This breathing problem happens often in otherwise healthy young people. The symptoms are general and include things like a high body temperature, a hacking cough, tightness in the chest, and trouble breathing (dyspnoea). Myalgia (muscle pain), weariness, joint pains, and stomach discomfort or pain are some less common symptoms of AEP. When there is a large decline in the amount of oxygen in the blood (hypoxaemia), AEP may rapidly worsen and produce acute respiratory failure. Within weeks or months, this might cause AEP patients to have life-threatening respiratory issues. Mechanical ventilation may be necessary for around two-thirds of people with AEP (Tsigkaropoulou et al. 2011).

1.3.2 Causes

IAEP is termed idiopathic since its etiology is uncertain. It is hypothesised that AEP is caused by a generic trigger that causes the body to manufacture eosinophils and attracts them to the lungs. But the precise cause of AEP's eosinophil overproduction and accumulation remains unclear. Sensitivity to environmental irritants including dust and smoke, as well as occupational variables, has been connected to AEP. However, it is very unlikely that AEP can be attributed to a single environmental component. It is more probable that a combination of circumstances, including a susceptible person's exposure to a triggering event, is necessary for the illness to emerge. The exact reason why some people get AEP and others do not is unknown (Miller et al. 2010).

Exposure to environmental elements including dust and smoke, both in and out of the workplace, has been linked to AEP. However, it is very unlikely that AEP can be attributed to a particular environmental aspect. More likely, the disease requires a confluence of circumstances, one of which is the presence of a trigger in an otherwise susceptible person. The precipitating reason for AEP may vary from patient to patient. People who have started smoking around the past 3 months prior to the onset of the disorder, who are still smoking after a temporary cessation, or who have recently increased the number of cigarettes smoked daily are especially at risk for developing the disorder. Smoking may be linked to "idiopathic" AEP in certain patients, according to findings in the medical literature. The exact part that smoking plays in the onset of AEP in such instances is, however, not well understood (Jeong et al. 2007).

Occupational factors that can trigger AEP are typically associated with exposure to inhaled dusts. AEP may be triggered by a broad variety of airborne contaminants and agents. There are studies in the medical literature associating AEP to the use of various medicines, in addition to occupational variables. Anti-depressants venlafaxine and daptomycin, as well as the antibiotic minocycline, have all been associated with AEP. A comprehensive list of drugs associated with AEP can be found on the website www.pneumotox.com. AEP may be triggered by a broad variety of airborne contaminants and agents. There are studies in the medical literature associating AEP to the use of various medicines, in addition to occupational variables. Anti-depressants venlafaxine and daptomycin, as well as the antibiotic minocycline, have all been associated with AEP (Shorr et al. 2004).

1.3.3 Diagnosis

AEP is diagnosed by considering the patient's symptoms, medical history, and physical exam in addition to doing tests such as bronchoalveolar lavage (BAL).

Investigating the possible causes of pulmonary eosinophilia and ruling them out are essential. Additional possible causes include parasite infections and medication exposure (Allen et al. 2002). Despite the lack of specificity, imaging tools such as chest X-rays may help confirm a diagnosis of AEP. Chest X-rays of people with AEP often show infiltrates or cloudy areas in the lungs. Chest CT scan findings indicative of the illness include interlobular septal thickening, mild-to-moderate bilateral pleural effusion, and bilateral alveolar consolidation. In the early stages of the illness, pulmonary function tests often show a confined pattern (Poletti et al. 2004).

1.3.4 Treatment

High dosages of corticosteroids are useful in treating AEP, and a response is often seen within days. However, before starting corticosteroid therapy, it is important to rule out any infectious causes of pulmonary eosinophilia. Some studies indicate that treatment with corticosteroids for AEP for just 2 weeks is adequate; however, this finding is not universally accepted in the medical literature. Most patients with AEP are given intravenous corticosteroids first, followed by oral administration, since there is no established dosage for corticosteroid treatment in AEP. Spontaneous remission, without any treatment, has also been reported in some cases. Once treated with corticosteroids, there is no risk of relapse. The long-term prognosis for AEP is excellent (Philit et al. 2002).

1.3.5 Banti Syndrome

Banti syndrome is an uncommon disorder that is characterised by hallmarks such as enlarged spleen (splenomegaly) and anaemia. Banti syndrome manifests initially with exhaustion, weakness, and anaemia, with anaemia becoming more severe as the condition develops. It is also known as Banti's disease, hypersplenism, idiopathic congestive splenomegaly, and idiopathic portal hypertension (Ravenna 1940).

1.3.6 Signs and Symptoms

Early symptoms of Banti syndrome include weariness, weakness, anaemia, and spleen enlargement. As the condition progresses, the anaemia worsens and may be compounded by bleeding in the oesophagus, leading to vomiting of blood and dark stools containing decomposed blood. When the liver develops and gets segmented by fibrous tissue, cirrhosis may occur. But spleen enlargement is the most noticeable sign of Banti syndrome. Individuals with Banti syndrome may tend to bruise easily and are at an increased risk of bacterial infections, often accompanied by prolonged fever. In addition to these, further symptoms might include gastrointestinal bleeding (haemorrhage), anaemia, leukopenia, thrombocytopenia, and an abnormal accumulation of fluid in the abdomen (ascites) (Waqar et al. 2004).

1.3.7 Causes

Banti syndrome can be caused by various factors that obstruct certain veins in the liver or spleen and cause abnormally high blood pressure within them. Cirrhosis and other underlying diseases may induce inflammation and blockage of hepatic veins, but other causes include congenital malformations of the veins or blood clots. In some cases, increased intake of arsenic may also be a factor. Additionally, reports have been made of instances of individuals using azathioprine for extended periods of time, particularly following kidney transplantation (McCormick 2003). Males and females are equally affected by Banti syndrome. It is rather frequent in some parts of India and Japan, but much less so in the Western region. The incidence may vary in different regions due to factors such as increased levels of arsenic in drinking water (Beers and Berkow 1999).

1.3.8 Diagnosis

Banti syndrome may be diagnosed with a thorough clinical evaluation and specialised testing such as splenic venography and magnetic resonance imaging (MRI). In magnetic resonance imaging (MRI), magnetic fields and radio waves are used to create precise cross-sectional pictures of particular parts of the body (Pickhardt and Balfe 1998).

1.3.9 Treatment

Banti syndrome treatment strategies are condition specific. If arsenic or azathioprine exposure is shown to be the cause of the disease, avoiding those substances is essential. Bleeding from the expanded blood vessels (varices) in the oesophagus or stomach is the key clinical issue. Vasoconstrictor medications or other methods used to treat portal hypertension may be used to stop active bleeding. In the case of persistent bleeding, surgical shunt treatments may be required to reroute blood flow (Behrman et al. 1996).

1.4 Buerger's Disease

Buerger's disease is an incredibly rare disorder that causes damage to the body's micro- and medium-sized blood vessels. Blood clots develop in the arteries involved, leading to discomfort, tissue damage, and even gangrene in extreme situations. Tobacco usage is closely linked to Buerger's disease, though the specific etiology is uncertain. Infections, genetics, and autoimmune diseases are other possible contributing causes. TAO, thromboangiitis obliterans, and inflammatory occlusive peripheral vascular disease are also additional forms for Buerger's disease (Arkkila 2006).

1.4.1 Signs and Symptoms

Buerger's disease is a type of peripheral vascular disease characterised by the constriction or blocking of small and intermediate-sized arteries and veins in the limbs, resulting in decreased blood flow to those regions. The illness usually comes suddenly and returns repeatedly over the course of a few weeks. Pain in the lower extremities, especially while at rest, is a hallmark of Buerger's disease. A symptom of claudication is the involuntary limping that might result from leg cramps when walking. Other symptoms include a lack of normal blood flow to the fingers and toes, especially in cold weather (known as Raynaud's phenomenon), hand discoloration, numbness, and tingling. Inflammation and clotting of specific veins, known as thrombophlebitis, are also possible. Finger and toe ulcers tend to be dry and black, and they may be very painful. Pain may be exacerbated by elevating the afflicted region. Extremity amputation may be required if the condition is severe enough to cause tissue death (gangrene) in the afflicted limbs (Saito et al. 2007).

1.4.2 Causes

Tobacco smoking is closely linked to the development of Buerger's disease, albeit the precise aetiology is yet unknown. Tobacco usage, either in the past or now, is considered by many scientists to be necessary for a diagnosis of Buerger's disease. The precise connection between cigarette smoking and onset of the illness is unclear. Since the frequency of Buerger's disease varies greatly across various ethnic groups, it is possible that genetic factors have a role in either its development or its severity. To determine whether genetics have a significant influence on the onset of Buerger's disease, further study is required. Ischaemia, or decreased blood flow, is the underlying cause of the limb pain and weakness characteristic with Buerger's illness. Pain, numbness, and even tissue death (gangrene) or amputation may result from inadequate blood supply. The United States and Europe have low incidences of Buerger's disease compared to other areas of the globe, especially sections of Asia and the Middle East. The estimated incidence rate of Buerger's disease in the United States is between 12.6 and 20 per 100,000 individuals. Buerger's disease is more common in nations where smoking is widespread. Tobacco consumption rates may vary greatly from country to country and region to region, which may help explain why the condition is more prevalent in certain areas than others. Buerger's disease is more common in certain ethnic groups and places than in others, which may be due in part to hereditary factors (Watts and Scott 2003).

1.4.3 Diagnosis

A history of recent or current tobacco smoking is often required by clinicians in order to identify Buerger's disease, which is otherwise diagnosed based on the presence of particular clinical characteristics and symptoms. For a definitive diagnosis, medical professionals may employ angiography or other non-invasive imaging studies. With the use of a specific dye injected into a patient's veins, angiography allows for clear imaging of the circulatory system. In this way, clinicians may look for the constriction or blockage of blood arteries, both of which are characteristics of Buerger's disease. Blood flow and anomalies in the afflicted limbs may be assessed using other non-invasive methods, such as ultrasound or magnetic resonance imaging (MRI) (Evans and Ratchford 2014).

1.4.4 Treatment

Tobacco reduction is the first line of defence against Buerger's disease. This is essential for slowing the disease's course and lowering the likelihood of consequences. Quitting smoking has been shown to be effective in reducing or eliminating symptoms and, in some instances, has even been linked to full remission of illness. To control symptoms and forestall consequences, conservative treatments may be performed if the afflicted person does not quit smoking. This may include medications such as anticoagulants to prevent blood clotting, vasodilators to increase blood flow, anti-inflammatory drugs to reduce inflammation, antibiotics to prevent infection, and analgesics to relieve pain. In some cases, surgical intervention may be necessary. This may include procedures to remove or bypass affected blood vessels or to improve blood flow to the affected areas. However, surgical options should be carefully considered and used only when conservative treatments have been ineffective or when there is a risk of tissue death (gangrene) or other serious complications (Arkkila 2006).

1.5 Chronic Lymphocytic Leukaemia

The white blood cells known as lymphocytes are the target of chronic lymphocytic leukaemia (CLL), a kind of blood cancer. Accumulation of aberrant lymphocytes in the blood, bone marrow, and lymph nodes causes a slowly progressive illness. Other terms for CLL include small lymphocytic lymphoma (Podhorecka et al. 2016).

1.5.1 Signs and Symptoms

It is estimated that between 50% and 75% of those with chronic lymphocytic leukemia have no symptoms at all until the disease has progressed significantly. In many cases, the condition is identified during checkups and screenings. While the symptoms of CLL may be similar between the two subtypes, Ig-mutated and Ig-unmutated, the latter subtype tends to progress more quickly and may result in more severe symptoms. Fatigue, unexplained weight loss, lack of appetite, trouble breathing, low-grade fever, spleen enlargement leading to a sense of fullness in the belly, and night sweats are all possible signs of CLL. Bacterial infections such as skin infections, pneumonia, and sinusitis are frequent in CLL patients. As the disease advances, patients become more susceptible to viral infections, which can pose a significant threat to their health (Mellstedt 2007).

1.5.2 Causes

The precise cause of chronic lymphocytic leukaemia remains unclear, although it involves several genetic mutations that occur within the blood-forming cells. The abnormal cells produced as a result of these mutations are unable to effectively combat infections. It is typical for patients with CLL to have an abnormal chromosome, usually resulting from a deletion of a portion of a chromosome. The most common deletions include part of chromosome 13, as well as chromosome 11 and 17 deletions. Some individuals also show signs of having an extra copy of chromosome 12, a condition known as trisomy 12. In families where many individuals are diagnosed with leukaemia, CLL is by far the most common kind. The average age of onset is 72 years old, and men are twice as likely as women to be affected. CLL is more common as people gets older; current estimates place the prevalence of the disease in the United States at around 3 per 100,000 people (Döhner et al. 1999).

1.5.3 Diagnosis

CLL is frequently detected through routine blood work that reveals an abnormally high count of white blood cells. To confirm a diagnosis, several tests can be conducted, including a complete blood cell count, flow cytometry, bone marrow biopsy, and lymph node biopsy. These tests help determine the presence and extent of the disease, as well as differentiate CLL from other related conditions. In addition to these tests, specialised tests may be recommended by your doctor to predict the likely progression of CLL and its response to treatment. However, the decision to treat the disease is primarily based on clinical factors, such as symptoms, blood counts, and presence of lymph nodes, among others (Cheson et al. 1996).

1.5.4 Treatment

The choice of treatment for CLL depends on symptoms, disease's progression, and prospects. Patients with CLL can remain asymptomatic for years and may not require any specific treatment. However, in advanced stages, chemotherapy is commonly used as a treatment option. Monoclonal antibody therapy is another option, where proteins are attached to cancer cells, initiating a mechanism that leads to their destruction. The use of both of these therapies together has been found to produce the highest treatment response. In 2010, the FDA approved the use of the anti-cancer drug Rituxan (rituximab) in combination with the chemotherapy drugs Fludara (fludarabine) and Cytoxan (cyclophosphamide) for first-line treatment of CLL. Another FDA-approved drug, Treanda (bendamustine hydrochloride), has shown effectiveness in treating CLL when administered once every 4 weeks, like other chemotherapies (Wu et al. 2012).

1.6 Cicatricial Alopecias

Cicatricial alopecias, also known as scarring alopecia, alopecia cicatrisata, and scarring hair loss, are a type of hair loss disease characterised by hair follicle destruction and replacement with scar tissue. This results in permanent hair loss and the inability of hair to regrow in affected areas (Price and Mirmirani 2011).

1.6.1 Signs and Symptoms

The scalp may display symptoms such as redness, scaling, altered pigmentation, pustules, or draining sinuses in areas affected by cicatricial alopecia. However, some cases may exhibit minimal signs of inflammation. The underlying cause of the condition is inflammation, which kills the hair follicle and leaves the scalp completely hairless and devoid of the regular pore marks (Beers et al. 2006). In the active stage of the disease, follicle destruction is predominantly caused by inflammatory cells, and this information is used to classify the cicatricial alopecias. In lichen planopilaris, frontofacial fibrosing alopecia, central centrifugal alopecia, pseudopelade (Brocq), and lymphocytic inflammation predominate. Tufted folliculitis and folliculitis decalvans are both caused by neutrophilic inflammation. Changes from a neutrophilic to a lymphocytic inflammatory response may occur. Mixed inflammatory infiltrates are seen in cicatricial alopecias such as dissecting cellulitis and folliculitis keloidalis (Bergfeld and Elston 2003).

1.6.2 Causes

Unfortunately, it has yet to be determined what causes the various forms of cicatricial alopecia. However, the inflammation responsible for cicatricial alopecia always manifests itself at the hair follicle's apex, just where the stem cells and sebaceous gland reside. Damage to the hair follicle's stem cells and sebaceous gland by inflammation precludes any possibility of regrowth. Alopecia caused by scarring is not communicable, so do not worry about spreading it (Wu et al. 2008). Primary cicatricial alopecia is rare in youngsters, although it may afflict healthy men and women of any age. These alopecias are seen all over, although there is no data regarding their prevalence from epidemiological investigations. The scarring hair loss caused by scar tissue is severe. Few families have been found to have more than one member affected by cicatricial alopecia; however, this is not the case for the vast majority of sufferers. Central centrifugal alopecia is most common in women of African heritage, and it may afflict many members of the same family (Blackwell and Rawnsley 2008).

1.6.3 Diagnosis

Cicatricial alopecia is first identified by a biopsy of the scalp. The diagnosis, evaluation, and treatment of cicatricial alopecia depend on the results of a scalp biopsy, which may provide information about the kind of inflammation, its location and severity, and other alterations to the scalp. After numbing the region, a biopsy punch is used to extract a tiny piece of skin (about the size of a rubber) for analysis. After horizontal and vertical sectioning of the skin samples, a biopsy or two are typically collected for analysis (Ochoa et al. 2008). In addition to the biopsy, a thorough clinical examination of the scalp is required. Inflammation is often characterised by the appearance of symptoms like itching, burning, pain, or soreness. Redness, scaling, and pustules on the scalp are all symptoms of inflammation. However, in certain situations, inflammation is still present while showing little outward symptoms or indications, and this is only revealed by a scalp biopsy. In addition to the results of the biopsy, the dermatologist considers the pattern and severity of hair loss in order to identify the exact kind of cicatricial alopecia that is present (Whiting 2001).

1.6.4 Treatment

Neutrophilic cicatricial alopecias, such as folliculitis decalvans and tufted folliculitis, are treated with antibiotics such as clindamycin, rifampicin, and doxycycline. If the patient does not respond to antibiotics, isotretinoin or oral prednisone may be used. It is important to note that cicatricial alopecias are chronic diseases and treatment is not always effective. It is essential for patients to maintain ongoing communication with their dermatologist and to follow the treatment plan as directed. In some cases, a combination of treatments may be necessary to achieve the best results. Early diagnosis and treatment can help prevent further hair loss and potentially improve hair regrowth (Olsen 2005).

1.7 Encephalocele

The birth defect encephalocele causes brain tissue to protrude through a skull defect. Birth defects of the brain and spinal cord result from an improper closure of the neural tube during embryonic development. As a result, the brain tissue protrudes out of the skull and is covered by a sac-like membrane. Other names are cephalocele, craniocele, and cranium bifidum (Al-Tubaikh and Reiser 2009).

1.7.1 Signs and Symptoms

The symptoms of encephalocele, a birth defect in which brain tissue protrudes through a hole in the skull, vary according to a number of circumstances. When considering clinical implications for therapy and prognosis, the location of the encephalocele is a key consideration. Because they do not include brain tissue, anterior encephaloceles have a better prognosis than their posterior counterparts, which are often linked to neurological issues. Therefore, the location of the encephalocele is essential in assessing the severity of the condition and determining appropriate treatment. Most encephaloceles originate in the top part of the skull and expand back towards the occipital bone from the front of the head. The base of the skull, the sinuses, the forehead, and the nose are all potential sites for encephaloceles. The location of the encephalocele can impact the severity of symptoms and treatment options, so it is important to identify the location of the encephalocele for proper diagnosis and management (Mahapatra 2007).

1.7.2 Causes

The majority of cases of encephalocele arise for basically no reason in particular. Genetic and environmental variables are both speculated to have a role in the onset of encephalocele. There is a genetic susceptibility for encephalocele among people who have a history of other neural tube disorders, such as spina bifida or anencephaly, in their families. Carrying one or more disease-causing genes does not always indicate that the condition will manifest in a person. Therefore, encephalocele may arise as a result of a combination of hereditary and environmental causes. Researchers have shown that the prevalence of encephaloceles varies by gender and geographical area. Encephaloceles are more common in females than men, and they tend to occur in the occipital region of the skull in females but the front of the head in males. In addition, encephaloceles are often seen in the occipital area of the skull in Western people but the front of the head in Southeast Asian groups. The necessity of taking demographic parameters into account while analysing encephalocele cases is highlighted by these discrepancies, which may be attributable to genetic, environmental, or other reasons (Moore et al. 1997).

1.7.3 Diagnosis

Encephaloceles are typically diagnosed either during a routine prenatal ultrasound or immediately after birth. However, small encephaloceles near the nose or forehead may not be detected at first. Encephaloceles can be identified as cysts on prenatal ultrasound scans because reflected sound waves are used to construct pictures of the growing fetus. Prenatal detection of an encephalocele may prompt evaluation for other abnormalities. In order to further assess the encephalocele and any related problems, prenatal care providers often employ magnetic resonance imaging (MRI) (Stoll et al. 2011).

1.7.4 Treatment

Children with encephalocele typically require surgical intervention, which is usually performed within the first 4 months of life, depending on factors such as the size and location of the encephalocele, any associated complications, and whether the encephalocele is covered by a thin layer of skin. Assuming that a protective skin layer is already in place, surgery may be delayed for several months. However, if the encephalocele is not protected by a layer of skin, surgery may be necessary shortly after birth (De Wals et al. 2007). During surgery, the protruding brain tissue is carefully placed back inside the skull. This is achieved through a craniotomy, in which a portion of the skull is cut and removed to allow contact with the brain. The neurosurgeon next makes an incision into the dura mater, the brain's protective outer barrier, to reach the encephalocele and correct any underlying problems. Encephalocele therapy must be individualised based on the patient's unique set of symptoms and anomalies. Surgical correction may be required in situations of craniofacial or other cranial anomalies. In addition, a shunt may be surgically implanted to treat hydrocephalus, a disorder characterised by an abnormal buildup of cerebrospinal fluid in the brain. This allows for the excess fluid to be drained and redirected to another part of the body, typically the abdominal cavity, where it can be reabsorbed by the body. The specific treatment approach will vary based on the individual case and the nature and severity of the associated conditions (Siffel et al. 2003).

1.7.4.1 Adrenoleukodystrophy

This is a rare hereditary condition affecting the adrenal glands and nerve system, sometimes known as adrenoleukodystrophy (ALD). Very-long-chain fatty acids (VLCFAs) build up in the body due to a mutation in the ABCD1 gene. Although both sexes are equally susceptible to ALD, men are disproportionately affected. Age of start and illness severity both have a role in how ALD manifests itself. The most severe type occurs in males between the ages of 4 and 10 and is called childhood cerebral form. Cognitive impairment, behavioural abnormalities, and loss of muscular control are all possible outcomes of this variant of the disease's fast progression. Although there is presently no cure for ALD, there are therapies that can alleviate its symptoms. Haematopoietic stem cell transplantation (HSCT) is an effective therapy that, when performed early on, can prevent the course of the illness (Moser and Raymond 2007).

Research into the pathogenesis and treatment of ALD is ongoing. The impact of very-long-chain fatty acids (VLCFAs) on disease progression has been the subject of recent research and on the use of gene therapy as a potential treatment option. One study, published in the journal *Neurotherapeutics*, investigated the use of gene therapy to correct the genetic defect responsible for ALD. The study used a viral vector to deliver a corrected version of the ABCD1 gene to cells from patients with ALD. The results showed that the corrected gene was able to produce functional proteins and reduce the accumulation of VLCFAs in the cells. The researchers concluded that gene therapy could be a promising treatment option for ALD in the future (Eichler and Aubourg 2006).

Another study, published in the journal *Brain*, investigated the use of a drug called Lorenzo's oil. In Lorenzo's oil, a combination of two fatty acids is used in the treatment of ALD that has been shown to reduce the levels of VLCFAs in the blood. In children with the cerebral type of ALD, Lorenzo's oil was shown to stabilise the illness, according to the study, but it did not reverse the neurological damage that had already occurred. In conclusion, ALD is a rare genetic disorder that can cause severe neurological symptoms. However, there are medicines that can alleviate symptoms and reduce the disease's course, so it is not hopeless just yet. Ongoing research into pathogenesis and treatment of ALD is providing new insights into the disease and offering hope for future treatments (Cartier and Aubourg 2018; Engelen et al. 2012).

1.7.4.2 Ehlers-Danlos Syndrome

Hypermobility of the joints, skin that is both flexible and brittle, and easy bruising are just some of the symptoms of Ehlers-Danlos syndrome (EDS), a collection of hereditary illnesses that affect the connective tissues in the body. There are currently 13 subtypes of EDS, each resulting from a unique set of gene mutations that alter collagen synthesis and structure. In recent years, there has been growing interest in EDS among medical researchers, as advances in genetics have allowed for a better understanding of the underlying causes of the disorder. This has led to improved diagnostic criteria and development of new treatments and therapies for patients with EDS.

One of the key challenges in studying EDS is the wide range of symptoms that can occur, which can make diagnosis difficult. In addition, there is often overlap between different subtypes of the disorder, which can further complicate diagnosis and treatment. Despite these challenges, significant progress has been made in understanding the molecular and genetic basis of EDS, which has led to the development of new diagnostic tools and potential treatments. For example, recent research has focused on the role of specific collagen genes in EDS and identified potential targets for drug therapies that could help alleviate symptoms and improve quality of life for patients. As research on EDS continues to advance, it is likely that new insights into the disorder will emerge, leading to further improvements in diagnosis and treatment. With increased awareness and understanding of the disorder, patients with EDS can receive better care and support and ultimately live healthier, more fulfilling lives (Malfait et al. 2017).

1.7.4.3 Clinical Manifestation

Hypermobility: excessive joint mobility or loose joints that can lead to frequent dislocations and sprains.

Skin: fragile, thin, and easily bruised skin that can be stretchy or saggy, and slow healing wounds.

Pain: chronic pain, especially in the joints or muscles, often present with a variety of other symptoms.

Gastrointestinal: irritable bowel syndrome, constipation, gastroparesis, and reflux.

Cardiovascular: heart palpitations, valve prolapse, and mitral regurgitation.

Other manifestations: dysautonomia, headaches, anxiety, and depression (Castori et al. 2017).

1.7.4.4 Diagnosis

Clinical evaluation and physical examination: A doctor will assess for hypermobility, skin abnormalities, and other characteristic signs of EDS. A geneticist can perform tests to identify specific gene mutations associated with different types of EDS. X-rays, MRIs, and CT scans may be done to identify joint dislocations, degeneration, or other abnormalities.

1.7.4.5 Treatment

Symptomatic management: EDS management typically involves treating the symptoms that arise from the condition. Physical therapy can help strengthen muscles, improve joint stability, and relieve aching joints. Non-steroidal anti-inflammatory drugs (NSAIDs), pain medications, and medications for gastrointestinal symptoms may be used. Joint stabilisation procedures may be necessary in some cases, particularly for those with joint dislocations and subluxations. The complexity of EDS cannot be overstated, and management and treatment will vary depending on the individual's symptoms and type of EDS. As a result, those who suffer from EDS need to create a tailored treatment strategy in close collaboration with their doctors (Colombi et al. 2015).

1.7.5 Usher Syndrome

Deafness and blindness are the results of Usher syndrome, a rare genetic condition. Mutations in genes involved in sensory cell formation and maintenance cause this disorder in the inner ear and the retina. Usher syndrome is thought to affect between 3 and 6 out of every 100,000 people, making it one of the leading causes of combined deafness and blindness. The symptoms of Usher syndrome typically present in early childhood, with hearing loss and balance problems being the first signs. Vision loss, which is caused by progressive retinal degeneration, usually becomes apparent in adolescence or early adulthood. There are currently no cures for Usher syndrome, and treatment options are limited to hearing aids, cochlear implants, and low-vision aids (Mathur and Yang 2015).

Research into Usher syndrome has made significant strides in recent years, with a particular focus on gene therapy and stem cell treatments. Results from animal studies and preliminary clinical tests are encouraging, and trials are underway to test these treatments in humans. Additionally, advances in assistive technologies, such as tactile and audio-based navigation systems, are helping individuals with Usher syndrome maintain their independence and quality of life. One of the challenges in studying Usher syndrome is the heterogeneity of the condition. There are three types of Usher syndrome, each resulting from a unique set of gene mutations. Additionally, severity and progression of symptoms can vary widely between individuals with the same type of Usher syndrome. As such, understanding the underlying genetic and molecular processes of the illness and developing personalised treatment solutions need additional investigation.

Usher syndrome is a kind of congenital oto-visual impairment. Gene mutations that disrupt signalling between the developing inner ear and the retina lead to this disorder. So far, more than ten genes have been associated with Usher syndrome, and different mutations in these genes can cause different subtypes of the disorder (Kimberling et al. 2010; Liu et al. 2017).

The three subtypes of Usher syndrome are the following:

1. Usher syndrome type 1 (USH1): In this most severe type of Usher syndrome, both hearing and vision are severely compromised or lost by early adolescence. Mutations in any of the six USH1-related genes can result in the disease.
2. Usher syndrome type 2 (USH2): Moderate to profound hearing loss at birth and progressive vision loss, typically starting in adolescence or early adulthood, characterise this variant of Usher syndrome. Any one of the three genes linked to USH2 can be mutated to produce the disease.
3. Usher syndrome type 3 (USH3): Progressive hearing loss and visual loss manifest in early infancy or adolescence in this variant of Usher syndrome. USH3 is brought on by mutations in the CLRN1 gene.

Inheritance of Usher syndrome follows an autosomal recessive pattern, which means that an affected individual inherits two copies of a mutated gene (one from each parent). Carriers of one mutated gene are usually asymptomatic but can pass the mutation on to their children. When two carriers have children, each child has a 25% chance of inheriting two mutated genes and therefore developing Usher syndrome, a 50% chance of inheriting one mutated gene and being a carrier like their parents, and a 25% chance of inheriting two normal genes. Genetic testing and counselling can be helpful for families affected by Usher syndrome to understand the threat of passing it on to future generations and to explore the options for managing the disorder (Geng et al. 2009; Fujinami et al. 2015).

1.7.6 Alpha-Hydroxylase Deficiency

Steroid hormone production is negatively impacted by alpha-hydroxylase deficiency, a rare genetic condition. Mutations in the CYP17A1 gene, which provides instructions for making cytochrome P450c17, are the root cause of this disorder. The adrenal gland and gonads rely heavily on this enzyme for the manufacture of both cortisol and androgens. Both a simple virilising version and a salt-losing form of alpha-hydroxylase deficiency exist. The early beginning of androgen excess and virilisation characterises the simple virilising type, whereas the extra loss of salt and water in urine characterises the salt-losing variant (Auchus 2017). In the simple virilising form, patients may experience ambiguous genitalia at birth, early onset of pubic and axillary hair growth, and accelerated growth and skeletal maturation. In females, clitoromegaly and infertility may occur, while males may experience precocious puberty and testicular enlargement. In the salt-losing form, patients may also experience dehydration, vomiting, and electrolyte imbalances, which can be life-threatening if left untreated (Kim et al. 2019). Diagnosis of alpha-hydroxylase deficiency is typically made through hormone testing, which may reveal low cortisol levels, high levels of adrenal androgens, and elevated levels of 17-hydroxyprogesterone. CYP17A1 mutations can also be confirmed by genetic testing. Treatment of alpha-hydroxylase deficiency typically involves hormone replacement therapy to replace the deficient hormones, as well as measures to manage salt and water balance in salt-losing patients. Long-term management may also involve regular monitoring of hormone levels and growth, as well as surgical interventions to correct any anatomical abnormalities (Turcu et al. 2015).

1.7.7 17-Beta-Hydroxysteroid Dehydrogenase X (17β-HSDX) Deficiency

17-Beta-hydroxysteroid dehydrogenase X (17β-HSDX) deficiency is a very uncommon genetic condition that disrupts the metabolism of female and male hormones. Mutations in the HSD17B10 gene, which supplies instructions for manufacturing the 17-HSDX enzyme, are responsible for the condition.

There are three types of 17β-HSDX deficiency, which vary in severity:

Type I: It is the most extreme manifestation of the illness; affected individuals typically have intellectual disability, developmental delays, seizures, and other neurological problems.

Type II: This form of the disorder is less severe than type I, and affected individuals will have cognitive and/or emotional challenges that range from mild to moderate.

Type III: There may be no outward signs or just modest developmental abnormalities in those with this variant of the illness (Wu and Chen 2019).

1.7.8 Beta-Hydroxysteroid Dehydrogenase X (17-HSDX) Deficiency

This is a rare autosomal recessive disease affecting the conversion of androgens and oestrogens, 17-beta-hydroxysteroid dehydrogenase X (17-HSDX) deficiency. This disorder results from a lack of or dysfunctional 17-HSDX, an enzyme encoded by the X-chromosomal gene HSD17B10. Androstenedione, a weak androgen, is converted to testosterone, a powerful androgen, by 17-HSDX in the testes, while estrone, a weak oestrogen, is converted to oestradiol, a powerful oestrogen, in the ovaries. In the absence of 17β-HSDX, these conversions are impaired, leading to an accumulation of androstenedione and estrone in the body and a decrease in testosterone and oestradiol levels. This imbalance in sex hormones can result in a range of clinical features depending on the severity of the deficiency and the sex of the affected individual (Çakır et al. 2019).

In males, 17β-HSDX deficiency can lead to underdeveloped testes (hypogonadism) and ambiguous genitalia, which can make it difficult to determine the sex of the individual at birth. In females, the condition can cause early onset of puberty, irregular menstrual cycles, and infertility. Although the specific mechanism by which 17-HSDX deficiency causes the aforementioned clinical symptoms is not yet known, it is thought to entail changes in the hypothalamic-pituitary-gonadal axis, which regulates the body's sex hormone synthesis. In addition, the accumulation of androstenedione and estrone in the body can have direct effects on various tissues, leading to abnormal development and function. There is currently no cure for 17β-HSDX deficiency, and treatment is aimed at managing the symptoms of the condition. In males, this may involve hormone replacement therapy to increase testosterone levels and promote masculinisation, while in females, treatment may involve hormonal contraceptives to regulate menstrual cycles and prevent unintended pregnancies (Al-Mutairi et al. 2016; Mehta et al. 2018).

Hypermobility of the joints, skin that is both flexible and brittle, and easy bruising are just some of the symptoms of Ehlers-Danlos syndrome (EDS), a collection of hereditary illnesses that affect the connective tissues in the body. There are presently 13 distinct forms of EDS, all of which are linked to mutations in genes involved in collagen synthesis and/or structure. In recent years, there has been growing interest in EDS among medical researchers, as advances in genetics have allowed for a better understanding of the underlying causes of the disorder. This has led to improved diagnostic criteria and development of new treatments and therapies for patients with EDS (Malfait et al. 2017).

There are 13 different forms of EDS, all with their own genetic roots and symptoms. The pathophysiology of EDS is characterised by defects in the synthesis, processing, or structure of collagen, a fundamental part of the extracellular matrix that gives connective tissues their structure and pliancy. Collagen is made up of three polypeptide chains, called alpha chains, which are wound together to form a triple helix. In EDS, the genetic mutations affect the formation, stability, or cross-linking of collagen fibres, leading to weakened or abnormal collagen. As a result, the affected tissues are prone to stretching, tearing, and other forms of damage. The severity and pattern of tissue involvement vary depending on the subtype of EDS (Grahame et al. 2000; Castori et al. 2017). For example, in classic EDS (cEDS), triggered by either COL5A1 or COL5A2 gene mutations, the production of type V collagen is reduced, leading to thin, fragile skin and hypermobile joints. In vascular EDS (vEDS), deficiencies in type III collagen due to COL3A1 gene mutations can result in arterial and organ rupture, gastrointestinal perforation, and other life-threatening complications. Other hallmarks of EDS are abnormal wound healing and scar formation, which may result from impaired collagen deposition and remodelling. In addition, EDS patients may have altered immune responses, chronic pain, and autonomic dysfunction, which can further contribute to their clinical features. Overall, the underlying genetic, molecular, and cellular pathways that contribute to EDS pathophysiology are intricate. In order to improve the quality of life for those who suffer from Ehlers-Danlos syndrome, however, it is imperative that we gain a deeper insight into the underlying abnormalities in collagen metabolism and tissue homeostasis (Brady et al. 2017).

One of the key challenges in studying EDS is the wide range of symptoms that can occur, which can make diagnosis difficult. In addition, there is often overlap between different subtypes of the disorder, which can further complicate diagnosis and treatment. Despite the difficulties, researchers have made great strides in understanding the molecular and genetic underpinnings of EDS, leading to the creation of novel diagnostic tools and potential therapies. Recent studies, for instance, have zeroed in on the function of individual collagen genes in EDS, pinpointing therapeutic targets that may reduce symptoms and enhance patients' quality of life. As research on EDS continues to advance, it is likely that new insights into the disorder will emerge, leading to further improvements in diagnosis and treatment. With increased awareness and understanding of the disorder, patients with EDS can receive better care and support and ultimately live healthier, more fulfilling lives (Kirschner and Sutcliffe 2005).

1.7.9 Achondrogenesis

Abnormal bone growth and development are the results of a rare genetic condition known as achondrogenesis (ACG). Type I and type II achondrogenesis are the two most common forms. Subtypes IA and IB are additional breakdowns of type I. Defects in the production or structure of type II collagen, a crucial component of cartilage, are at the heart of the pathophysiology of achondrogenesis. Chondrocytes, which are specialised cells that compose the cartilage matrix, generate type II collagen. In achondrogenesis, mutations in genes encoding type II collagen or other factors required for its proper formation and function result in abnormal chondrogenesis. This leads to the formation of defective cartilage, which cannot support the developing skeleton. The lack of structural support for the developing fetus can result in limb and spine deformities, as well as respiratory insufficiency and perinatal death (Bonafe et al. 2015).

Achondrogenesis is classified into three subtypes based on the genetic mutations involved:

1. Mutations in the TRIP11 gene, which codes for Golgi microtubule-associated protein 210 (GMAP-210), a key regulator of vesicular trafficking and secretion, cause achondroplasia type 1A (ACG1A).
2. Mutations in SLC26A2, which codes for the sulphate transporter diastrophic dysplasia sulphate transporter (DTDST), are the cause of achondrogenesis type 1B (ACG1B). When DTDST is absent or dysfunctional, proteoglycans, critical components of the cartilage matrix, do not receive enough sulphate to properly function.
3. The alpha-1 chain of type II collagen is encoded by the COL2A1 gene, which is mutated in patients with achondrogenesis type 2 (ACG2). Defects in cartilage development are caused by mutations in this gene, which affect collagen synthesis, processing, and secretion.

In summary, the pathophysiology of achondrogenesis involves mutations in genes that are essential for the development and upkeep of skeletal tissue. These mutations lead to severe skeletal abnormalities and can be fatal in the most severe cases (Kornak and Mundlos 2003; Forlino and Marini 2016; Lee et al. 2012).

1.7.10 RPC1B Deficiency

Rare genetic condition known as RPC1B deficiency compromises the immune system's capacity to fend off infections. Mutations in the ARPC1B gene, which codes for a protein involved in a complex known as the Arp2/3 complex, cause this disorder. Important for the health of many cell types, including immune system cells, this complex is involved in the process of actin polymerisation (Kahr et al. 2017; Burns et al. 2013).

Involved in the control of actin dynamics, ARPC1B belongs to the family of actin-related proteins (ARPs). Branching actin networks are required for numerous cellular functions, including cell motility, cell division, and phagocytosis, and ARPC1B is a key component of the Arp2/3 complex required for their formation. The Arp2/3 complex is essential for T cell and NK cell activity by contributing to the development of the immunological synapse. Affected individuals inherit two mutant copies of the ARPC1B gene, one from each parent, making ARPC1B deficiency an extremely rare autosomal recessive illness. The disorder was first described in 2017 in a cohort of patients with platelet abnormalities and susceptibility to inflammatory disease. Since then, several additional cases have been reported, including individuals with recurrent infections, severe viral infections, and autoimmune disorders (Bigley et al. 2018; Kahr et al. 2017).

Even in the same family, ARPC1B deficiency might present with a wide range of symptoms. It is possible that some people will experience just minimal symptoms, such as recurrent infections, while others may have more severe complications, such as lymphoproliferative disorders or autoimmune diseases. The severity of the disease may also depend on the specific mutation in the ARPC1B gene. Diagnosis of ARPC1B deficiency typically involves clinical observation, genetic analysis, and immunological research. The existence of ARPC1B gene mutations can be verified by genetic testing, while immunological studies can reveal defects in T cell and NK cell function, as well as abnormalities in neutrophil migration and platelet formation (Burns et al. 2013; Bigley et al. 2018).

Treatment of ARPC1B deficiency is mainly supportive and involves the use of antibiotics and antiviral medications to manage infections, as well as immunoglobulin replacement therapy to boost immune function. Haematopoietic stem cell transplantation (HSCT) has been studied as a possible therapy for some diseases. Yet, it is not yet well established that HSCT is successful in ARPC1B deficiency, and thus the risks and advantages of this operation should be carefully considered on an individual basis. The pathophysiology of ARPC1B deficiency involves defects in several aspects of immune system function, including impaired T cell and natural killer (NK) cell activity, defective neutrophil migration, and abnormal platelet formation. These defects result in an increased susceptibility to infections, particularly by certain types of bacteria and viruses. Studies have shown that ARPC1B deficiency can cause a range of clinical manifestations, including recurrent infections, severe viral infections, autoimmune disorders, and lymphoproliferative disorders. The severity of these manifestations can vary widely among affected individuals, even within the same family, suggesting that other genetic and environmental factors may also play a role. Diagnosis of ARPC1B deficiency is typically based on clinical evaluation, genetic testing, and immunological studies. Treatment may involve the use of antibiotics and antiviral medications to manage infections, as well as immunoglobulin replacement therapy to boost immune function (Al-Saud and Al-Mousa 1993; Simon and Seger 2019).

1.7.11 Bamforth-Lazarus Syndrome (BLS)

Congenital hypothyroidism is the result of Bamforth-Lazarus syndrome (BLS), a rare autosomal recessive condition that disrupts thyroid gland development. It is also known as thyroid dysgenesis type 2, as it is one of the two main types of thyroid dysgenesis, the other being thyroid dysgenesis type 1. Although the specific genetic origin of BLS is unknown at this time, numerous genes have been identified as possible contributors. BLS is typically caused by mutations in the gene NKX2–1 (sometimes called TTF-1). Thyroid transcription factor 1 (TTF-1) is a protein that is necessary for proper thyroid development and is encoded by the NKX2–1 gene. The mutation of this gene causes BLS by interfering with the normal growth and function of the thyroid gland (Castanet et al. 2020). Aside from PAX8 and FOXE1, NKX2–5 and FOXE1 have also been linked to BLS. Mutations in these genes can also disrupt thyroid gland development and lead to congenital hypothyroidism. However, mutations in these genes are less common causes of BLS compared to NKX2–1. The pathophysiology of BLS involves the failure of the thyroid gland to develop properly during embryonic development. The thyroid gland normally forms from a structure called the thyroid diverticulum, which develops from the pharyngeal floor. The thyroid diverticulum then migrates downward and fuses with the laryngeal cartilage to form the thyroid gland (Schmitt et al. 2012).

In BLS, the thyroid diverticulum fails to migrate or fuse properly, resulting in a small or absent thyroid gland. This leads to a deficiency of thyroid hormones, which are essential for normal growth and development. Without treatment, congenital hypothyroidism can lead to intellectual disability and other developmental problems. Treatment for BLS involves lifelong thyroid hormone replacement therapy, which replaces the deficient thyroid hormones and prevents the development of the serious complications of hypothyroidism. Congenital hypothyroidism is the result of Bamforth-Lazarus syndrome (BLS), a rare autosomal recessive condition that disrupts thyroid gland development. There are two main types of thyroid dysgenesis, which include BLS (Park and Chatterjee 2005).

Thyroid dysgenesis type 1: The absence or underdevelopment of the thyroid gland characterises this kind of thyroid dysgenesis. Approximately 80% of all occurrences of thyroid dysgenesis are due to this kind. A lack of proper thyroid gland development is the root cause of this disorder. Thyroid dysgenesis type 2 (Bamforth-Lazarus syndrome): A tiny or non-existent thyroid gland and a failure to properly construct the larynx and trachea characterise this kind of thyroid dysgenesis. A lack of proper thyroid gland maturation is the root cause of this condition. Congenital hypothyroidism, a disorder in which the thyroid gland does not generate enough of the thyroid hormones necessary for proper growth and development, can originate from any form of thyroid dysgenesis (Muzza et al. 2014).

1.7.12 Bare Lymphocyte Syndrome (BLS) Type I

There are two main types of bare lymphocyte syndrome (BLS), each classified based on the specific genetic defect that causes the disorder: BLS type I: Mutations in CIITA, RFXANK, or RFX5, which code for transcription factors that govern the production of major histocompatibility complex class II (MHC II) molecules, are the causative factors in this form of BLS. Reduced antigen presentation and a reduced immune response characterise type I Behcet's disease, which is caused by a shortage of MHC II molecules on the surfaces of immune cells. Type II BLS is caused by structural or functional defects in the MHC II molecule, which are encoded by mutations in the MHC II genes. In BLS type II, MHC II molecules are either absent or non-functional, leading to a similar impairment in antigen presentation and immune response as seen in BLS type I (Reith et al. 2005).

There is also a third type of BLS, known as "atypical BLS", which has been reported in a few individuals. Mutations in other genes involved in the MHC II pathway can produce atypical BLS, which is characterised by a partial lack of MHC II molecules. Recurrent infections, persistent diarrhoea, and other immune-related illnesses are all possible outcomes of BLS, a rare genetic condition that impairs the immune system. Symptom management and infection prevention are the primary goals of treatment, with stem cell transplantation being considered in rare patients (Günther et al. 2020).

1.7.13 Regenerate Response

Major histocompatibility complex class II (MHC II) deficiency, or bare lymphocyte syndrome (BLS) type I, is a rare inherited immunological illness. Important immune cells including B cells, T cells, and antigen-presenting cells (APCs) have no or extremely few MHC II molecules on their surfaces, making them less effective in recognising and eliminating infections. The genetic basis of BLS type I is primarily attributed to mutations in three genes: CIITA, RFXANK, and RFX5. The expression of MHC II molecules is controlled by transcription factors that are encoded by these genes. Mutations in these genes reduce MHC II expression, leaving the body less able to fight against infections.

The pathophysiology of BLS type I is primarily related to the inability of affected immune cells to present antigens to T cells. Antigens are the proteins and other tiny molecules that the immune system recognises as foreign and attacks. These antigens are presented to T lymphocytes via MHC II molecules, which then activate B cells and other immune cells to mount a response. In individuals with BLS type I, the absence or very low levels of MHC II molecules on immune cells prevent the presentation of antigens to T cells, leading to a weakened immune response and increased susceptibility to infections. This condition is usually diagnosed in childhood, and affected individuals often suffer from recurrent infections, chronic diarrhea, and other immune-related disorders. Treatment for BLS type I focuses on symptom management and infection prevention in the absence of a cure. Although stem cell transplantation has shown promise in some situations, it is not appropriate for every patient due to its high-risk profile (Bousfiha et al. 2015).

1.7.14 Cantú Syndrome

Cantú syndrome is a rare genetic condition that has far-reaching consequences for many bodily functions. Mutations in the ABCC9 gene on chromosome 12 are responsible for this disorder. The protein SUR2, which regulates the transport of potassium ions across cell membranes, is encoded by the ABCC9 gene. Cantú syndrome is caused by mutations in ABCC9, which cause the SUR2 protein to become hyperactive. This, in turn, causes intracellular potassium levels to rise. The body's organs and systems may be negatively impacted by this imbalance. Cantú syndrome has been linked to mutations in the ABCC9 gene; however, this is the only known form of the disorder. However, even among people who share a genetic mutation, there is often a huge range in how severely they are impacted. This variability may be due to differences in the specific mutation, as well as other genetic and environmental factors (Harakalova et al. 2012).

Some researchers have suggested that there may be different subtypes or variations of Cantú syndrome based on the specific symptoms and features that individuals exhibit. However, this is not yet widely accepted or recognised in the medical community, and further research is needed to explore the possibility of subtypes. Cantú syndrome is a multisystem illness caused by mutations in the ABCC9 gene, and it is extremely uncommon. Individuals afflicted may have a wide range of symptoms and degrees of severity (Kaur et al. 2018). Facial characteristics such as a broad nose, huge eyes, and a wide head circumference are hallmarks of Cantú syndrome. Heart problems, including enlarged chambers, irregular beats, and valve anomalies, are also possible in people with Cantú syndrome. Other possible symptoms include joint hypermobility, muscle weakness, and intellectual disability. There is presently no treatment or cure for Cantú syndrome, despite continuous research. Managing symptoms and consequences as they develop is the main focus of treatment (Cooper and Shendure 2011).

1.7.15 Combined Oxidative Phosphorylation Deficiency (COXPD)

The mitochondria, the cells' power plants, are compromised in a series of extremely uncommon genetic illnesses known as combined oxidative phosphorylation deficit (COXPD). COXPD is characterised by a defect in the electron transport chain (ETC) of mitochondria, which results in a decreased ability to produce ATP, the main source of energy for cellular processes. Mutations in genes encoding ETC-related proteins or in mitochondrial DNA (mtDNA) have been linked to COXPD. Several subtypes of COXPD may be identified, each of which results from mutations in a different gene. COXPD comes in a variety of forms, some of which are as follows:

Leigh syndrome: In most cases, the onset of this severe neurological condition occurs during infancy or early childhood. Mutations in genes encoding proteins essential for mitochondrial function, such as the ETC, cause Leigh syndrome. Symptoms of Leigh syndrome can include developmental delay, muscle weakness, seizures, and respiratory problems (Lake et al. 2016). MELAS, or mitochondrial encephalomyopathy, lactic acidosis, and stroke-like events, is a condition of the mitochondria that often manifests in young adults or children. It is caused by mutations in mtDNA and can result in a range of symptoms, including muscle weakness, epileptic attacks, and stroke-like symptoms. Deficiency in cytochrome oxidase: COX is the final enzyme in the ETC, and mutations in genes that encode COX can result in COXPD. Symptoms of COX deficiency can include muscle weakness, developmental delay, and seizures (Finsterer 2018).

The pathophysiology of COXPD is complex and involves a defect in the ETC that results in decreased ATP production. Several protein complexes work together to create an electrochemical gradient of protons across the inner mitochondrial membrane; this process is known as the electron transport chain (ETC). As a result of this proton gradient, ATP is synthesised by ATP synthase. If the genes encoding ETC-related proteins are mutated in COXPD, electron transport and ATP generation may be impaired. Secondary consequences of COXPD include increased generation of reactive oxygen species (ROS) and altered mitochondrial dynamics, in addition to the basic deficiency in the ETC. DNA, proteins, and lipids are all vulnerable to destruction at the hands of reactive oxygen species (ROS). The increased generation of reactive oxygen species (ROS) by COXPD may contribute to mitochondrial dysfunction and disease progression (Nogueira et al. 2019). The clinical manifestations of combined oxidative phosphorylation deficiency (COXPD) might differ in terms of age of start, severity of disease, and precise genetic mutation responsible for the condition. However, there are some general symptoms that may be present in individuals with COXPD, including the following:

Muscle weakness: Many types of COXPD can cause weakness in the muscles, which can affect mobility and the ability to perform everyday activities.
Developmental delay: Intellectual impairment or developmental delays have been linked to COXPD because of its effect on brain and nervous system maturation.
Seizures: Seizures are periods of aberrant electrical activity in the brain that can result in a wide variety of symptoms, and they have been linked to some forms of COXPD (Rodenburg 2016).
Respiratory problems: COXPD can affect the muscles involved in breathing, leading to respiratory problems such as difficulty breathing or rapid breathing.
Vision and hearing problems: Some types of COXPD can cause vision or hearing problems, which can range from mild to severe.
Gastrointestinal symptoms: COXPD can affect the muscles of the gastrointestinal tract, leading to symptoms such as nausea, vomiting, and diarrhoea.
Cardiac symptoms: Some types of COXPD can affect the function of the heart, leading to symptoms such as arrhythmias, heart failure, or sudden cardiac arrest.

It is worth noting that not everyone with COXPD has these symptoms and that the intensity with which they manifest varies greatly from person to person. Getting a professional medical opinion is crucial if you or a loved one is suffering from any of these symptoms (Alston et al. 2017).

1.8 Conclusion

Rare genetic disorders are a complex and challenging group of diseases. However, there is hope for the future. With ongoing research, new treatments are being developed that can help people with these disorders live longer, healthier lives. In addition, there are a number of resources available to help people with rare genetic disorders and their families. It is important to remember that rare genetic disorders are not rare in terms of the impact they have on people's lives. Millions of people around the world are affected by these disorders, and they can have a profound impact on individuals, families, and communities. By raising awareness of rare genetic disorders, we can help to ensure that people with these disorders have access to the resources they need to live full and productive lives. Early diagnosis and intervention are essential for people with rare genetic disorders. Early diagnosis can help to ensure that people receive the correct treatment and support, and it can also help to prevent or delay the onset of complications. Intervention can help to improve the quality of life for people with rare genetic disorders, and it can also help to maximise their potential. People with rare genetic disorders and their families often face unique challenges, such as a lack of awareness and understanding, difficulty accessing specialised medical care, and financial burden of caring for a child with a rare genetic disorder. Support can help these individuals and families cope with these challenges and live full and productive lives.

References

Allen JN, Magro CM, King MA (2002) The eosinophilic pneumonias. Semin Respir Crit Care Med 23:127–134. http://www.ncbi.nlm.nih.gov/pubmed/16088605

Al-Mutairi N, Al-Shaikh A, Nampoothiri N (2016) 17β-Hydroxysteroid dehydrogenase type 10 deficiency presenting as male pseudohermaphroditism. Saudi Med J 37(5):573–576

Al-Saud B, Al-Mousa H (1993–2022) ARPC1B deficiency. 2021 Sep 28. In: Adam MP, Ardinger HH, Pagon RA, et al. (eds) GeneReviews®. University of Washington, Seattle, WA. Available from: https://www.ncbi.nlm.nih.gov/books/NBK518710/

Al-Semari A, Wakil SM, Al-Muhaizea MA et al (2013) Novel FGD1 variant underlying Aarskog-Scott syndrome with myopathy and distal arthropathy. Clin Dysmorphol 22:13–17

Alston CL, Rocha MC, Lax NZ, Turnbull DM, Taylor RW (2017) The genetics and pathology of mitochondrial disease. J Pathol 241(2):236–250. https://doi.org/10.1002/path.4809. Epub 2016 Nov 22

Al-Tubaikh JA, Reiser MF (2009) Encephalocele. In: Congenital diseases and syndromes—an illustrated radiological guide. Springer-Verlag, Berlin, Heidelberg, pp 3–4

Al-Yahyaee SA, Al-Kindi MN, Habbal O, Kumar DS (2003) Clinical and molecular analysis of Grebe acromesomelic dysplasia in an Omani family. Am J Med Genet A 121:9–14

Arkkila PE (2006) Thromboangiitis obliterans (Buerger's disease). Orphanet J Rare Dis 1(1):1–5

Auchus RJ (2017) The genetics, pathophysiology, and management of human deficiencies of P450c17. Endocrinol Metab Clin 46(2):375–393

Bartels CF, Bukulmez H, Padayatti P et al (2004) Mutations in the transmembrane natriuretic peptide receptor NPR-B impair skeletal growth and cause acromesomelic dysplasia type Maroteaux. Am J Hum Genet 75:27–34

Beers MH, Berkow R (eds) (1999) The Merck manual, 17th edn. Merck Research Laboratories, Whitehouse Station, NJ, p 973

Beers MH, Porter RS, Jones TV (2006) The Merck manual of diagnosis and therapy. Merck Research Laboratories, Whitehouse Station, NJ

Behrman RE, Kliegman RM, Arvin AM (eds) (1996) Nelson textbook of pediatrics, 15th edn. W.B. Saunder Company, Philadelphia, PA, p 1439

Bergfeld WF, Elston DM (2003) Cicatricial Alopecia. In: Olsen EA (ed) Disorders of hair growth: diagnosis and treatment. McGraw-Hill, New York, NY

Bigley V, Maisuria S, Cytlak U, Jardine L, Care MA, Green K, Gunawan M, Milne P, Dickinson R, Harrison RE, Haniffa M, Collin M (2018) Biallelic interferon regulatory factor 8 mutation: a complex immunodeficiency syndrome with dendritic cell deficiency, monocytopenia, and immune dysregulation. J Allergy Clin Immunol 141(6):2234–2248

Blackwell KE, Rawnsley JD (2008) Aesthetic considerations in scalp reconstruction. Facial Plast Surg 24(1):11–21

Bonafe L, Cormier-Daire V, Hall C, Lachman R, Mortier G, Mundlos S, Nishimura G, Sangiorgi L, Savarirayan R, Sillence D, Spranger J, Superti-Furga A, Warman M, Unger S (2015)

Nosology and classification of genetic skeletal disorders: 2015 revision. Am J Med Genet A 167A:2869–2892

Bottani A, Orrico A, Galli L et al (2007) Unilateral focal polymicrogyria in a patient with classical Aarskog-Scott Syndrome due to a novel missense variant in an evolutionary conserved RhoGEF domain of the faciogenital dysplasia gene FGD1. Am J Med Genet A 143A:2334–2338

Bousfiha A, Jeddane L, Al-Herz W et al (2015) The 2015 IUIS phenotypic classification for primary immunodeficiencies. J Clin Immunol 35(8):727–738. https://doi.org/10.1007/s10875-015-0201-1. Epub 2015 Oct 23

Brady AF, Demirdas S, Fournel-Gigleux S et al (2017) The Ehlers-Danlos syndromes, rare types. Am J Med Genet C Semin Med Genet 175(1):70–115. https://doi.org/10.1002/ajmg.c.31550

Burns SO, Zenner HL, Plagnol V, Curtis J, Mok K, Eisenhut M, Kumararatne D, Doffinger R, Thrasher AJ (2013) Investigation of primary immunodeficiencies in patients with chronic mucocutaneous candidiasis. Immunology 138(3):235–245

Çakır E, Çamtosun F, Gürkan F et al (2019) 17β-Hydroxysteroid dehydrogenase type 10 deficiency: a novel case with a potential homozygous founder mutation. J Pediatr Endocrinol Metab 32(6):661–665

Cartier N, Aubourg P (2018) Gene therapy for X-linked adrenoleukodystrophy. Neurotherapeutics 15(3):587–597

Castanet M, Park SM, Smith A, Bost M, Léger J (2020) Genetics of thyroid dysgenesis: evolving insights. Semin Cell Dev Biol 98:155–162. https://doi.org/10.1016/j.semcdb.2019.11.011. Epub 2019 Nov 20

Castori M, Tinkle B, Levy H, Grahame R, Malfait F, Hakim A (2017) A framework for the classification of joint hypermobility and related conditions. Am J Med Genet C Semin Med Genet 175(1):148–157

Cheson BD, Bennett JM, Grever M, Kay N, Keating MJ, O'Brien S, Rai KR (1996) National Cancer Institute-sponsored Working Group guidelines for chronic lymphocytic leukemia: revised guidelines for diagnosis and treatment. Blood 87(12):4990–4997

Colombi M, Dordoni C, Chiarelli N, Ritelli M (2015) Differential diagnosis and diagnostic flow chart of joint hypermobility syndrome/Ehlers-Danlos syndrome hypermobility type compared to other heritable connective tissue disorders. Am J Med Genet C Semin Med Genet 169(1):6–22

Cooper GM, Shendure J (2011) Needles in stacks of needles: finding disease-causal variants in a wealth of genomic data. Nat Rev Genet 12(9):628–640

De Wals P, Tairou F, Van Allen MI et al (2007) Reduction in neural-tube defects after folic acid fortification in Canada. N Engl J Med 357:135–142. http://www.ncbi.nlm.nih.gov/pubmed/17625125

Döhner H, Stilgenbauer S, Döhner K, Bentz M, Lichter P (1999) Chromosome aberrations in B-cell chronic lymphocytic leukemia: reassessment based on molecular cytogenetic analysis. J Mol Med 77:266–281

Eichler FS, Aubourg P (2006) Therapy for X-linked adrenoleukodystrophy. Curr Opin Neurol 19(6):606–610

Engelen M, Kemp S, Poll-The BT, van Geel BM, Wanders RJ, Aubourg P (2012) X-linked adrenoleukodystrophy (X-ALD): clinical presentation and guidelines for diagnosis, follow-up and management. Orphanet J Rare Dis 7(1):1–21

Evans NS, Ratchford EV (2014) Vascular disease patient information page: venous thromboembolism (deep vein thrombosis and pulmonary embolism). Vasc Med 19(2):148–150

Finsterer J (2018) Mitochondrial encephalomyopathies. J Neurol Sci 395:4–9. https://doi.org/10.1016/j.jns.2018.10.012. Epub 2018 Oct 10

Forlino A, Marini JC (2016) Osteogenesis imperfecta. Lancet 387:1657–1671

Fujinami K, Zernant J, Chana RK et al (2015) Clinical and molecular characteristics of childhood-onset Usher syndrome. Am J Ophthalmol 159(5):987–996. https://doi.org/10.1016/j.ajo.2015.02.005

Geng R, Geller SF, Hayashi T et al (2009) Usher syndrome IIIA gene clarin-1 is essential for hair cell function and associated neural activation. Hum Mol Genet 18(14):2748–2760. https://doi.org/10.1093/hmg/ddp211
Gorski JL (2003) Aarskog syndrome. In: NORD guide to rare disorders. Lippincott Williams & Wilkins, Philadelphia, PA, p 142
Grahame R, Bird HA, Child A (2000) The revised (Brighton 1998) criteria for the diagnosis of benign joint hypermobility syndrome (BJHS). J Rheumatol 27(7):1777–1779. PMID: 10914845
Günther S, Schlaf G, Sticht H, Schwarz K (2020) Transcriptional regulation of MHC class II genes: lessons from a disease. Immunogenetics 72(1–2):1–14. https://doi.org/10.1007/s00251-019-01131-1. Epub 2019 Nov 16
Harakalova M, van der Smagt JJ, de Kovel CGF, Van't Slot R, Poot M, Nijman I, Houweling AC (2012) Incomplete segregation of MYH11 variants with thoracic aortic aneurysms and dissections and patent ductus arteriosus. Eur J Hum Genet 20(6):581–584
Jeong YJ, Kim KL, Seo IJ et al (2007) Eosinophilic lung diseases: a clinical, radiologic, and pathologic overview. Radiographics 27:617–637. http://www.ncbi.nlm.nih.gov/pubmed/17495282
Jones KL (ed) (1997) Smith's recognizable patterns of human malformation, 5th edn. W. B. Saunders Co., Philadelphia, PA, pp 128–129
Kahr WHA, Pluthero FG, Elkadri A, Warner N, Drobac M, Chen CH, Lo RW, Li L, Li R, Li Q, Thoeni C, Pan J, Leung G, Lara-Corrales I, Murchie R, Cutz E et al (2017) Loss of the Arp2/3 complex component ARPC1B causes platelet abnormalities and predisposes to inflammatory disease. Nat Commun 8:14816
Kaname T, Yanagi K, Okamoto N, Naritomi K (2006) Neurobehavioral disorders in patients with Aarskog-Scott Syndrome affected by novel FGD1 variants. Am J Med Genet A 140:1331–1332
Kaur S, Smith HS, Mathew S (2018) Cantu syndrome: a rare genetic disorder. J Genet Couns 27(5):1237–1242
Khan S, Basit S, Khan MA, Muhammad N, Ahmad W (2016) Genetics of human isolated acromesomelic dysplasia. Eur J Med Genet 59(4):198–203
Kim YJ, Kim GH, Cheon CK, Cho HJ (2019) Clinical and genetic analysis of 4 Korean patients with congenital adrenal hyperplasia due to 17α-hydroxylase/17, 20-lyase deficiency. Ann Pediatr Endocrinol Metab 24(4):247–252
Kimberling WJ, Hildebrand MS, Shearer AE et al (2010) Frequency of Usher syndrome in two pediatric populations: implications for genetic screening of deaf and hard of hearing children. Genet Med 12(9):512–516. https://doi.org/10.1097/GIM.0b013e3181ec44f8
Kirschner RE, Sutcliffe AJ (2005) The Ehlers-Danlos syndromes. Clin Plast Surg 32(4):627–647. https://doi.org/10.1016/j.cps.2005.05.004
Kornak U, Mundlos S (2003) Genetic disorders of the skeleton: a developmental approach. Am J Hum Genet 73:447–474
Lake NJ, Compton AG, Rahman S, Thorburn DR (2016) Leigh syndrome: One disorder, more than 75 monogenic causes. Ann Neurol 79(2):190–203. https://doi.org/10.1002/ana.24532. Epub 2015 Dec 19
Lebel RR, May M, Pouls S, Lubs HA, Stevenson RE, Schwartz CE (2002) Non-syndromic X-linked mental retardation associated with a missense variant (P312L) in the FGD1 gene. Clin Genet 61:139–145
Lee BH, Kim YM, Kim OH, Choi JH, Yoo WJ, Cho TJ (2012) Update on the molecular pathology of skeletal dysplasias. J Pathol Transl Med 46:561–571
Liu XZ, Angeli S, Liu W et al (2017) Cochlear implantation in individuals with Usher type 1 syndrome. Int J Pediatr Otorhinolaryngol 101:45–50. https://doi.org/10.1016/j.ijporl.2017.08.009
Mahapatra AK (2007) Management of encephaloceles. In: Ramamurthi R, Sridhar K, Vasudevan MC (eds) Textbooks of operative neurosurgery. BI Publications, New Delhi, pp 279–290
Malfait F, Francomano C, Byers P et al (2017) The 2017 international classification of the Ehlers-Danlos syndromes. Am J Med Genet C Semin Med Genet 175(1):8–26

Mathur P, Yang J (2015) Usher syndrome: hearing loss, retinal degeneration and associated abnormalities. Biochim Biophys Acta 1852(3):406–420. https://doi.org/10.1016/j.bbadis.2014.05.015

McCormick PA (2003) Banti syndrome. In: NORD guide to rare disorders. Lippincott Williams & Wilkins, Philadelphia, PA, pp 377–378

Mehta A, Sundaram D, Chaudhary S et al (2018) 17β-Hydroxysteroid dehydrogenase type 10 deficiency: a rare cause of hyperandrogenism in females. J Pediatr Adolesc Gynecol 31(3):250–253

Mellstedt H (2007) Clinical signs and symptoms at diagnosis and its differential diagnosis. Ann Oncol 18:i14–i21

Miller BA, Gray A, Leblanc TW et al (2010) Acute eosinophilic pneumonia secondary to daptomycin: a report of three cases. Clin Infect Dis 50:63–68. http://www.ncbi.nlm.nih.gov/pubmed/20420515

Moore CA, Li S, Li Z, Hong SX, Gu HQ, Berry RJ, Mulinare J, Erickson JD (1997) Elevated rates of severe neural tube defects in a high-prevalence area in northern China. Am J Med Genet 73(2):113–118

Moser HW, Raymond GV (2007) X-linked adrenoleukodystrophy. Lancet Neurol 6(11):1044–1055

Muzza M, Rabbiosi S, Vigone MC, Zamproni I, Cirello V, Maffini MA, Bonomi M, Colombo C, Fugazzola L (2014) Congenital hypothyroidism with eutopic thyroid gland: analysis of clinical and biochemical features at diagnosis and after re-evaluation. J Clin Endocrinol Metab 99(12):4333–4340. https://doi.org/10.1210/jc.2014-2349. Epub 2014 Sep 10

Nogueira C, Uziel G, Perez MJ et al (2019) Impaired complex III activity in peripheral blood cells of patients with Parkinson's disease. Eur Neurol 81(1–2):1–7. https://doi.org/10.1159/000496424. Epub 2019 Jan 9

Ochoa BE, King LE Jr, Price VH (2008) Lichen planopilaris: annual incidence in four hair referral centers in the United States. J Am Acad Dermatol 58(2):352–353

Olney RC, Bukulmez H, Bartels CF, Prickett TC, Espiner EA, Potter LR, Warman ML (2006) Heterozygous mutations in natriuretic peptide receptor-B (NPR2) are associated with short stature. J Clin Endocrinol Metab 91:1229–1232

Olsen EA (2005) Female pattern hair loss and its relationship to permanent/cicatricial alopecia: a new perspective. J Investig Dermatol Symp Proc 10:217–221

Orrico A, Galli L, Cavaliere ML et al (2004) Phenotypic and molecular characterisation of the Aarskog syndrome: a survey of the clinical variability considering FDG1 variant analysis in 46 patients. Eur J Hum Genet 12:16–23

Orrico A, Galli L, Buoni S et al (2005) Attention-deficit/hyperactivity disorder (ADHD) and variable clinical expression of Aarskog-Scott Syndrome due to a novel FGD1 gene variant (R408Q). Am J Med Genet A 135:99–102

Orrico A, Galli L, Obregón MG, de Castro Perez MF, Falciani M, Sorrentino V (2007) Unusually severe expression of craniofacial features in Aarskog-Scott syndrome due to a novel truncating variant of the FDG1 gene. Am J Med Genet A 143:58–63

Orrico A, Galli L, Faivre L et al (2010) Aarskog-Scott syndrome: clinical update and report of nine novel variants of the FGD1 gene. Am J Med Genet A 152A:313–318

Orrico A, Galli L, Clayton-Smith J, Fryns J-P (2015) Clinical utility gene card for: Aarskog–Scott syndrome (faciogenital dysplasia)—update 2015. Eur J Hum Genet 23(4):558. https://doi.org/10.1038/ejhg.2014.178

Park SM, Chatterjee VK (2005) Genetics of congenital hypothyroidism. J Med Genet 42(10):379–389. https://doi.org/10.1136/jmg.2004.024646

Philit F, Etienne-Mastroianni B, Parrot A et al (2002) Idiopathic acute eosinophilic pneumonia. A study of 22 patients. Am J Respir Crit Car Med 166:1235–1239. http://www.ncbi.nlm.nih.gov/pubmed/12403693

Pickhardt PJ, Balfe DM (1998) Portal vein calcification and associated biliary stricture in idiopathic portal hypertension (Banti's syndrome). Abdom Imaging 23:180–182

Pilozzi-Edmonds L, Maher TA, Basran RK et al (2011) Fraternal twins with Aarskog-Scott Syndrome due to maternal germline mosaicism. Am J Med Genet A 55A:1987–1990

Podhorecka M, Halicka D, Szymczyk A, Macheta A, Chocholska S, Hus M, Darzynkiewicz Z (2016) Assessment of red blood cell distribution width as a prognostic marker in chronic lymphocytic leukemia. Oncotarget 7(22):32846

Poletti V, Costabel U, Casoni GL et al (2004) Rare infiltrative lung diseases: a challenge for clinicians. Respiration 71:431–443. http://www.ncbi.nlm.nih.gov/pubmed/15467318

Potter LR, Abbey-Hosch S, Dickey DM (2006) Natriuretic peptides, their receptors, and cyclic guanosine monophosphate-dependent signaling functions. Endocr Rev 27:47–72

Price V, Mirmirani P (2011) Cicatricial alopecia: an approach to diagnosis and management. Springer, New York Dordrecht Heidelberg London

Ravenna P (1940) Banti syndrome (fibrocongestive splenomegaly): definition, classification and pathogenesis. Arch Intern Med 66(4):879–892

Reith W, LeibundGut-Landmann S, Waldburger JM (2005) Regulation of MHC class II gene expression by the class II transactivator. Nat Rev Immunol 5(1):793–806. https://doi.org/10.1038/nri1707

Rhee CK, Min KH, Yim NY, Lee JE, Lee NR, Chung MP, Jeon K (2013) Clinical characteristics and corticosteroid treatment of acute eosinophilic pneumonia. Eur Respir J 41:402–409. http://www.ncbi.nlm.nih.gov/pubmed/22599359

Rodenburg RJ (2016) Mitochondrial complex I-linked disease. Biochim Biophys Acta 1857(5):938–945. https://doi.org/10.1016/j.bbabio.2016.01.005. Epub 2016 Jan 8

Saito Y, Sasaki KI, Katsuda Y, Murohara T, Takeshita Y, Okazaki T, Arima K, Katsuki Y, Shintani S, Shimada T, Akashi H (2007) Effect of autologous bone-marrow cell transplantation on ischemic ulcer in patients with Buerger's disease. Circ J 71(8):1187–1192

Satoh M, Yokoya S (2006) Anabolic steroid and gonadotropin releasing hormone analog combined treatment increased pubertal height gain and adult height in two children who entered puberty with short stature. J Pediatr Endocrinol Metab 19:1125–1131

Savarirayan R, White SM, Goodman FR et al (2003) Broad phenotypic spectrum caused by an identical heterozygous CDMP-1 mutation in three unrelated families. Am J Med Genet A 117:136–142

Schmitt TL, Parikh P, Crary SE (2012) Bamforth-Lazarus syndrome: report of a case with a novel mutation in the NKX2-1 gene and review of the literature. Pediatr Dev Pathol 15(4):318–322. https://doi.org/10.2350/11-08-1066-CR.1. Epub 2012 Mar 29

Schwartz CE, Gillessen-Kaesbach G, May M et al (2000) Two novel variants confirm FDG1 is responsible for the Aarskog syndrome. Eur J Hum Genet 8:869–874

Shalev SA, Chervinski E, Weiner E, Mazor G, Friez MJ, Schwartz CE (2006) Clinical variation of Aarskog syndrome in a large family with 2189delA in the FGD1 gene. Am J Med Genet A 140:162–165

Shorr AF, Scoville SL, Cersovsky SB et al (2004) Acute eosinophilic pneumonia among US military personnel deployed in or near Iraq. JAMA 292:2997–3005. http://jama.ama-assn.org/content/292/24/2997.long

Siffel C, Wong LY, Olney RS, Correa A (2003) Survival of infants diagnosed with encephalocele in Atlanta, 1979–98. Paediatr Perinat Epidemiol 17:40–48. http://www.ncbi.nlm.nih.gov/pubmed/12562471

Simon HU, Seger R (2019) ARPC1B deficiency—from bedside to bench and back again. Immunol Rev 287(1):62–72

Stoll C, Dott B, Alembik Y, Roth MP (2011) Associated malformations among infants with neural tube defects. Am J Med Genet A 155A:565–568. http://www.ncbi.nlm.nih.gov/pubmed/21337695

Szczaluba K, Hilbert K, Obersztyn E, Zabel B, Mazurczak T, Kozlowski K (2005) Du Pan syndrome phenotype caused by heterozygous pathogenic mutations in CDMP1 gene. Am J Med Genet A 138(4):379–383

Tsigkaropoulou E, Hatzilia D, Rizos E et al (2011) Venlafaxine-induced acute eosinophilic pneumonia. Gen Hosp Psychiatry 33:e7–e9. http://www.ncbi.nlm.nih.gov/pubmed/21762842

Turcu AF, Auchus RJ, Rainey WE (2015) Adrenal androgens and androgen precursors—defining their role in human physiology and disease. Endocr Rev 36(3):348–373

Verhoeven WM, Egger JI, Hoogeboom AJ (2012) X-linked Aarskog syndrome: report on a novel FGD1 gene variant. Executive dysfunction as part of the behavioural phenotype. Genet Couns 23:157–167
Waqar SN, Jindani S, Baig NS et al (2004) Banti's syndrome: case report and review of the literature. J Pak Med Assoc 54:99–101
Watts RA, Scott DG (2003) Epidemiology of the vasculitides. Curr Opin Rheumatol 15(1):11–16
Whiting DA (2001) Cicatricial Alopecia: clinico-pathological findings and treatment. Clin Dermatol 19:211–225
Wu H, Chen CL (2019) 17β-Hydroxysteroid dehydrogenase type 10 and its role in human disease. Mol Cell Endocrinol 489:34–42
Wu WY, Otberg N, McElwee KJ, Shapiro J (2008) Diagnosis and management of primary cicatricial Alopecia: Part II. Skinmed 7(2):78–83
Wu F, Ding H, Zhang Z (2012) Antibody-mediated drug delivery system for lymphatic targeting treatment. In: Antibody-mediated drug delivery systems: concepts technology, and applications, pp 169–190. https://doi.org/10.1002/9781118229019.ch9

Techniques for the Diagnosis of Rare Genetic Disorders

2

Iyshwarya B K and Ramakrishnan Veerabathiran

Abstract

Rare genetic disorders are hereditary conditions brought on by changes or mutations in a person's DNA. A few people are affected by these illnesses, often inherited from parents with mutant genes. Rare diseases afflict more than 300–400 million people worldwide, including 30 million in the United States, and often result in chronic illness, disability, and premature death. Using heuristic approaches to diagnose rare diseases is a common practice based on clinical experience and medical literature. Diagnosing rare diseases remains challenging, regardless of their prevalence in clinical practice. Patients with rare diseases are often left undiagnosed for years, and many do not receive a correct diagnosis before they die. But still, improvements in genetic testing and research are improving the understanding of these disorders and raising the potential for identifying effective therapies and remedies. Molecular identification of rare and undiagnosed diseases has been made possible by gene panels, microarrays, and exome sequencing in recent years. New diagnostic approaches based on next-generation sequencing (NGS) technology have made it possible to diagnose genetically heterogeneous disorders, even when clinical diagnostic hypotheses are unclear. While new technologies are used extensively in many health facilities and health systems, their use is significantly different. This chapter provides clinicians and researchers with strategies for a group of rare hereditary diseases brought on by genetic abnormalities, which only affect a minor portion of the population. The rarity of these conditions and the wide variety of symptoms they might present make things more complicated. The precision and rapidity of diagnosis have increased due to genetic testing technology developments.

I. B. K · R. Veerabathiran (✉)
Human Cytogenetics and Genomics Laboratory, Faculty of Allied Health Sciences, Chettinad Hospital and Research Institute, Chettinad Academy of Research and Education, Kelambakkam, Chennai, Tamil Nadu, India

M. Umair et al. (eds.), *Rare Genetic Disorders*,
https://doi.org/10.1007/978-981-99-9323-9_2

Keywords
Rare disease · Genetics testing · Diagnosis · Cytogenetics · Molecular testing

2.1 Introduction

Rare diseases (RDs) are chronic, debilitating, and often deadly disorders that only impact a small proportion of the general population (Richter et al. 2015; Danese and Lippi 2018). However, they are increasingly recognized as a severe global public health hazard. RDs are estimated to impact a disproportionately higher proportion of the world's population (Schieppati et al. 2008). Several issues hinder research on RDs, making diagnosing and treating RDs more difficult. Even though several international projects are being undertaken to address the problems caused by RDs, more work still needs to be done to address this underutilized area of health. The scarcity of appropriate clinical resources and widely dispersed genetic service centers in underdeveloped nations such as India has impeded RD research (Aggarwal and Phadke 2015; Kasthuri 2018). It has been proposed that the primary reasons for India's genetic diversity are its long history of migration and its highly diversified population structure, characterized by sociocultural, geographic, linguistic, and religious isolation (Basu et al. 2016). However, a comparatively greater frequency of genetic diseases in India may have been caused by the biological isolation of multiple endogamous population groupings (Pradhan et al. 2011).

Jammu and Kashmir (J&K), one of India's most diversified conglomerations of distinct inbred population groupings, is predicted to represent an undiscovered reservoir of genetic diseases. J&K's population is organized into many endogamous groups with different marital relationships, such as consanguinity. Moreover, the different environments separating various races may give J&K a distinct genetic architecture and illness profile. Regrettably, the area has gotten little attention in terms of genetic research. This may lead to a general lack of understanding of the great majority of genetic diseases, which are probably limited to certain extended families or communities. Nevertheless, genetic research in J&K and high-throughput genomics-based methods have lagged behind the proper identification of certain monogenic illnesses, which are typically uncommon internationally but are expected to become more common in the region due to increased inbreeding rates (Rai et al. 2016; Kuchay et al. 2019). Rare genetic illnesses may have a profound effect on both individuals and their relatives. Physical, emotional, and financial difficulties, social isolation, and stigma affect many persons with these disorders. Nonetheless, genetic and healthcare technology developments increase the potential for better diagnosis, treatment, and management of uncommon genetic illnesses.

2.2 Rare Genetic Disease

In the 1980s and 1990s, linkage analysis and fine mapping within large multiplex pedigrees were predominantly employed to map disease genes, focusing on rare, monogenic, and syndromic disorders. Sanger sequencing of the genes found to map into the linked locus after genetic signal localization was often used to detect disease-causing alleles. Confirmatory functional studies typically follow pathogenicity assessments in cellular and animal models based on associating a potentially causal variant with the illness when there is no evidence that the risk genotype was present in healthy individuals across multiple families. By 2000, about 1000 of the 7000 single-gene hereditary disorders had been found, although it had taken a long time. Several diseases, including cystic fibrosis and Huntington's disease, have a substantial biological impact. Sequencing the whole human genome has resulted in a fourfold increase in the number of genes connected to rare, single-gene diseases. This work relied mainly on exome and genome sequencing, microarray-based structural variation detection, in silico analysis, and prioritization of found genetic variations. The growing reference datasets representing population genetic diversity across diverse ethnic origins have enabled causal solid inference (Vissers et al. 2003; Ng et al. 2009; Wheeler et al. 2008; 1000 Genomes Project Consortium 2015). The aptitude to detect all sorts of causal genetic variation, from minor structural rearrangements to single mutations, in a single test has recently been made feasible by high-throughput sequencing technology. These innovations have transitioned from the laboratory to the clinic, allowing earlier and more rapid diagnosis of genetic diseases (Lek et al. 2016; Karczewski et al. 2019). There is currently continuing research into rare genetic disorders, and developments in genetic testing and gene therapy provide promise for future treatments and perhaps a cure for some diseases. So far, managing these diseases might need ongoing research from experts, health researchers, and the larger scientific community due to their rarity and complexity.

2.3 Standard Classifications of Genetic Testing

A genetic test examines a person's DNA (genetic material) to identify any alterations or mutations that might be linked to a specific genetic illness or disease. Usually, a sample of blood, saliva, or other tissues is used for the testing. Standard classification of genetic tests can be used for many different purposes; here are lists of some of the significant uses discussed below. A schematic representing common classifications of genetic techniques is shown in Fig. 2.1.

2.3.1 Newborn Screening Test

Newborn screening is the most popular use for genetic testing (NBS). Every year, up to 50 genetic and metabolic diseases are examined in almost ten million

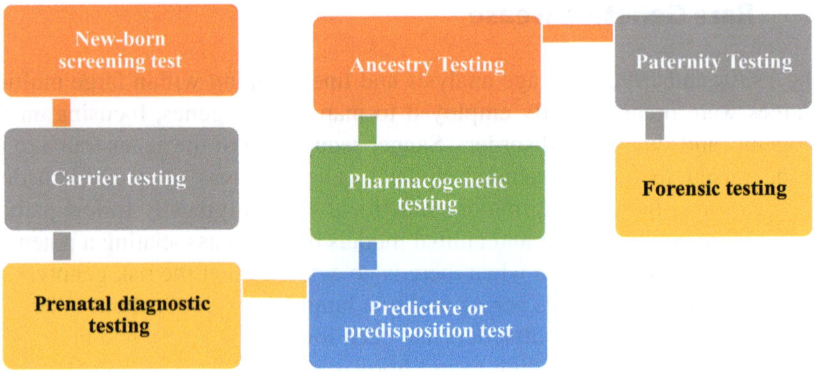

Fig. 2.1 Common classifications of genetic techniques

newborns globally. In newborn screening, infants are identified as at risk for these illnesses and may need additional testing to diagnose them definitively. It is imperative to complete the analysis and report immediately to avoid delaying potentially lifesaving treatment. It is not uncommon for a typical NBS laboratory to examine and assess more than 1000 samples for 50 diseases daily. The three subtypes of metabolic illnesses include protein metabolism disorders, organic acidemias, and fatty acid oxidation disorders. Tandem mass spectrometry may detect these three types of metabolic diseases (MSs). A wet-chemical method and spectrophotometric or fluorescence detection can help diagnose galactosemia and biotinidase deficiency. Screening samples are collected from the baby's heel with a tiny skin puncture (CLSI 2013). A few blood clots from the heel stick are collected on filter paper. The sample is commonly called the Guthrie spot in honor of Robert Guthrie's contributions to NBS. The exact form that the filter paper sample is connected to also provides demographic data about the patient. Significant cultural, budgetary, and medical ramifications precede the publication of an aberrant test result. Enhanced newborn screening for more than 20 diseases using tandem mass spectrometry had a cumulative frequency of around 1 case per 2400 neonates, a false-positive rate of 0.33%, and a false-positive rate per 2400 newborns (Schulze et al. 2003). A false-positive result impacts around 13,000 of the approximately four million infants born in the United States yearly. Confirmatory testing often takes 2–7 days, and it may strain parent-child ties and make parents concerned about their children's future and health (Gurian et al. 2006; Hewlett and Waisbren 2006; Morrison and Clayton 2011). It is possible that the insurance company will not always pay for confirmatory tests, and there may be a substantial out-of-pocket cost. Others may not exhibit clinical signs until much later in life, even though certain diseases may need hospitalization of the kid for clinical and biochemical testing. Early detection of these disorders may lead to interventions that lower the severity of the condition or postpone the onset of symptoms. It is essential to remember that neonatal screening cannot identify all genetic diseases, and even though a baby screens harmful, it might still be at risk for specific issues in the future. Hence, newborn screening is just one of the various tools that can be employed to assist a child's long-term health and well-being.

2.3.2 Carrier Testing

It might help couples decide whether their kids are at risk of inheriting recessive genes that cause genetic illnesses, sickle cell illness, cystic fibrosis, and Tay-Sachs disease, for example. Individuals with a family history of a genetic condition and members of ethnic groups at increased risk of specific congenital abnormalities are generally qualified for this testing. The test may reveal facts regarding a couple's probability of delivering a baby with a genetic condition if both parents are checked. It is used to identify patients who have a recessive autosomal or X-linked genomic variant for a disorder. The carrier, in this case, is aware of possible reproductive concerns but frequently displays no symptoms. Couples who are contemplating having children should seek carrier testing. Pre- and post-test counseling is required owing to the possible social and personal consequences of test results, notwithstanding deviations from predictive testing. Based on the genetic condition, the doctor and carrier should consider different options, such as preimplantation diagnosis and pregnancy without prenatal testing (Pagon et al. 2001; Bioethics 2001). To assist people and couples who wish to have children and to make informed decisions concerning their potential reproductive options, carrier testing may be a valuable tool. In addition, if both couples have the same gene mutation, they can prefer to use different biological approaches, such as in vitro fertilization with preimplantation genetic screening, to decrease the likelihood that the modification would be passed on to their offspring.

2.3.3 Prenatal Diagnostic Testing

It is used to detect chromosomal or genetic alterations in fetuses. This testing is available to couples more likely to become pregnant with a child with a genetic or chromosomal abnormality. Two approaches for acquiring a tissue sample for testing are amniocentesis and chorionic villus sampling. Prenatal testing is done to establish the pregnancy's genetic condition and detect any potential hazards for a genetic problem. Risk factors include ethnicity, maternal age, and family history. A few invasive prenatal alternatives include chorionic villus sampling, amniocentesis, and maternal serum screening.

Additionally, placental biopsies and periumbilical blood collection are more advanced, highly skilled methods. However, there may be dangers to the unborn child from prenatal testing; for this therapy, genetic counseling and informed consent are necessary (Carlson and Vora 2017). Prenatal testing is commonly done for screening reasons. These diagnostics, which target women with pregnancies at high risk of chromosomal abnormalities or congenital disabilities, include blood screening, carrier screening, and ultrasound. Although ultrasonography may identify specific problems, such as an open neural tube anomaly, serum screening mainly aims to aid women with high-risk kids. Different alternatives for serum screening have varying test criteria and employment time available (Anon 2016). These tests may help to determine the status of several diseases, including sickle cell disease, Down syndrome, and cystic fibrosis. Prenatal diagnostic testing has certain complications, such as a slight chance of miscarriage or other issues, which should be considered during the process. It is crucial to consider the ethical implications of genetic testing and, if necessary, seek counseling or support.

2.3.4 Predictive or Predisposition Test

This genetic testing may identify people at risk of illness before symptoms appear. These tests may be relevant if a person has a family history of a particular sickness and there is a treatment option that may postpone the development of the problem or lessen its severity. Predictive testing may identify mutations that raise an individual's likelihood of having diseases with a genetic basis, such as some cancers. Predictive testing establishes an asymptomatic genetic condition's genetic status and family history. Finding a particular disease mutation in a sick family member is necessary for predictive testing. Presymptomatic testing and predisposition testing are two types of predictive testing. Presymptomatic testing is often used to identify disorders such as Huntington's disease. If there is a family history of the genetic illness for which the faulty gene is known, a predictive test is necessary to prove the hereditary nature of this mutant gene. It also indicates if the problem may be handled successfully, safely, or both. It also demonstrates if genetic data may be leveraged to gain an edge. The employment of a growing arsenal of screening and preventative techniques may be guided by predictive genetic testing, which offers a great deal of promise for accurate risk assessment.

Nevertheless, depending on the level of risk, the precision of risk prediction, the solutions available to lower the risk, an individual's prior experience, and the requirements and understanding of family members, the usefulness of testing varies greatly. Also, when knowledge advances, new preventive measures are created, and prices shift, it is conceivable that the benefit of a specific predictive genetic test may alter over time. Due to the intricacy of these variables, discussion regarding testing must be carefully personalized to the testing set as well as the requirements and preferences of the individual (Evans et al. 2001). It is essential to remember that genetic testing is not always 100% accurate and can only provide probabilities of future health effects. Before undertaking a predictive or predisposition genetic test, it is essential to speak with a genetic counselor to understand the potential risks and benefits of testing fully.

2.3.5 Pharmacogenetic Testing

According to estimates made by Classen et al. in 1997 and Lazarou et al. in 1998, more than 770,000 hospital patients suffer harm or pass away due to adverse drug events (ADEs) each year, costing the healthcare system millions of dollars. One possibility for lowering healthcare expenses related to ADEs and inadequate response to medication is the area of genetic medicine (Classen et al. 1997; Lazarou et al. 1998). Pharmacogenetics, in particular, uses a patient's genetic profile in conjunction with other clinical data to develop a tailored pharmaceutical regimen that is more effective and secure for the specific patient. Pharmacogenetic information is available to help in proper dose or selection of many presently administered drugs. Pharmacogenetic assessments are also often carried out throughout medication development. It is likely that including genetic data from pharmacogenetic testing would enhance treatment decision-making by increasing effectiveness and minimizing adverse effects (Liou et al. 2012). One of the most promising clinical uses of personalized medicine is pharmacogenetic (PGx) testing, which has the potential to enhance treatment outcomes by decreasing adverse drug reactions and increasing the chance of response. PGx testing typically has fewer ethical and societal consequences than disease-based genetic tests (Roses 2000). Because pharmacogenetic testing is still a novel subject, it is essential to remember that there is ongoing research to better understand these tests' clinical efficacy and cost-effectiveness. As a result, pharmacogenetic testing in medical practice may differ depending on the patient and medical professional.

2.3.6 Ancestry Testing

Modern scientific developments have given this discussion a new context. Researchers may now determine genetic or biogeographic heritage using DNA testing. In conclusion, genetic testing enables the statistically precise determination and quantification of an individual's ancestry. This corpus of work's scientific foundation is the cornerstone of television drama and forensic inquiry. Applications of ancestry testing have just lately made their way into clinical settings and biological research (Nalls et al. 2008; Salari et al. 2005; Kao et al. 2008). Understanding that ancestral DNA testing is not a medical procedure and cannot determine individual risk for certain illnesses is crucial. It is a tool for family history research and learning more about our genetic ancestry. A significant step toward the biomedical research future we envision is the routine use of gene ancestry assessment. This involves forecasting overall (genetic and environmental) risk for common diseases, sequencing an individual's whole genome, social and ecological factors, tailored meals, lifestyle recommendations, and medication treatment.

2.3.7 Paternity Testing

These same factors were combined throughout the twentieth century to interpret fingerprint data in paternity testing, another forensic practice field. It is generally known that a sequence of scientific advancements in serology, human genetics, and later molecular biology led to a revolution in paternity testing techniques in the twentieth century (Patzelt 2004). Knowing whether a child is or is not the biological child of confident parents has significant implications. According to the historian Milanich (2019), contemporary concepts of fatherhood developed at the junction of several political, legal, social, and cultural challenges in the Americas and Europe (Milanich 2019). Moreover, by integrating experts from various academic backgrounds, paternity was established as a biological fact that could be studied using scientific methods. Several legal (and political) systems have distinct regulations regarding paternity testing techniques' practical use and legal acceptability. Yet, determining biological paternity is a topic of empirical study with implications for a wide range of social issues, including those connected to law, governance, and culture. This testing aims to determine if the child and the alleged father are biologically related. The child's DNA matches that of the alleged father, and the study's results may establish paternity beyond a shadow of a doubt. It is important to note that paternity genetic testing is often used in judicial cases involving disputed child support, custody, and inheritance difficulties. Those who are having testing should carefully consider the potential consequences since it may also have emotional effects.

2.3.8 Forensic Testing

Forensic genetics uses genetic technologies and scientific methods to resolve criminal and civil disputes (Bukyya et al. 2021). Biological evidence may include DNA or cellular material extracted from crime sites. Genetic techniques were expanded as technology developed to include human and nonhuman forensic investigations. While any genome may benefit from these techniques, human DNA typing has become the industry standard due to the extensive use of databases and established procedures. Since its introduction into the criminal justice system, forensic DNA testing was first utilized in the mid-1980s to help convict the guilty and exonerate the innocent (Gill et al. 1985). The remains of people who have gone missing or died are significant. By matching reference samples to retrieved remains, disasters have been reassociated and recognized (Clayton et al. 1995). New techniques are often offered and validated to improve the skills of laboratories trying to obtain DNA results with enhanced sensitivity and informativeness. Forensic laboratories have automated adequate planning and data gathering to meet growing throughput demands (Butler 2012, 2015). Forensic genetic testing, a vital tool in investigating crimes, has altered criminal investigations. Nonetheless, using these techniques ethically and responsibly is essential since they can affect people's lives and rights.

Genetic testing results can be used to confirm a diagnosis in a symptomatic individual or monitor the prognosis and response to treatment for a given disease. Tests are conducted to determine whether symptomatic individuals have a genetic condition that is known or suspected. In contrast to DNA testing, which is less costly and more reliable and causes less patient injury, the creatine kinase assay and muscle biopsy are equivocal in detecting myotonic dystrophy (McPherson 2006; Botkin 2016). Moreover, genetic testing may affect a person's family members, who may become more likely to experience certain conditions due to the test's results. The test's outcome might be a report detailing the particular genetic variations found and information about their potential effects.

2.4 Cytogenetic Testing

Chromosomes play an essential role in cell function, which is the focus of cytogenetics. Chromosomes are found in all tissues and cells. In biology and medicine, cytogenetics is widely used to understand chromosomes and their role in hereditary disorders. Pharmacogenomics-targeted cancer treatment and personalized medicine are examples of medical concepts that have been significantly influenced by molecular cytogenetics, a new branch of cytogenetics (Pickard et al. 2005). A schematic representing possible cytogenetic techniques is shown in Fig. 2.2.

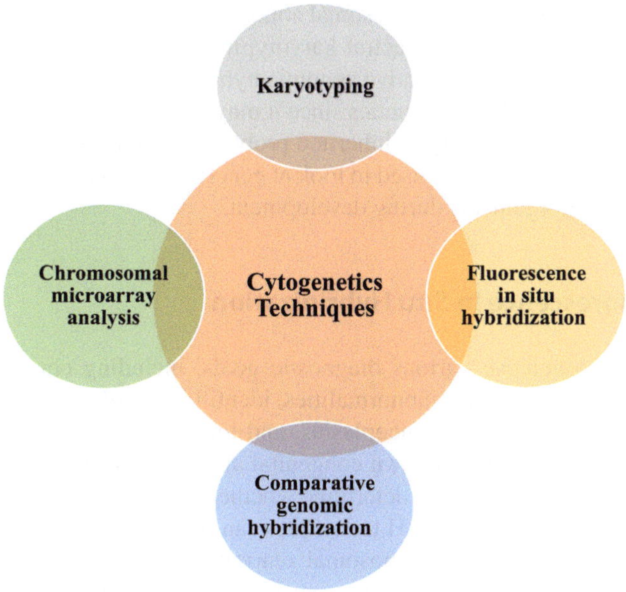

Fig. 2.2 Methods of cytogenetics techniques

2.4.1 Karyotyping

This method is used to identify chromosomal anomalies in babies that are numerical and structural to foretell the emergence of a genetic disorder that may impede growth and progress. There are different banding approaches, the most common of which is G-banding for cytogenetic research. While this strategy has been around for a long time, new techniques to give more accurate and relevant findings while simultaneously decreasing the time between diagnosis and treatment have emerged. Even though it is suggested that it be used with methods such as chromosomal microarrays (CMAs) to identify additional clinically significant cytogenetic information, such as aneuploidies and unbalanced rearrangements, it is still used in clinical practice today (Wapner et al. 2012a). Nevertheless, karyotyping has lower diagnostic rates for prenatal diagnosis than techniques like comparative genomic hybridization; diagnostic rates ranged from 2.5% to 4.2% in a study comparing different approaches for assessing chromosomal alteration in high-risk individuals (Saldarriaga et al. 2015). It is used to examine specific chromosomes (Rooney and Czepulkowski 1992). The karyotyping procedure is illustrated in Fig. 2.3.

Congenital abnormalities' newborn findings and prenatal diagnosis (Yu et al. 2013), chromosomal breakage, translocation, deletion, inversion, ring chromosomes, isochromosomes, mosaicism, and chimerism are structural abnormalities of rearrangements. Diagnose neuropsychiatric disorder (ReichalCp 2020) and mental retardation (Schaeffer et al. 2004). Fluorescent in situ hybridization may detect chromosomal abnormalities in nondividing cells and detect the rate of miscarriage (Cremer et al. 1986). Minor chromosomal abnormalities, such as telomeric chromatin translocation, are seen by spectral karyotyping but not by fluorescence in situ hybridization (FISH) or comparative genomic hybridization (CGH). Karyotyping is a crucial technique in clinical genetics since it may detect genetic abnormalities and provide data on the propensity of inherited problems to be passed on to future generations. In addition, it may be used to look at genetic variations and the interactions between different organisms during development.

2.4.2 Fluorescence In Situ Hybridization

FISH is used to achieve various diagnostic goals, including chromosomal gene mapping, detection of genetic abnormalities, identification of genetic abnormalities linked to hereditary disease or neoplasms, and location of viral genomes (Bishop 2010). FISH can only detect known congenital abnormalities by hybridizing a particular probe into the samples to determine whether or not that specific genetic aberration is present. Since most FISH techniques can only identify known imbalances, they cannot be utilized as chromosomal rearrangement screening assays. When locus-specific probes or chromosome-specific DNA libraries are employed, the analysis can only be performed on a particular chromosome or chromosomal subregion. Cytogenetics still makes extensive use of FISH. In the battle against cancer, genetic testing has also developed into a crucial diagnostic and research tool.

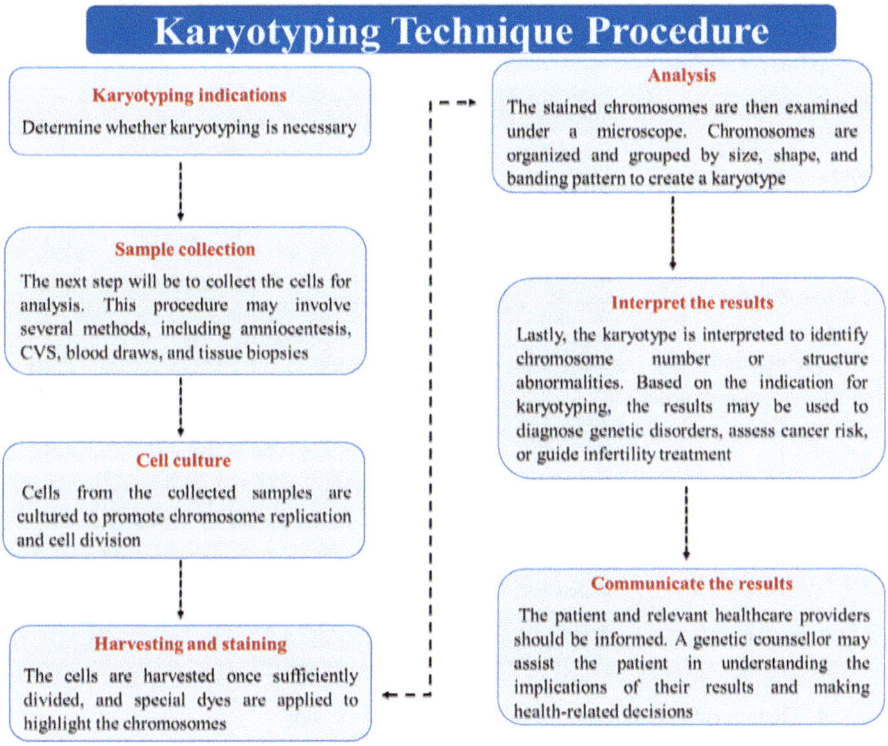

Fig. 2.3 Procedure of chromosome preparation from peripheral blood

Multiplex FISH and spectral karyotyping are research techniques. They might be utilized to look at changes in the intranuclear genomic organization and cytoplasmic RNA profiling (Lai et al. 2016). The FISH approach moved cell-based analysis from metaphases to interphases, allowing researchers to discover a broader spectrum of genetic disorders, from chromosomal abnormalities to submicroscopic copy number variations (CNVs). With this technology, the human genome may be physically mapped to determine gene and polymorphism positions within the chromosomes. Because of its high analytical resolution to a single-gene level, specificity, and susceptibility, it is also used for genetic diagnosis of relatively frequent aneuploidies such as microdeletion/microduplication syndromes and subtelomeric rearrangements (Cui et al. 2016; Ratan et al. 2017). The widely used FISH method has several uses in biology and medicine. It is an essential tool for genetic analysis and diagnosis because of its sensitivity and specificity in detecting and visualizing particular DNA sequences (Fig. 2.4).

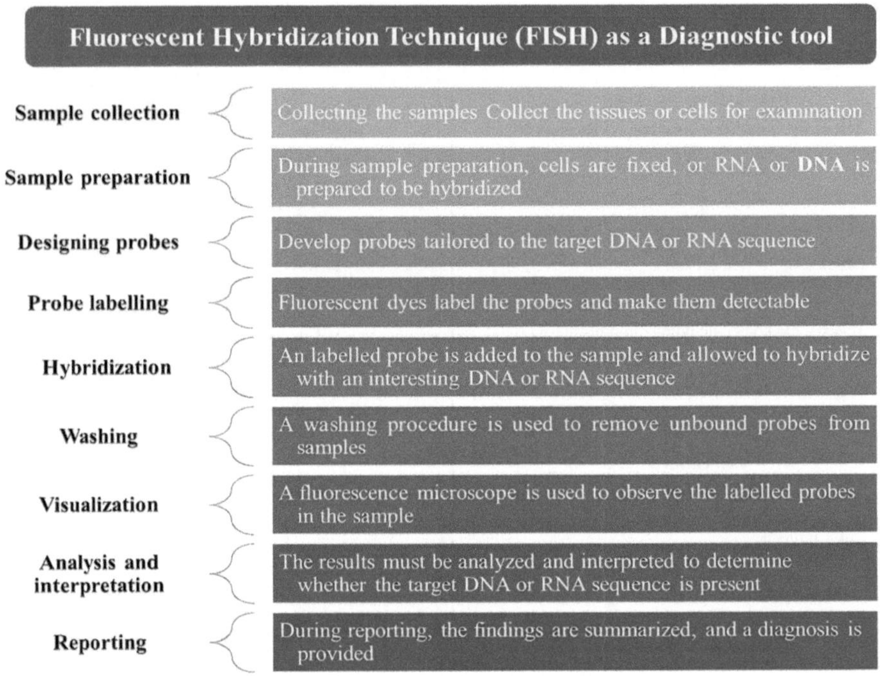

Fig. 2.4 Fluorescent hybridization technique as a diagnostic tool

2.4.3 Comparative Genomic Hybridization

The first successful approach was comparative genomic hybridization (CGH), which examines whole genomes for changes in DNA copy number (du Manoir et al. 1993; Kallioniemi et al. 1992). It has several applications for pediatric patients, adults, and prenatal diagnostics. Comparative genomic hybridization is a molecular method used to analyze chromosomes. CGH can detect microdeletions and microduplications with a base pair count of more than 500, but karyotypes cannot. CGH should be the primary diagnostic test for adults with neurological diseases, autism, cognitive impairments, and neonates with congenital abnormalities of unknown cause, according to a 2010 consensus statement and economic assessment. This would be used in place of karyotyping. However, prenatal diagnosis often relies on aCGH despite its high cost. The additional costs are typically covered by health insurance and nations or governments that permit abortion (Katsanis and Katsanis 2013). Array CGH is one of several top-down techniques that can provide complete information on biological features or disorders. These methods may soon provide foundational knowledge about essential characteristics of genome structure and correlational information that is important for therapeutic applications in cancer and medical genetics. Clinical applications of medical genetics are incredibly appealing and are anticipated to become widely used soon. Much care must be taken to preserve acceptable levels of false-positive abnormality signals and to guarantee that

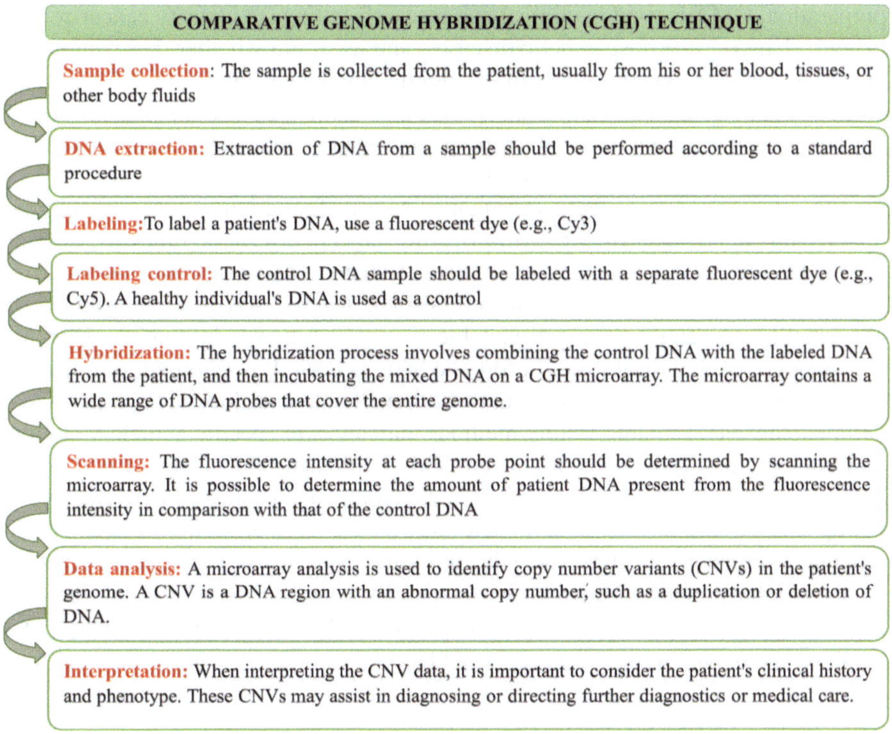

Fig. 2.5 Flowchart for comparative genomic hybridization (CGH) diagnosis

the detected aberrations are rigorously assessed. With this information, it is possible to investigate how the genomes of diverse species have evolved through time and find genetic abnormalities connected to diseases and pathologies (Fig. 2.5).

2.4.4 Chromosomal Microarray Analysis (CMA)

There is a promise for increased diagnostic accuracy with the advent of more contemporary molecular cytogenetic approaches, such as chromosomal microarray analysis (CMA). CMA easily outperforms traditional karyotyping, which can only identify imbalances bigger than 7–10 million bases due to its ability to detect imbalances in the kilobase range. In postnatal investigations of babies with congenital abnormalities, developmental delays, or intellectual impairment, CMA will have an extra diagnostic yield of 12–15% of clinically critical subchromosomal abnormalities (Manning et al. 2010; Miller et al. 2010). In 2013, the publication of a large, multicenter NICHD-funded study by Wapner and associates demonstrated the therapeutic benefit of CMA in prenatal diagnosis (Wapner et al. 2012b). CMA offers additional diagnostic benefits by identifying copy number variations (CNVs) or submicroscopic imbalances undetectable on a regular G-banded chromosomal preparation. CMA is used in clinical and research settings to identify cancer-related genetic changes, diagnose genetic diseases, and ascertain whether these anomalies are the underlying genetic causes of developmental delays, intellectual impairment, autism spectrum disorder, and other problems. In the medical genomics sector, it has become an essential tool that supports improved genetic counseling for patients and their families, as well as diagnosis and treatment (Fig. 2.6).

The outcomes of cytogenetic testing may be used for various purposes, including identifying genetic abnormalities, discovering carriers of genetic illnesses, estimating the likelihood of a genetic problem being transmitted to offspring, and assessing the efficacy of cancer therapies. It may also be used to screen for chromosomal abnormalities in fetuses during prenatal testing, which may help find hereditary illnesses or congenital disabilities before the baby is born. It is essential to remember that although cytogenetic testing may be beneficial, it cannot always provide a definitive answer. Existing testing methods may uncover specific genetic mutations that are too small to be detected or may only be present in a subset of body cells. Typical signs or health problems could not accompany specific genetic changes. It is important to discuss the benefits and drawbacks of cytogenetic testing with the clinician, similar to what we do with any medical test. New cytogenetic testing techniques have made it much easier to identify genetic defects and provide individualized treatment plans. Much more advanced techniques should emerge as a result of future technological advancements.

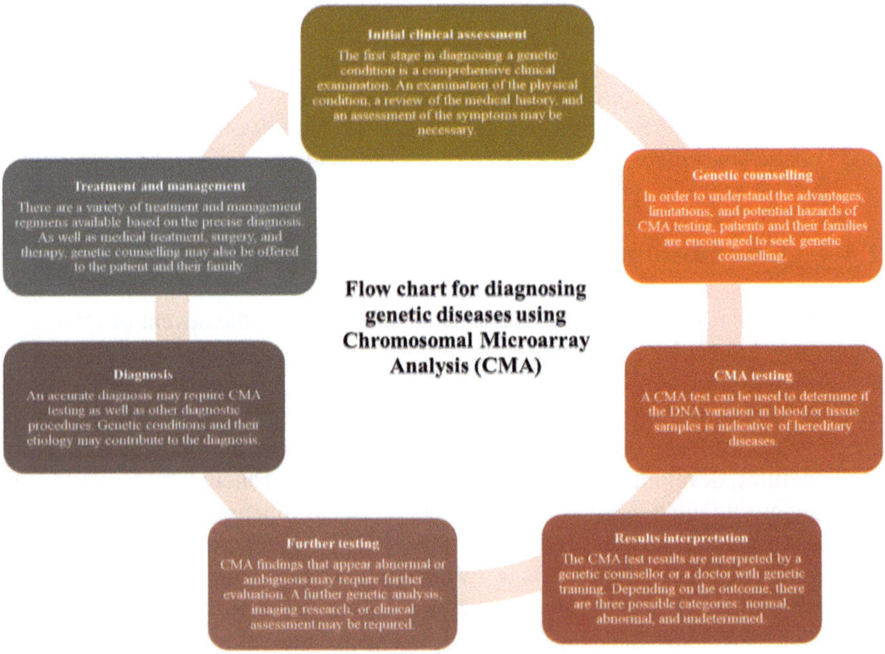

Fig. 2.6 Flowchart for diagnosing genetic diseases using CMA

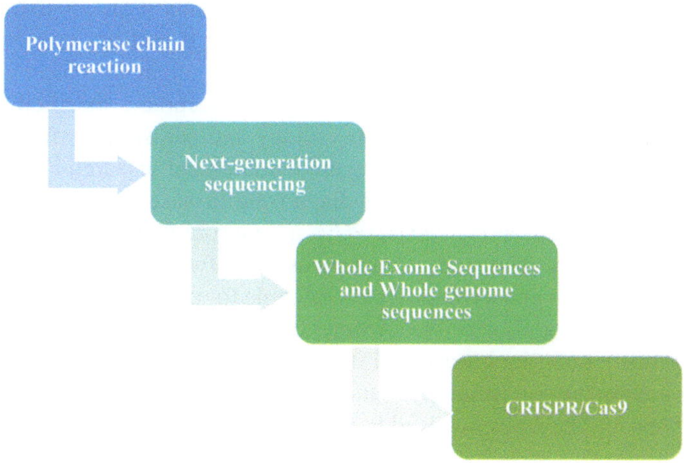

Fig. 2.7 Methods of molecular techniques

2.5 Molecular Testing

There has been an increase in the sophistication of molecular diagnostic tests performed on symptomatic individuals. Up until recently, it used to be that such tests could only be performed on a few loci. Large-insert clone and oligonucleotide arrays, which allow higher resolutions for querying patients' whole genomes and enable the identification of medium- to extensive genetic alterations, have altered the landscape. This can now be done with single-nucleotide accuracy utilizing whole-exome and whole-genome sequencing (WGS) (Pasche and Absher 2011). In addition to its niche applications for rare disorders, genetic testing has become a tool for personal use and complex diseases (Sequeiros et al. 2012; Kiezun et al. 2012). A graphic representing potential molecular techniques is shown in Fig. 2.7.

2.5.1 Polymerase Chain Reaction (PCR)

The discovery of the PCR enabled scientific advances, including genetic sequence analysis and expression levels in recombinant systems, molecular genetics research, fast paternity determination, and infectious disease diagnostics (Novais et al. 2004a; Speers et al. 2003). With in vitro nucleic acid synthesis made possible by PCR, a DNA fragment may be selectively reproduced semiconservatively. Typically, it demonstrates excellent detection limitations (Cortelli et al. 2003; Spolidorio and Spolidorio 2005). Real-time PCR, a technological advancement that may provide quantitative data, has grown significantly in clinical diagnostics and research facilities. In contrast to traditional PCR, which only shows the qualitative findings, this method accompanies the reaction and presentation of data quicker and more precisely (Kubista et al. 2006; Morillo et al. 2003; Novais et al. 2004b). Molecular biology-related techniques have significantly advanced and are now valuable in many medical research areas. Identification of microorganisms necessitates the development of PCR, and this enhances diagnostic accuracy, specificity, and efficiency. The use of PCR technology in mycology and parasitology increases early microorganism identification, enhancing epidemiological research and fungal and parasite diagnosis, which is critical for diagnosing diseases. By performing diagnostic tests to find microorganisms, PCR in microbiology enables significant, instantaneous findings. The study of microbes like *Mycobacterium tuberculosis* is made possible through genotyping. This is valuable for public health since it encourages early detection and effective treatment. The invention of the PCR method marked a turning point in science and medicine. In dentistry, it is now a standard diagnostic and research technology that enables the early detection of disorders, as mentioned above. The steps involved in PCR are illustrated in Fig. 2.8.

2.5.2 Next-Generation Sequencing

The advancement of NGS technology has helped several life sciences sectors, including functional genomics, transcriptomics, cancer, evolutionary biology, forensic sciences, and medicine. Another advantage of NGS technologies is their capacity to identify genetic changes in patients accurately and inexpensively. This has the potential to transform medicine radically. NGS may generate 100 times more data than the most advanced Sanger technique capillary sequencers (Pareek et al. 2011). With NGS technology, the entire genome of interest can be sequenced simultaneously, which is significantly more scalable than earlier sequencing procedures. This is accomplished by segmenting the genome, which is accomplished by randomly selecting a piece and then sequencing it using one of many techniques. The widespread use of sequencing in clinical labs has dramatically improved the genomic diagnosis of many inherited illnesses. Various molecular tests are available depending on the genetic condition, including single-gene testing, exome or genome sequencing, and NGS panels (Di Resta et al. 2018). However, the effectiveness of genetic testing and the difficulty of choosing the proper genetic test have substantial limits. As a result, combining various laboratory methods to obtain reliable results is strongly advised.

Rapid clinical testing based on NGS technology is fundamentally altering clinical diagnosis. It is critical to emphasize that comparative genomic hybridization and other auxiliary methods often complement NGS-based gene panel testing, resulting in a comprehensive and valuable approach for various illnesses (Xue et al. 2015). Single tests are often used because they are suited for disorders with recognized clinical symptoms and limited site heterogeneity, even though these data are constantly changing due to the fast development of innovative genetic testing technology. As a result, doctors must make tough decisions about which diagnostic instrument or technique is appropriate for their patients suffering from a range of genetic disorders. Twelve human exomes were collected and massively parallel

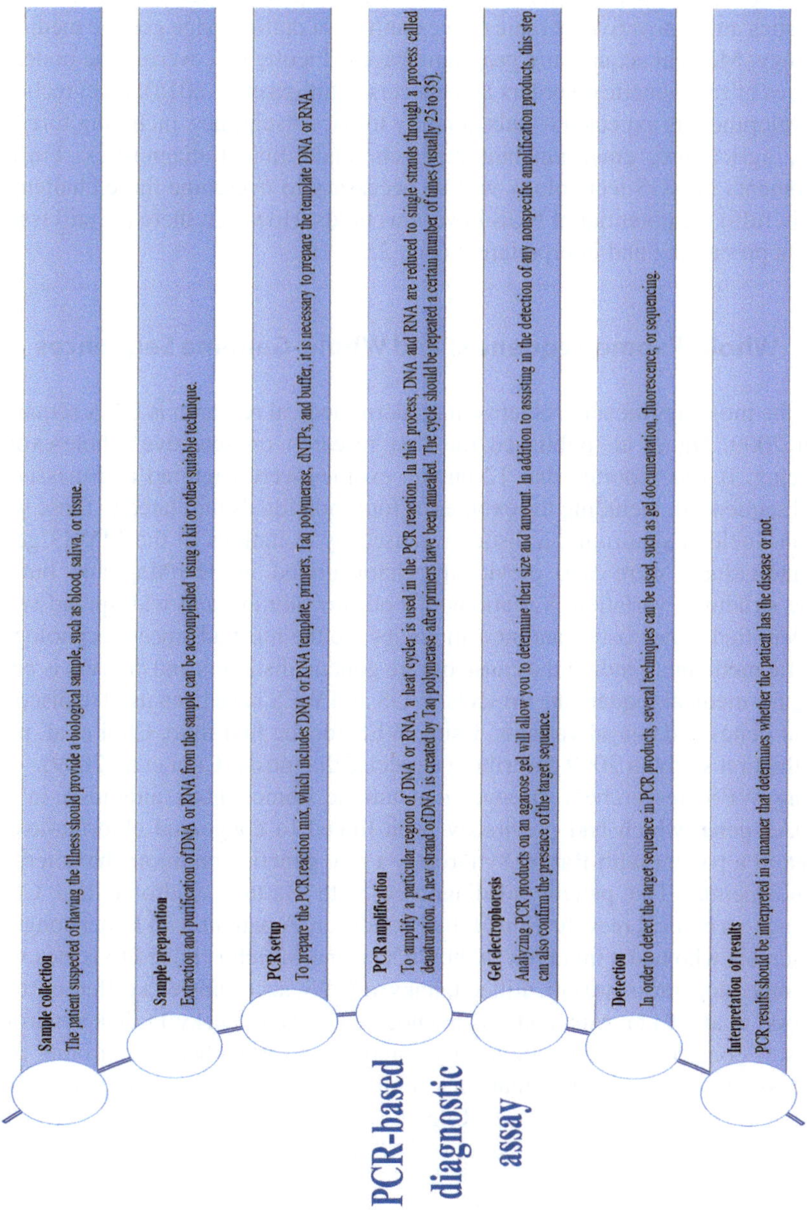

Fig. 2.8 Basic flowchart for a PCR-based diagnostic assay

sequenced, including the exomes of four people unrelated to those with Freeman-Sheldon syndrome, a disease caused by mutations in the MYH3 gene. Previously, HapMap and human genome structural variation were utilized to characterize eight people. Variations exist in both uncommon and common types (Ng et al. 2009). Diagnostics and innovative therapies are examples of cutting-edge genetic medicine technology. Medical sequencing programs have difficulty discovering the underlying causes of rare genetic disorders and cancers (Phillips et al. 2018). NGS technology development is expected to substantially impact disciplines, including forensic science, agriculture, environmental research, and clinical diagnostics. Further developments in NGS technology will be necessary to overcome these challenges and fully fulfil the potential of NGS in various fields. However, there are still issues with data processing and interpretation (Fig. 2.9).

2.5.3 Whole-Exome Sequences and Whole-Genome Sequences

One of the most significant problems in modern medical research is DNA sequencing. In 2009, Ng et al. published the first research on selective whole-exome sequencing. It was reported that 12 human exomes were captured and massively parallel sequenced, including the exomes of four individuals unrelated to those with Freeman-Sheldon syndrome, a disorder caused by mutations in the MYH3 gene. Eight individuals were also previously characterized by HapMap and human genome structural variation. Variations of both uncommon and widespread types were identified. Diagnostics and new therapies are the most advanced technologies in genetic medicine. Finding the cause of rare genetic disorders and cancers is challenging for medical sequencing projects. WGS and WES have been used to discover different genes and causal variants. Using WES for the first time, Choi et al. published the first study in 2009 describing a precise diagnosis (Choi et al. 2009).

Using WES, researchers discovered a unique homozygous alteration in the SLC26A3 gene, which had previously been linked to congenital chloride-losing diarrhea in a patient with Bartter syndrome, a rare genetic condition characterized by hypokalemia. The patient misdiagnosed with Bartter syndrome had CLD, according to a clinical reevaluation of their condition. A patient with Leber congenital amaurosis who had a mutation in the PEX1 gene, which is linked to issues with peroxisome biogenesis, was identified thanks to WES, according to a different study (Majewski et al. 2011). A patient misdiagnosed with Hermansky-Pudlak syndrome type 2 was considered for homozygosity mapping and WES because of other phenotypic symptoms like oculocutaneous albinism (OCA) and neutropenia. After WES, two disease-causing genes, SLC45A2 (related to OCA) and G6PC3 (related to neutropenia), were found. The associated AP3B1 gene was not mutated (Cullinane et al. 2011).

WES makes prenatal diagnosis, disease screening, and treatment more accessible. Once exome sequencing identified a mutation in the X-linked inhibitor of the apoptosis gene, a 15-month-old immunodeficient kid was diagnosed with Crohn's disease. A hematopoietic progenitor cell transplant effectively treated the patient

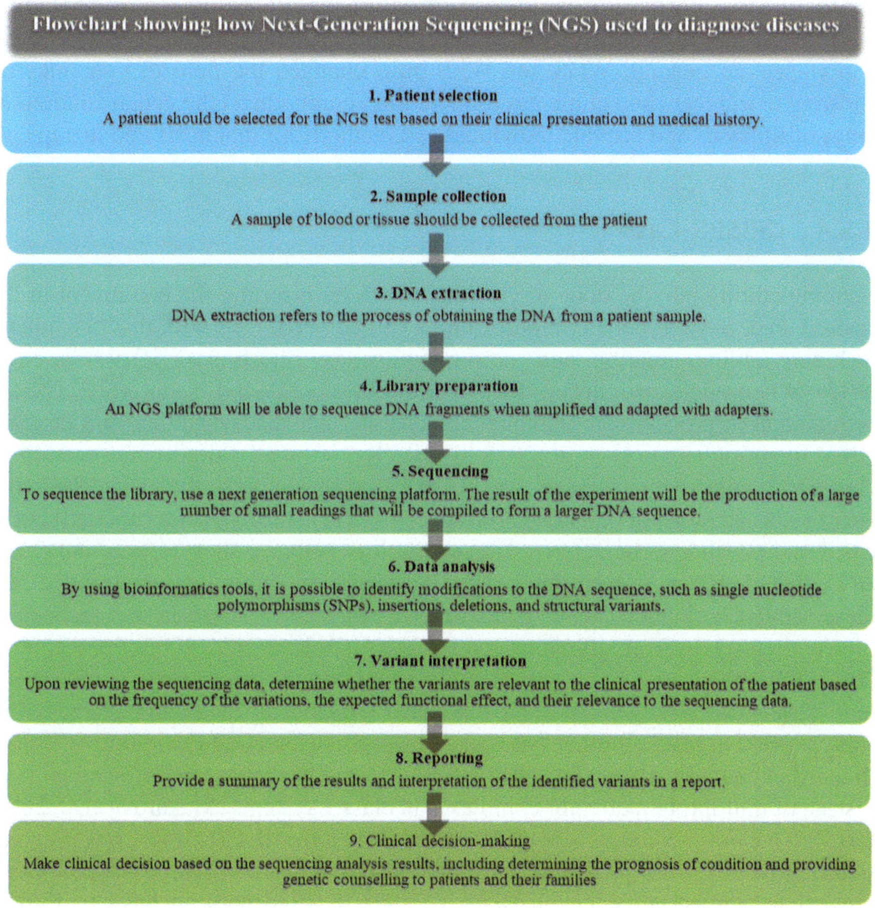

Fig. 2.9 Flowchart showing how next-generation sequencing can be used to diagnose diseases

(Worthey et al. 2011). In individuals with mutations in the ABCC8 or KCNJ11 genes, insulin is replaced with oral sulfonylurea. The utility of WES for neonatal diabetes mellitus (NDM) screening has been examined (Bonnefond et al. 2010). In their study, Bell et al. employed NGS to do carrier testing for 448 severe recessive pediatric illnesses. The use of fetal DNA in maternal blood during prenatal diagnosis revealed the efficacy of WES in noninvasively identifying aneuploidies (Chiu et al. 2008; Chen et al. 2011; Sehnert et al. 2011). This screening approach for carriers is less expensive than treating and caring for unwell newborns (Bell et al. 2011). The preclinical use of WES was identified by looking at gene abnormalities generating disorders with similar symptoms. Jiao and colleagues evaluated ten non-familial pancreatic neuroendocrine tumors using WES. The prognosis was based on the prevalent gene mutations (Jiao et al. 2011). WGS and WES include sequencing a person's DNA from their cells, which may be done using different materials such as blood, saliva, or tissue. Bioinformatics tools may examine the output data to find genetic variations, annotate their functional effects, and determine their clinical importance. In general, WGS and WES have changed the field of genomics by allowing researchers and clinicians to more fully comprehend the genetic underpinnings of illnesses and provide individualized methods for diagnosis and therapy.

2.5.4 CRISPR-Cas9

Genome editing may be done in vivo or in vitro by inserting the equipment in the desired area. Several genetic modifications can be performed with this machinery, including adding, ablation, and "correcting" genes (Ghosh et al. 2019; Gaj et al. 2013). When nuclease-induced double-strand breaks (DSBs) occur, very efficient recombination processes in mammalian cells are activated, resulting in a targeted DNA alteration (Kosicki et al. 2018). CRISPRs, or small palindromic repetitions clustered regularly interspersed, were discovered in *E. coli* in the early 1980s and later in a broad range of other bacterial species (Li et al. 2020). It was not until 2005, when several studies characterized the similarities between short repeats and phage DNA that their function became clear. A subsequent investigation revealed that these sequences induced RNA-guided DNA cleavage by bacteria and archaea to protect themselves from offending foreign DNA (Bolotin et al. 2005; Pourcel et al. 2005). CRISPR-Cas systems are often classified into two kinds based on the structural variety and arrangement of the Cas genes (Jinek et al. 2012). Six CRISPR-Cas subtypes and at least 29 other types have been discovered, and the number is continually expanding. Class 2 CRISPR-Cas systems have just one effector protein instead of multiprotein effector complexes in class 1 systems (Makarova et al. 2011, 2015). The most common CRISPR systems are type II CRISPR-Cas9 systems, which utilize the Cas9 protein from *Streptococcus pyogenes* to target specific DNA sequences (SpCas9) (Jiang et al. 2013). CRISPR-Cas9 technology has transformed genetics and is used in various fields, such as gene therapy, agriculture, and drug development. Yet there are also safety and ethical issues with the technology, which researchers and decision-makers are aggressively tackling (Fig. 2.10).

2.6 Future Prospectives and Conclusion

RDs are a significant public health issue that must be addressed. Figuring out the source of the illness, identifying brand-new therapeutic targets and diagnostic biomarkers, and generating tailored treatment programs are required for each patient with RDs. The internationalization of RD-related R&D has advanced significantly in recent years, and several chances to expand on productive projects, programs, and partnerships have emerged. Because of groundbreaking fundamental research into the disease process, we now have a greater understanding of the pathophysiology of many RDs, which has also prompted the development of helpful orphan medications, healthcare improvements, and treatment techniques. There are various

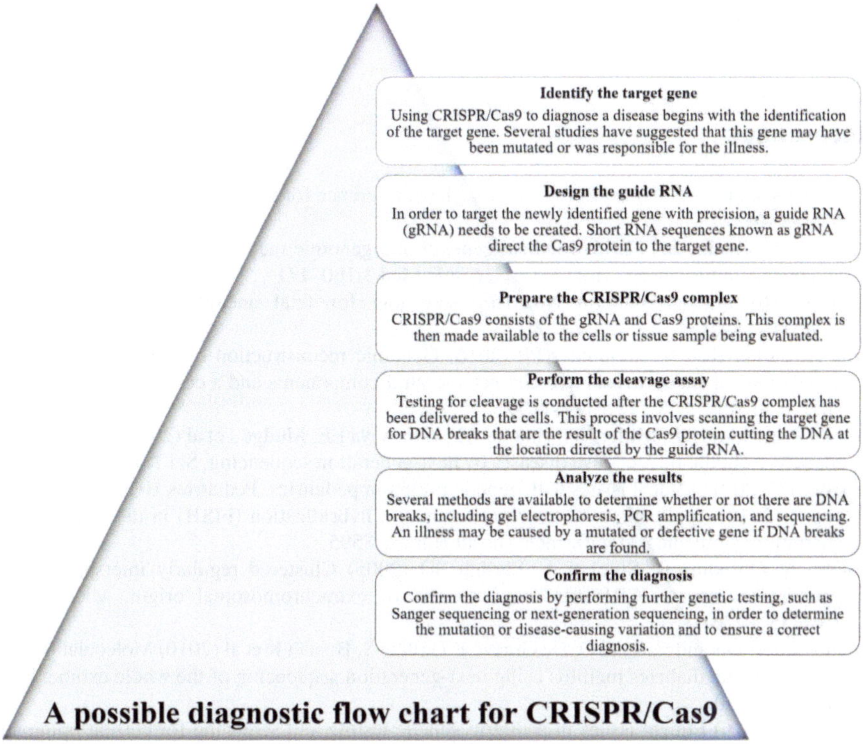

Fig. 2.10 Possible diagnostic flowchart for CRISPR-Cas9

barriers to RD research and healthcare, and more has to be done to support appropriate RD-related R&D and legislative activities within individual nations so that all patients worldwide have equitable access to treatment alternatives. Although the internet is the primary resource for patients and their families seeking information on support and research organizations, certain nations still need national RD support websites. A wide range of clinical information sources and infrastructure must be updated often to meet the clinical problems related to RDs. For medical students to understand the concepts of genetics and how they apply to human health, the academic program must contain a particular clinical genetics course. Considering the limitations of WES, combining WES with WGS and other "omics" platforms is necessary to establish the complicated molecular etiology of RDs. Novel techniques that may sustain and accelerate the pace of clinical and genetic developments while also supplying information for future therapeutic breakthroughs should be developed in parallel. In summary, genetic testing has the potential to revolutionize healthcare in the future. Genetic testing may enhance patient outcomes and prevent illness by identifying genetic abnormalities and creating individualized treatment options. Privacy and discrimination issues must be addressed to ensure that genetic testing is applied ethically.

Acknowledgements The authors are grateful to the Chettinad Academy of Research Education for its continued support and encouragement.

References

1000 Genomes Project Consortium (2015) A global reference for human genetic variation. Nature 526:68–74

Aggarwal S, Phadke SR (2015) Medical genetics and genomic medicine in India: current status and opportunities ahead. Mol Genet Genomic Med 3:160–171

Anon (2016) Practice bulletin No. 163: screening for fetal aneuploidy. Obstet Gynecol 127(5):e123–e137

Basu A, Sarkar-Roy N, Majumder PP (2016) Genomic reconstruction of the history of extant populations of India reveals five distinct ancestral components and a complex structure. Proc Natl Acad Sci U S A 113:1594–1599

Bell CJ, Dinwiddie DL, Miller NA, Hateley SL, Ganusova EE, Mudge J et al (2011) Carrier testing for severe childhood recessive diseases by next-generation sequencing. Sci Transl Med 3:65ra4

Bioethics C (2001) Ethical issues with genetic testing in pediatrics. Pediatrics 107:14511455

Bishop R (2010) Applications of fluorescence in situ hybridization (FISH) in detecting genetic aberrations of medical significance. Biosci Horiz 3:8595

Bolotin A, Quinquis B, Sorokin A, Ehrlich SD (2005) Clustered regularly interspaced short palindrome repeats (CRISPRs) have spacers of extrachromosomal origin. Microbiology 151:2551–2561

Bonnefond A, Durand E, Sand O, De Graeve F, Gallina S, Busiah K et al (2010) Molecular diagnosis of neonatal diabetes mellitus using next-generation sequencing of the whole exome. PLoS One 5:e13630

Botkin JR (2016) Ethical issues in pediatric genetic testing and screening for current opinion in pediatrics. Curr Opin Pediatr 28:700704

Bukyya JL, Tejasvi MLA, Avinash A, Chanchala HP, Talwade P, Afroz MM, Pokala A, Neela PK, Shyamilee TK, Srisha V (2021) DNA profiling in forensic science: a review. Glob Med Genet 8(4):135–143

Butler JM (2012) Advanced topics in forensic DNA typing: methodology. Elsevier Academic Press, San Diego, CA

Butler JM (2015) Advanced topics in forensic DNA typing: interpretation. Elsevier Academic Press, San Diego, CA

Carlson LM, Vora NL (2017) Prenatal diagnosis: screening and diagnostic tools. Obstetr Gynecol Clin 44(2):245–256

Chen EZ, Chiu RW, Sun H, Akolekar R, Chan KC, Leung TY et al (2011) Noninvasive prenatal diagnosis of fetal trisomy 18 and trisomy 13 by maternal plasma DNA sequencing. PLoS One 6:e21791

Chiu RW, Chan KC, Gao Y, Lau VY, Zheng W, Leung TY et al (2008) Noninvasive prenatal diagnosis of fetal chromosomal aneuploidy by massively parallel genomic DNA sequencing in maternal plasma. Proc Natl Acad Sci U S A 105:20458–20463

Choi M, Scholl UI, Ji W, Liu T, Tikhonova IR, Zumbo P et al (2009) Genetic diagnosis by whole exome capture and massively parallel DNA sequencing. Proc Natl Acad Sci U S A 106:19096–19101

Classen D, Pestotnik S, Evans R, Lloyd J, Burke J (1997) Adverse drug events in hospitalized patients. excess length of stay, extra costs, and attributable mortality. JAMA 277:301–306

Clayton TM, Whitaker JP, Maguire CN (1995) Identification of bodies from the scene of a mass disaster using DNA amplification of short tandem repeat (STR) loci. Forensic Sci Int 76:7–15

CLSI (2013) Blood collection on filter paper for newborn screening programs; approved standard, CLSI Document NBS01-A6, vol 29, No. 25, 6th edn. Clinical and Laboratory Standards Institute, Wayne, PA

Cortelli SC, Jorge AOC, Querido SMR, Cortelli JR (2003) PCR e cultura na detecção subgengival de Actinobacillus actinomycetemcomitans: estudo comparativo. Cien Odontol Bras 6(2):58–64
Cremer T, Landegent J, Brückner A, Scholl HP, Schardin M, Hager HD et al (1986) Detection of chromosome aberrations in the human interphase nucleus by visualization of specific target DNAs with radioactive and non-radioactive in situ hybridization techniques: diagnosis of trisomy 18 with probe L1.84. Hum Genet 74:346–352
Cui C, Shu W, Li P (2016) Fluorescence in situ hybridization: cell-based genetic diagnostic and research applications. Front Cell Dev Biol 4:89
Cullinane AR, Vilboux T, O'Brien K, Curry JA, Maynard DM, Carlson-Donohoe H et al (2011) Homozygosity mapping and whole-exome sequencing to detect SLC45A2 and G6PC3 mutations in a single patient with oculocutaneous albinism and neutropenia. J Invest Dermatol 131:2017–2025
Danese E, Lippi G (2018) Rare diseases: the paradox of an emerging challenge. Ann Transl Med 6:329
Di Resta C, Galbiati S, Carrera P, Ferrari M (2018) Next-generation sequencing approach for the diagnosis of human diseases: open challenges and new opportunities. EJIFCC 29:414
du Manoir S, Speicher MR, Joos S, Schrock E, Popp S et al (1993) Detection of complete and partial chromosome gains and losses by comparative genomic in situ hybridization. Hum Genet 90:590–610
Evans JP, Skrzynia C, Burke W (2001) The complexities of predictive genetic testing. BMJ 322(7293):1052–1056
Gaj T, Gersbach CA, Barbas CF (2013) ZFN, TALEN, and CRISPR/Cas-based methods for genome engineering. Trends Biotechnol 31:397–405
Ghosh D, Venkataramani P, Nandi S, Bhattacharjee S (2019) CRISPR-Cas9 a boon or bane: the bumpy road ahead to cancer therapeutics. Cancer Cell Int 19:12
Gill P, Jeffreys AJ, Werrett DJ (1985) Forensic application of DNA 'fingerprints'. Nature 318:577–579
Gurian EA, Kinnamon DD, Henry JJ, Waisbren SE (2006) Expanded newborn screening for biochemical disorders: the effect of a false-positive result. Pediatrics 117:1915–1921
Hewlett J, Waisbren SE (2006) A review of the psychosocial effects of false-positive results on parents and current communication practices in newborn screening. J Inherit Metab Dis 29:677–682
Jiang W, Bikard D, Cox D, Zhang F, Marraffini LA (2013) RNA-guided editing of bacterial genomes using CRISPR-Cas systems. Nat Biotechnol 31:233–239
Jiao Y, Shi C, Edil BH, de Wilde RF, Klimstra DS, Maitra A et al (2011) DAXX/ATRX, MEN1, and mTOR pathway genes are frequently altered in pancreatic neuroendocrine tumors. Science 331:1199–1203
Jinek M et al (2012) A programmable dual-RNA-guided DNA endonuclease in adaptive bacterial immunity. Science 337:816–821
Kallioniemi A, Kallioniemi OP, Sudar D, Rutovitz D, Gray JW et al (1992) Comparative genomic hybridization for molecular cytogenetic analysis of solid tumors. Science 258:818–821
Kao WH, Klag MJ, Meoni LA et al (2008) MYH9 is associated with nondiabetic end-stage renal disease in African Americans. Nat Genet 40(10):1185–1192
Karczewski KJ et al (2019) Variation across 141,456 human exomes and genomes reveals the spectrum of loss-of-function intolerance across human protein-coding genes. bioRxiv. https://doi.org/10.1101/531210
Kasthuri A (2018) Challenges to healthcare in India—the five A's. Indian J Community Med 43:141–143
Katsanis SH, Katsanis N (2013) Molecular genetic testing and the future of clinical genomics. Nat Rev Genet 14:415426
Kiezun A et al (2012) Exome sequencing and the genetic basis of complex traits. Nature Genet 44:623–630
Kosicki M, Tomberg K, Bradley A (2018) Repair of double-strand breaks induced by CRISPR-Cas9 leads to large deletions and complex rearrangements. Nat Biotechnol 36:765–771

Kubista M, Andrade JM, Bengtsson M, Forootan A, Jonák J, Lind K, Sindelka R, Sjoback R, Sjogreen B, Strombom L, Stahlberg A, Zoric N (2006) The real-time polymerase chain reaction. Mol Asp Med 27(2–3):95–115

Kuchay RAH, Mir YR, Zeng X, Hassan A, Musarrat J, Parwez I et al (2019) ARSACS as a worldwide disease: novel SACS mutations identified in a consanguineous family from the remote tribal Jammu and Kashmir Region in India. Cerebellum 18:807–812

Lai L-T, Meng Z, Shao F, Zhang L-F (2016) Simultaneous RNA-DNA FISH. Methods Mol Biol 1402:135–145

Lazarou J, Pomeranz B, Corey P (1998) Incidence of adverse drug reactions in hospitalized patients: a meta-analysis of prospective studies. JAMA 279:1200–1205

Lek M et al (2016) Analysis of protein-coding genetic variation in 60,706 humans. Nature 536:285–291

Li H, Yang Y, Hong W, Huang M, Wu M, Zhao X (2020) Applications of genome editing technology in the targeted therapy of human diseases: mechanisms, advances and prospects. Signal Transduct Target Ther 5(1):1

Liou S, Stringer F, Hirayama M (2012) The impact of pharmacogenomics research on drug development. Drug Metab Pharmacokinet 27:2–8

Majewski J, Wang Z, Lopez I, Al Humaid S, Ren H, Racine J et al (2011) A new ocular phenotype associated with an unexpected but known systemic disorder and mutation: novel use of genomic diagnostics and exome sequencing. J Med Genet 48:593–596

Makarova KS et al (2011) Evolution and classification of the CRISPR-Cas systems. Nat Rev Microbiol 9:467–477

Makarova KS et al (2015) An updated evolutionary classification of CRISPR-Cas systems. Nat Rev Microbiol 13:722–736

Manning M, Hudgins L, Professional P, Guidelines C (2010) Array-based technology and recommendations for utilization in medical genetics practice for detection of chromosomal abnormalities. Genet Med 12:742–745

McPherson E (2006) Genetic diagnosis and testing in clinical practice. Clin Med Res 4:123129

Milanich NB (2019) Paternity: the elusive quest for the father. Harvard University Press

Miller DT, Adam MP, Aradhya S, Biesecker LG, Brothman AR, Carter NP et al (2010) Consensus statement: chromosomal microarray is a first-tier clinical diagnostic test for individuals with developmental disabilities or congenital anomalies. Am J Hum Genet 86:749–764

Morillo JM, Lau L, Sanz M, Herrera D, Silva A (2003) Quantitative real time PCR based on single copy gene sequence for detection of Actinobacillus actinomycetemcomitans and Porphyromonas gingivalis. J Periodontal Res 38(5):518–524

Morrison DR, Clayton EW (2011) False positive newborn screening results are not always benign. Public Health Genom 14:173–177

Nalls MA, Wilson JG, Patterson NJ et al (2008) Admixture mapping of white cell count: genetic locus responsible for lower white blood cell count in the Health ABC and Jackson Heart studies. Am J Hum Genet 82(1):81–87

Ng SB et al (2009) Targeted capture and massively parallel sequencing of 12 human exomes. Nature 461:272–276

Novais CM, Pires-Alves M, Silva FF (2004a) PCR em tempo real. Rev Biotecnol Cienc Des 33:10–14

Novais CM, Pires-Alves M, Silva FF (2004b) PCR em tempo real. Rev Biotecnol Cienc Des 33:10–13

Pagon RA, Hanson NB, Neufeld-Kaiser W, Covington ML (2001) Genetic testing. West J Med 174:344347

Pareek CS, Smoczynski R, Tretyn A (2011) Sequencing technologies and genome sequencing. J Appl Genet 52:413435

Pasche B, Absher D (2011) Whole-genome sequencing: a step closer to personalized medicine. JAMA 305:1596–1597

Patzelt D (2004) History of forensic serology and molecular genetics in the sphere of activity of the German Society for Forensic Medicine. Forensic Sci Int 144:185–191

Phillips KA, Deverka PA, Hooker GW, Douglas MP (2018) Genetic test availability and spending: where are we now? Where are we going? Health Aff (Millwood) 37:710716
Pickard BS, Millar JK, Porteous DJ, Muir WJ, Blackwood DHR (2005) Cytogenetics and gene discovery in psychiatric disorders. Pharmacogenomics J 5:81
Pourcel C, Salvignol G, Vergnaud G (2005) CRISPR elements in Yersinia pestis acquire new repeats by preferential uptake of bacteriophage DNA, and provide additional tools for evolutionary studies. Microbiology 151:653–663
Pradhan S, Sengupta M, Dutta A, Bhattacharyya K, Bag SK, Dutta C et al (2011) Indian genetic disease database. Nucleic Acids Res 39:D933–D938
Rai E, Mahajan A, Kumar P, Angural A, Dhar MK, Razdan S et al (2016) Whole exome screening identifies novel and recurrent WISP3 mutations causing progressive Pseudorheumatoid Dysplasia in Jammu and Kashmir. India Sci Rep 6:27684
Ratan ZA, Zaman SB, Mehta V, Haidere MF, Runa NJ, Akter N (2017) Application of fluorescence in situ hybridization (FISH) technique for the detection of genetic aberration in medical science. Cureus 9:e1325
ReichalCp RS (2020) Occurrence of behavioral changes and its management in persons with mental illness due to lunar effects. Drug Invention Today. competitive strategy model and its impact on micro business unit of local development banksin jawa pjaee, 17 (7) (2020) 642
Richter T, Nestler-Parr S, Babela R, Khan ZM, Tesoro T, Molsen E et al (2015) Rare disease terminology and definitions-a systematic global review: report of the ISPOR rare disease special interest group. Value Health 18:906–914
Rooney DE, Czepulkowski BH (1992) Human cytogenetics: malignancy and acquired abnormalities. IRL Press, Oxford
Roses AD (2000) Pharmacogenetics and the practice of medicine. Nature 405:857–865
Salari K, Choudhry S, Tang H et al (2005) Genetic admixture and asthma-related phenotypes in Mexican American and Puerto Rican asthmatics. Genet Epidemiol 29(1):76–86
Saldarriaga W, Garcı'a-Perdomo HA, Arango-Pineda J, Fonseca J (2015) Karyotype versus genomic hybridization for the prenatal diagnosis of chromosomal abnormalities: a meta-analysis. Am J Obstet Gynecol 212(330):e1–e10
Schaeffer AJ, Chung J, Heretis K, Wong A, Ledbetter DH, Lese MC (2004) Comparative genomic hybridization-array analysis enhances the detection of aneuploidies and submicroscopic imbalances in spontaneous miscarriages. Am J Hum Genet 74(6):1168–1174
Schieppati A, Henter JI, Daina E, Aperia A (2008) Why rare diseases are an important medical and social issue. Lancet 371:2039–2041
Schulze A, Lindner M, Kohlmuller D et al (2003) Expanded newborn screening for inborn errors of metabolism by electrospray ionization–tandem mass spectrometry: results, outcome, and implications. Pediatrics 111:1399–1406
Sehnert AJ, Rhees B, Comstock D, de Feo E, Heilek G, Burke J et al (2011) Optimal detection of fetal chromosomal abnormalities by massively parallel DNA sequencing of cell-free fetal DNA from maternal blood. Clin Chem 57:1042–1049
Sequeiros J et al (2012) The wide variation of definitions of genetic testing in international recommendations, guidelines and reports. J Commun Genet 3:113–124
Speers DJ, Ryan S, Harnett G, Chidlow G (2003) Diagnosis of malaria aided by polymerase chain reaction in two cases with low-level parasitaemia. Inter Med J 33(12):613–615
Spolidorio DMP, Spolidorio LC (2005) Técnicas básicas de biologia molecular. In: Estrela C (ed) Metodologia científica-Ciência-Ensino-Pesquisa. Artes Médicas, São Paulo
Vissers LE et al (2003) Array-based comparative genomic hybridization for the genome-wide detection of submicroscopic chromosomal abnormalities. Am J Hum Genet 73:1261–1270
Wapner RJ, Martin CL, Levy B et al (2012a) Chromosomal microarray versus karyotyping for prenatal diagnosis. N Engl J Med 367:21752184
Wapner RJ, Martin CL, Levy B, Ballif BC, Eng CM, Zachary JM et al (2012b) Chromosomal microarray versus karyotyping for prenatal diagnosis. N Engl J Med 367:2175–2184
Wheeler DA et al (2008) The complete genome of an individual by massively parallel DNA sequencing. Nature 452:872–876

Worthey EA, Mayer AN, Syverson GD, Helbling D, Bonacci BB, Decker B et al (2011) Making a definitive diagnosis: successful clinical application of whole exome sequencing in a child with intractable inflammatory bowel disease. Genet Med 13:255–262

Xue Y, Ankala A, Wilcox WR, Hegde MR (2015) Solving the molecular diagnostic testing conundrum for Mendelian disorders in the era of next-generation sequencing: single-gene, gene panel, or exome/genome sequencing. Genet Med 17:444451

Yu SCY, Jiang P, Choy KW, Chan KCA, Won H-S, Leung WC et al (2013) Noninvasive prenatal molecular karyotyping from maternal plasma. PLoS One 8:e60968

Prenatal Screening and Counseling for Rare Genetic Disorders

3

K. Kumaran, Suruthi Abirami, Abna Ajeesh, J. Hemarangan, T. L. Vasanth Kanth, P. Shriya, and K. N. Aruljothi

Abstract

Genetic counseling is an important component of healthcare that provides individuals and families affected by rare genetic illnesses with important assistance and direction. Genetic counselors work with patients and families to provide information on the disease's genetic basis and potential dangers and advantages of genetic testing and to assist families in making informed decisions. Genetic counselors can assist them in receiving the best possible care and support by collaborating with healthcare practitioners and researchers. Numerous genetic tests available can help families affected by rare genetic illnesses make educated reproductive decisions, allowing them to make more informed decisions about lowering the chance of passing on the condition to future generations. Several obstacles are still involved with genetic testing for uncommon diseases, including issues relating to result interpretation, testing access, and testing cost. Hence, we need to go deeper into the specific issues and factors involved in providing genetic counseling. In this chapter, we will look at the most recent technological advances in genetic testing and how they affect diagnosis and therapy. This chapter also discusses the psychological and emotional consequences of rare genetic illnesses, as well as the need to provide proper support and resources to individuals and families.

K. Kumaran and Suruthi Abirami contributed equally with all other contributors.

K. Kumaran · S. Abirami · A. Ajeesh · J. Hemarangan · T. L. Vasanth Kanth · P. Shriya · K. N. Aruljothi (✉)
Faculty of Science and Technology, Department of Genetic Engineering, SRM Institute of Science and Technology, Kattankulathur, Chengalpattu, Chennai, Tamil Nadu, India
e-mail: aruljotn@srmist.edu.in

M. Umair et al. (eds.), *Rare Genetic Disorders*,
https://doi.org/10.1007/978-981-99-9323-9_3

Keywords
Genetic counseling · Prenatal diagnosis · Genetic analysis · Pregnancy · Rare diseases

3.1 Introduction

Rare diseases are a group of medical conditions that affect a few people. Despite their low prevalence, these diseases can significantly impact the individuals who suffer from them and their families. Due to the rarity of these diseases, they often go undiagnosed or misdiagnosed for extended periods, causing unnecessary suffering and thereby delaying effective treatment. Genetic testing and sequencing technologies have greatly improved the ability to diagnose rare diseases and identify their genetic causes. This has led to the development of targeted therapies that aim to correct or compensate for the specific congenital disability causing the disease. However, many challenges remain, including the high cost and limited availability of genetic testing, the complexity of some genetic disorders, and the need for more research to better understand the underlying genetic mechanisms of rare diseases (Cornelis et al. 2022).

Globally, it is estimated that there are between 5000 and 7000 rare diseases, affecting an estimated 400 million people worldwide. Many of these rare diseases are chronic, debilitating, and life-threatening, and significant unmet medical needs exist in the diagnosis, treatment, and management of these conditions. Approximately 80% of rare diseases are estimated to have a genetic component, meaning they are caused by changes or mutations in a person's DNA. This includes monogenic rare diseases, which are caused by mutations in a single gene, and polygenic rare diseases, which are caused by changes in multiple genes. The remaining 20% of rare diseases may have non-genetic causes, such as environmental factors, infections, or other unknown factors. In this book chapter, we will delve into the specific challenges and considerations involved in providing genetic counseling for rare genetic diseases. We will explore the latest technological advancements in genetic testing and their implications for diagnosis and treatment. We will also address the psychological and emotional impact of rare genetic diseases and the importance of providing individuals and families with adequate support and resources (Dive et al. 2022).

3.1.1 Genetic Counseling

Genetic counseling is a vital component of healthcare that offers crucial support and guidance to individuals and families affected by rare genetic diseases. These conditions are often complex and challenging to diagnose and manage, with limited resources and treatment options available. Genetic counselors are critical in helping affected individuals make informed decisions about their care by providing them with accurate information, emotional support, and practical resources (Bamshad et al. 2018).

For patients and their families, receiving a rare disease diagnosis marks more of the beginning than the conclusion of their journey. Patients need assistance once a genetic diagnosis has been made to comprehend the results' consequences and the psychosocial issues at play. Additionally, as genetics runs in the family, it affects the closest relatives and increases the likelihood that they may possess a disease-causing mutation. Although uncommon disorders may also emerge de novo, with the patient being the first member of the family to be affected, these disease-causing genetic alterations can be passed down from one generation to the next. Due to the increased risk of passing on the disease to future generations due to the inheritance of these variants, some relatives who may be carriers will want support, assistance, and direction to manage and plan a future pregnancy to lower the likelihood of having an affected child. Additionally, certain genetic disorders have complicated inheritance patterns and disease mechanisms, necessitating the assistance of genetic professionals with strong communication skills to help families comprehend genetic etiology and its implications for the family.

3.2 The Role of Genetic Counseling for Rare Diseases

Genetic counseling is a communication process that aids individuals in understanding the implications of a genetic variant on their life and health and in adjusting to the medical, psychological, and familial repercussions of genetic disease. It may be provided by clinical geneticists or genetic counselors. A genetic counselor's job entails a variety of duties that must be carried out to guarantee that a person has access to enough individualized information and support to deal with their hereditary disease and to make educated decisions. Patients should ideally obtain support from a genetic counselor on their first appointment with a specialist (in collaboration with interdisciplinary genetics units), who can examine the patient's clinical and family history and provide a suitable clinical letter. Arming families with knowledge and directing them to specialists (neurologists, ophthalmologists, nephrologists, social services, etc.) for their follow-up may aid in the identification of potential inheritance patterns while also helping families cope with challenges along the way. In fact, a study assessing the economic impact of employing genome sequencing to diagnose RD revealed that parents perceive genetic counselors to be an invaluable resource for facilitating complex decisions, and several other studies emphasize the cost-effectiveness of offering genetic counseling. Finally, it is important to note the harm caused when RD patients do not receive appropriate genetic counseling, which has an influence on many levels. This unfavorable consequence could manifest as psychological repercussions, unsuitable genetic testing, incorrect result interpretation, or even insufficient disease therapy, all of which would cause patients distress and discomfort (Smith et al. 2023; Genoff Garzon et al. 2018).

The following definition of genetic counseling was created by the "National Society of Genetic Counselors" (NSGC) Genetic Counseling Definition Task Force and accepted by the NSGC Board of Directors in 2006: The practice of genetic counseling involves assisting individuals in comprehending and adjusting to the medical, psychological, and familial ramifications of hereditary influences on disease. This method incorporates the following: assessment of the likelihood that a disease would develop or recur using an interpretation of family and medical histories; education in the areas of management, prevention, testing, and resources; and counseling to encourage savvy decision-making and acclimatization to the danger or condition. In addition to interpreting genetic test results based on your personal and family history, genetic counselors provide patients wanting additional information about how inherited diseases and ailments may affect them or their families with guidance and assistance (Kaye 2023; Benn and Chapman 2010).

3.2.1 Prenatal Genetic Counseling

For parents, being pregnant may be a joyous time but also uncertain, especially if there are hereditary concerns about the pregnancy's health. Prenatal genetic counseling involves a professional evaluation of the couple's medical and pregnancy history and any potential risk factors in the couple's family. It can reveal underlying genetic causes for any anomalies in the pregnancy and establish risks for diseases that could be passed on to the child (Lamb et al. 2018; Sullivan-Pyke and Dokras 2018; Caceres et al. 2022; Dumars et al. 1976).

3.2.2 Preconception Genetic Counseling

There is much excitement and eagerness while planning a family. Parents-to-be prefer the most secure pregnancy possible for the mother and the unborn child. Before conceiving is one of the best times to get tested for common genetic illnesses and to learn whether a condition is likely to run in the family. Preconception genetic counseling can be educational and reassuring, regardless of whether a couple is dealing with a known genetic issue or has no known risks (Simpson and Rechitsky 2019; De Wert et al. 2012; Kihlbom 2016; Ren et al. 2023) (Fig. 3.1).

3.3 Genetic Testing for Rare Diseases

Genetic testing helps us understand the underlying genetic condition a fetus may contain (Bedinghaus 2002). If so, the parents may continue the pregnancy or seek alternative options. Prenatal genetic screening can be performed if a family history of genetic problems is known, if the first child has a genetic disorder, if there is a history of recurrent pregnancy loss, or if there are abnormal ultrasound results (Van den Veyver 2016). The position of genes on chromosomes was determined using genetic mapping or linkage analysis in the traditional study. Linkage analysis was performed utilizing chromosomal markers. The linkage analysis was used along with polymerase reactions, Sanger sequence, next-generation sequencing techniques, and microarray techniques for noteworthy results (Bailey-Wilson and Wilson 2011; Pulst 1999).

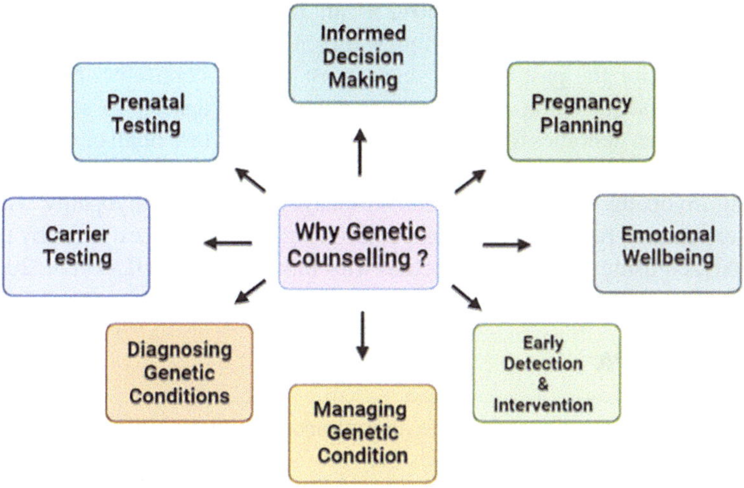

Fig. 3.1 The use of genetic counseling in the analysis of medical conditions

3.3.1 Sanger Sequencing

The "chain termination method," known as the Sanger sequencing method, determines the DNA nucleotide sequence (Sanger et al. 1977). During Sanger sequencing, target PCR templates are employed for chain termination PCR. Except for adding modified nucleotides (ddNTPs), also called dideoxy ribonucleotides, chain termination PCR is comparable to conventional PCR (dNTPs). DNA polymerases add dNTPs to a growing DNA strand during the extension stage of a conventional PCR. If DNA polymerase incorporates ddNTPs, elongation is halted. Four PCR reactions are constructed using a single type of ddNTP utilizing manual Sanger sequencing (ddATP, ddTTP, ddGTP, and ddCTP).

Each ddNTPs is fluorescently labeled uniquely, and are mixed in a single automated procedure. The samples separated through gel electrophoresis are read at the final step in DNA sequencing to determine the input DNA's sequence. Manual sequencing requires viewing all four gel lanes from bottom to top (3' to 5') to identify the terminal ddNTP in each route. In an automated procedure, a computer examines each capillary gel strip and uses fluorescence to identify each end. The result is a chromatogram that displays the fluorescent peaks for each nucleotide over the whole length of the DNA template (CD Genomics 2020).

3.3.2 Next-Generation Sequencing

The next-generation sequencing is a modern day technology to study uncommon mutations. The targeted exome or whole-exome sequencing analyzes mutations in either a specified region or the entire coding region of the exome (Hu et al. 2021). Using advanced platforms, these systems can execute billions of sequencing reactions with a read length of 150–250 nt. In recent decades, a few essential mutations have been detected in the intron region, causing concern among researchers. Therefore, for the past several years, most genetic counselors have opted for whole-genome sequencing, which corresponds to the capacity to sequence many individual human genomes to obtain a comprehensive picture of all clinically significant differences (Di Scipio et al. 2020; Vaz-Drago et al. 2017). A few rare diseases, including myopathy with areflexia and dysphagia, hereditary myopathy with early respiratory failure, respiratory distress, and limb-girdle muscular dystrophy type 1E, were analyzed using NGS in conjunction with linkage analysis (Liu et al. 2019).

3.3.3 MS-MLPA

MS-MLPA analyzes epigenetic gene alterations. This technology provides genetic counselors with a significant analysis for analyzing uncommon disorders induced by methylation (Moelans et al. 2018). Using probes unique to a gene in conjunction with restriction enzymes, the MS-MLPA is done to identify unmethylated sequences. Using the peak observed by comparing the probe sample to the control sample, the amount of methylation in the targeted gene is determined. Methylation-specific MLPA is used to evaluate rare disorders such as Angelman syndrome and Prader-Willi syndrome (Procter et al. 2006).

3.3.4 Chromosomal Microarray

CMA is the most recent technology used to diagnose chromosomal microdeletions and microduplications (Levy and Wapner 2018). It can detect microdeletions and microduplications with a resolution of up to 200 kb. It is a panel test based on chips. There are now two resolutions in use: 315 K, a low solution that can detect up to 2–1 mb, and 750 K, a high resolution that can detect up to 400–200 kb (Bignell et al. 2004).

CMA is superior to karyotyping because, unlike karyotyping, microarrays do not require living cells and can detect microdeletions and microduplications (Wapner et al. 2012). It is more likely to be used than karyotype analysis since noninvasive samples can be used. Microarray analysis is instrumental when a karyotype cannot be established from product of conception (POC) materials. Microarrays can detect unusual genetic disorders, variants of unclear importance, and imbalanced rearrangements, even in the prenatal stage. Along with prenatal microarray analysis, we always do maternal cell contamination (MCC) analysis to exclude the possibility of maternal cell contamination in fetal cells. This prevents erroneous positive and negative outcomes. DNA-based microarrays, gene expression microarrays, and protein microarrays are the three types of microarrays. We may select an approach based on our requirements.

3.3.5 Genetic Counselors and Genetic Testing

Genetic counselors rely on the results of genetic tests and diagnoses conducted on a patient's sample. Through analysis and interpretation of the results of the plethora of available tests, genetic counselors provide empathic and psychological assistance to patients to assist them in making informed, prudent decisions (Garber and Hixon 1990; Biesecker 2001). Testing and diagnosis are continuously refined and reinvented to further equip the genetic counseling sector with minimally invasive and noninvasive testing procedures that yield considerably more reliable and simply interpretable results (Smith et al. 2023; Farmer et al. 2019; Temel et al. 2014).

3.4 Carrier Screening

Since the early days of screening for uncommon diseases, newborn bloodspot screening (NBS) for phenylketonuria and carrier screening for conditions including Tay-Sachs disease and β-thalassemia have been used. Carrier screening is a genetic testing procedure performed on individuals who do not show any symptoms of a genetic disorder but may carry a variant allele associated with a particular condition. It is recommended that pregnant women receive information about carrier screening, and it is best to undergo the screening before pregnancy. Couples can then completely comprehend their reproductive risks and weigh all of their options as a result. The reproductive partner of a person who is a carrier of a particular illness should also be offered testing and counseling to help them understand the possible results of their pregnancy. To make educated decisions regarding their reproductive options, couples who are found to be carriers of the same genetic disease should obtain genetic counseling (Delatycki et al. 2020; Henneman et al. 2016).

If available, information about the patient's family history and their partner's history can serve as an effective screening tool to assess the potential inherited risks. Details on the ethnicity of family members, as well as any consanguineous relationships, should be included in the family history. If a genetic problem runs in the family, it is suggested that carrier screening and genetic counseling be provided. Knowing the exact mutation can be useful for testing and counseling purposes (Dive and Newson 2021).

Testing for genetic carriers can help identify couples who are most likely to give birth to a child who has a recessive condition. Over 1800 rare diseases can be inherited in this manner, with symptoms ranging from mild to severe. When both partners are carriers of the same autosomal recessive disorder, their offspring have a one-in-four chance of being affected by the disorder. There is a 50% likelihood that sons will be affected by an X-linked illness in cases where the woman is a carrier and a 50% chance that girls will be carriers. This risk affects approximately 1 in 100 couples worldwide, and certain subpopulations may have an even higher risk due to their geographic origin or ancestry. Most carriers are unaware of their status because it typically does not impact their health, leading to unexpected occurrences of affected children at birth (Richardson et al. 2022; van den Heuvel et al. 2022).

Preconception carrier screening is a valuable option for individuals who want to identify their risk of having a child with a genetic disorder. It offers more reproductive options than prenatal carrier screening, allowing individuals to make informed decisions before conception. However, preconception carrier screening may not always be readily available, as it may not be routinely offered as part of preconception care. It is important to note that carrier screening does not test for an existing disease but rather aims to detect healthy individuals who carry a genetic mutation, and if their partner is a carrier, they run the possibility of having a child who has the condition (Kingsmore et al. 2022; Schuurmans et al. 2019).

The recommendations and findings from the American College of Obstetricians and Gynecologists are as follows:

- It is important to offer all pregnant women information about genetic carrier screening. However, it is ultimately the patient's decision whether or not to undergo the screening after receiving counseling.
- Carriers should be screened and counseled, ideally before conception or early in pregnancy, to allow individuals to make informed decisions about their reproductive options.
- When a person is found to be a carrier of a particular genetic disorder, their reproductive partner should also be provided with testing to receive complete genetic counseling and decide how they want to proceed with their prospective pregnancy. The simultaneous screening of both individuals is advised if there is a deadline for making decisions on prenatal diagnostic evaluation.
- If both individuals are found to be carriers of an inheritable condition, it is crucial to provide genetic counseling to go over possible ways to minimize the likelihood of developing an affected child, such as prenatal diagnosis and cutting-edge reproductive technologies. Furthermore, if a person is discovered to have been a carrier of a genetic illness, it may give their family members a chance of developing the same mutation. As such, the patient should be encouraged to let them know about the risk and the possibility of carrier screening. Obstetrician-gynecologists and other healthcare professionals must respect patients' right to privacy and should not share this information without those patients' expressed consent.
- It is important to note that prenatal carrier and newborn screening serve different purposes and should not be considered interchangeable. Prenatal carrier screening is designed to identify potential genetic risks during pregnancy, while newborn screening is used to detect certain conditions that may not be apparent at birth. Therefore, prenatal carrier screening should not be considered a substitute for newborn screening, and vice versa. Both types of screening can provide valuable information for healthcare providers and families to make informed decisions about the health and well-being of the child.
- Patients who request carrier screening for a specific genetic condition, for which testing is readily available and may be considered in other screening strategies, should be provided with the option to undergo testing after receiving appropriate counseling on the potential risks, benefits, and limitations of screening. This

offer should be extended regardless of the patient's ethnicity or family history. Genetic counseling can help patients make informed decisions about carrier screening, considering their personal circumstances and preferences.
- It is crucial to remember that testing using commercially available broad carrier screening panels may be less expensive than carrier screening for a specific genetic disorder. The cost of each option to the patient and the healthcare system must be considered when choosing a carrier screening strategy. This information can help patients and providers make informed decisions about carrier screening that consider both the clinical benefits and the financial impact.

3.5 Prenatal Diagnosis to Prevent Rare Diseases

Two types of samples are used for prenatal testing: amniotic fluid and chorionic villi. After 12–13 weeks (about 3 months) of gestation, it is optimum to collect CVS from the fetus, whereas AF is typically obtained after 16+ weeks.

3.5.1 Amniotic Fluid

Amniotic fluid sampling, or amniocentesis, is a test performed between 14 and 20 weeks of pregnancy to diagnose genetic abnormalities, birth malformations, and other fetal anomalies. During this invasive operation, a tiny needle is introduced into the pregnant mother's abdomen to collect amniotic fluid from the amniotic sac. Amniotic fluid is a transparent, slightly yellowish fluid produced during pregnancy that surrounds the newborn infant and is present in the amniotic sac containing the baby (Jindal et al. 2022). After 12 days of pregnancy, it forms in the amniotic sac to cushion and protect the fetus. The outpatient is advised to refrain from strenuous physical activity and seek prompt medical attention if there are any noteworthy changes in the abdomen region, fevers or chills, or variations in the fetus' activity level.

3.5.2 Chorionic Villus Sampling (CVS)

Chorionic villus sampling, also known as chorionic villus biopsy, is a prenatal diagnostic method in which a tissue sample from the placenta is obtained and tested for chromosomal and genetic abnormalities. Two approaches are used for the collecting of tissue:

- Transabdominal technique.
- Transcervical technique.

A sufficient amount of chorionic villi (placental villi) from the placenta is collected and evaluated between the tenth and 12th weeks of pregnancy when a needle or catheter is inserted into the patient's abdomen with the aid of ultrasound and negative pressure (Vink and Quinn 2021). Women carrying twins or multiples will be required to give placenta samples for each child (Rao et al. 2004).

3.5.3 Cord Blood Genetic Evaluation

Blood that remains in the umbilical cord and placenta after birth is called cord blood. It comprises totipotent stem cells that can differentiate into any cell. Cordocentesis, often known as percutaneous umbilical cord blood sampling (PUBS), is a procedure used to collect cord blood. It is a diagnostic test that analyzes fetal blood to identify fetal abnormalities. The following are detected or diagnosed using cord blood:

1. Fetal anemia.
2. Number of fetal platelets in the mother.
3. Immunization test.
4. Pregnancy-related infections.
5. Congenital disabilities in the fetus.
6. Measurement of blood gases to determine if the concentrations of oxygen and other chemicals are normal.
7. Determination of bilirubin concentrations.
8. Examination of the infant for indications of exposure to nonprescription medicines that the mother would have consumed during pregnancy.
9. If these results are hazardous to the fetus, the healthcare or genetic counselor can recommend appropriate preventive measures or therapies.

3.5.4 When and How Should Cordocentesis Be Performed?

In most situations, cordocentesis is performed after the baby is born. However, it can also be conducted during mid- or late- pregnancy (after 17 weeks). The umbilical cord is cut shortly after birth to separate the newborn from the mother. Using ultrasonography, the umbilical cord is visualized in real time. The ultrasonography transducer is placed after determining the portion of the umbilical cord to be pierced. The plane of scanning the umbilical cord relies on the chord's visibility and the position of the fetus. Most medical professionals would instead puncture the placenta or umbilical cord and then use a loose loop that could easily slip away.

This technique is typically performed immediately after birth, although leading health organizations recommend cutting the umbilical cord 1 min after childbirth to increase the baby's blood circulation. The umbilical cord is then clamped to prevent bleeding. Blood is extracted by placing a syringe or needle into a vein. After sufficient blood is collected, the sample is packaged and sent to a laboratory for analysis (Schmidtke and Krawczak 2022).

3.5.5 Complications

PUBS/cordocentesis is an emerging technology conducted with exceptional care to assess fetal danger. Even though it is a highly modern technique, it carries significant hazards and should only be utilized when amniocentesis or CVS cannot provide the necessary information. Some dangers of cordocentesis include bleeding at the puncture site, umbilical cord hematoma, laceration, and fetal distress. Also of concern is the proximity to the puncture site. As the distance to the puncture site grows, the ultrasound's resolution decreases, making it more challenging to guide the needle tip. In some instances, the free loop is better than cord insertion.

3.5.6 Anomaly Scanning

Anomaly scanning, as its name suggests, is the practice of ultrasonographically scanning an expecting woman, often between 20 and 24 weeks, to detect any abnormalities or anomalies in the unborn (Anomaly Scan 2023). Ultrasonography is performed with an ultrasound instrument that emits and receives ultrasound waves (high-frequency sound waves). The frequency feedback data is then gathered, transformed into graphical form, processed, and displayed as 2D/3D black-and-white images on a monitor. Ultrasound gel is placed on the surface of the patient's abdomen to prevent picture noise and decrease impedance (Afzal et al. 2022). The typical ultrasonic gel consists of water and propylene glycol, giving it a thick consistency. Ultrasound gels continuously undergo research and development to make them more cost-effective and derived from natural substances. It is a transabdominal procedure, meaning that the ultrasound device is put over the pregnant woman's abdomen. It is generally recommended to fast for 10–12 h before a test since undigested food particles may hinder a clear image from being produced.

During the scan, the following are performed:

- Component of the fetal body.
- Placenta position.
- Amount of amniotic fluid.
- Fetal growth estimation.

Additionally, particular focus is placed on organs like the kidneys, face, spine, heart, stomach, colon, and limbs. The primary objective of the anomaly scan is to conduct a comprehensive analysis of the pregnancy and detect any abnormal conditions, such as those of the brain, spine, and heart. The procedure can also be used to determine the baby's gender. If any anomalies are detected, the significance of the results will be evaluated, and patients will be given the option to consult with a physician.

3.6 Genetic Counseling for Uncommon Diseases: Current Requirements and Future Directions

Genomics has advanced to the point where diagnosis and therapy are both practical and cost-effective. These treatments are targeted at specific sites in the body. In an era where genetic solutions are available for almost all diseases diagnosed, genetic counseling plays a vital role in advising patients on what treatments are available for a particular condition. Genetic counselors' abilities and knowledge are crucial in efficiently applying genomic medicine. Genetic counseling is a communication process that deals with genetic issues, according to the first official definition of the term released by the American Society of Human Genetics (ASHG) in 1975.

Genetic counseling aids in preparing women whose fetus has been diagnosed with specific disabilities. The counseling provides the mothers with adequate and accurate information about the disabilities in the fetus and, if required, suggests that they take more tests to confirm the condition. A genetic provider can aid in good counseling by learning about the parents' worries and needs related to the child's condition. It was hypothesized that the parent's perception of the genetic counselor could have a positive or negative impact on the usefulness of the counseling. India reports a significantly high rate of genetic disorders due to the rate of consanguineous marriages. Most couples in India report for prenatal testing in the later stages of the second trimester, and phenotypic diagnosis plays a part. Cordocentesis is only performed on such families. Due to the higher risk of fetal death associated with cordocentesis, when genetic counseling is given after diagnosis in the index case, it is crucial to underline the significance of reporting PND early in the second pregnancy (Fig. 3.2).

Expanded carrier screening (ECS), another stream, provides couples with information about their reproductive options, ideally before conception. In this type of counseling, genetic practitioners offer a population-based ECS test assessed by the degree of patient-informed choice. Genetic counselors work side by side with healthcare providers to address the majority of issues the patients face. The globe gradually moves towards the era of precision medicine as technology advances. In addition to focusing on early detection and individualized therapies like tumor profiling, genetic counselors are now emphasizing prevention. Depending on the context, though, complete integration of genomes into conventional medicine would jeopardize the existing trend of clinical specialization.

To ensure that genetic counselors provide high-quality and accurate genetic services, the quality has to be defined and uniform for all genetic counselors. The Research, Quality, and Outcomes Committee of the National Society of Genetic Counselors collaborated with Discern Health, a value-based healthcare policy consulting company, to create a genetic counseling care continuum model. The US healthcare system's currently available quality metrics for genetic counseling were evaluated using the suggested methodology, allowing for the identification of gaps and top areas for future development. Although few measures were specifically for genetic counseling or genetic diseases, a total of 560 quality measures were found to apply to various facets of the care continuum model across various clinical

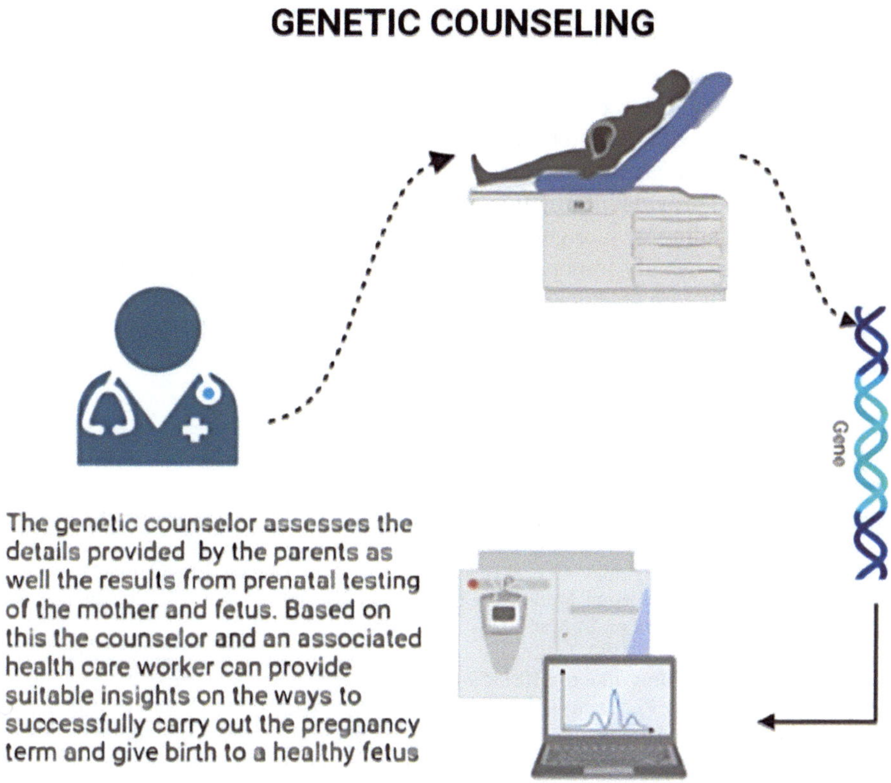

Fig. 3.2 Schematic representation showing the role of genetic counselor during pregnancy stages

specialist areas in genetic counseling. The attitudes towards genetic testing, family communication, stigma, and problems with justice, equity, diversity, and inclusion were some of the areas where quality metrics fell short.

3.7 Conclusion

Genetic testing is a valuable tool in diagnosing and managing rare genetic diseases. Sequencing technologies have made it increasingly possible to identify the genetic basis of these conditions, leading to earlier diagnosis, more targeted treatments, and improved outcomes for patients and families. Genetic testing can also inform reproductive decision-making for families affected by rare genetic diseases, allowing them to make informed choices about having children and reducing the risk of passing on the condition to future generations. However, there are still many challenges associated with genetic testing for rare diseases, including issues related to the interpretation of results, access to testing, and cost of testing. Genetic counselors are critical in helping patients and families navigate these challenges and make informed decisions about genetic testing. Genetic counseling plays a crucial role in the management of rare genetic diseases. Genetic counselors work with patients and families to provide information about the genetic basis of the disease, discuss the potential risks and benefits of genetic testing, and help families make informed decisions about their reproductive options. Genetic counseling can help individuals and families cope with the emotional and psychological impact of a rare congenital disease diagnosis. It can also facilitate access to appropriate medical care and support services and provide information about available treatments and clinical trials. As our understanding of rare genetic diseases continues to evolve, genetic testing will play an increasingly important role in diagnosing and managing these conditions. By working with healthcare providers and researchers, genetic counselors can help ensure that patients and families affected by rare genetic diseases receive the best possible care and support.

References

Afzal S, Zahid M, Rehan ZA, Fayzan Shakir HM, Javed H, Aljohani MMH, Mustafa SK, Ahmad M, Hassan MM (2022) Preparation and evaluation of polymer-based ultrasound gel and its application in ultrasonography. Gels 8(1):42. https://doi.org/10.3390/GELS8010042

Anomaly Scan (2023) The Fetal Medicine Centre. Available online: https://fetalmedicine.com/anomaly-scan/ (accessed on 24 Mar 2023)

Bailey-Wilson JE, Wilson AF (2011) Linkage analysis in the next-generation sequencing era. Hum Hered 72:228–236. https://doi.org/10.1159/000334381

Bamshad MJ, Magoulas PL, Dent KM (2018) Genetic counselors on the frontline of precision health. Am J Med Genet C Semin Med Genet 178:5–9. https://doi.org/10.1002/AJMG.C.31610

Bedinghaus JM (2002) Genetic testing. N Engl J Med:351–356. https://doi.org/10.1056/NEJMOA012113

Benn PA, Chapman AR (2010) Ethical challenges in providing noninvasive prenatal diagnosis. Curr Opin Obstet Gynecol 22:128–134. https://doi.org/10.1097/gco.0b013e3283372352

Biesecker BB (2001) Goals of genetic counseling. Clin Genet 60:323–330. https://doi.org/10.1034/J.1399-0004.2001.600501.X

Bignell GR, Huang J, Greshock J, Watt S, Butler A, West S, Grigorova M, Jones KW, Wei W, Stratton MR et al (2004) High-resolution analysis of DNA copy number using oligonucleotide microarrays. Genome Res 14:287–295. https://doi.org/10.1101/GR.2012304

Caceres V, Murray T, Myers C, Parbhoo K (2022) Prenatal genetic testing and screening: a focused review. Semin Pediatr Neurol 42:100976. https://doi.org/10.1016/J.SPEN.2022.100976

CD Genomics (2020) Sanger sequencing steps and method. Available online: https://www.sigmaaldrich.com/IN/en/technical-documents/protocol/genomics/sequencing/sanger-sequencing (accessed on 15 Mar 2023)

Cornelis SS, Runhart EH, Bauwens M, Corradi Z, De Baere E, Roosing S, Haer-Wigman L, Dhaenens CM, Vulto-van Silfhout AT, Cremers FPM (2022) Personalized genetic counseling for stargardt disease: offspring risk estimates based on variant severity. Am J Hum Genet 109:498–507. https://doi.org/10.1016/J.AJHG.2022.01.008

De Wert GMWR, Dondorp WJ, Knoppers BM (2012) Preconception care and genetic risk: ethical issues. J Community Genet 3:221–228. https://doi.org/10.1007/s12687-011-0074-9

Delatycki MB, Alkuraya F, Archibald A, Castellani C, Cornel M, Grody WW, Henneman L, Ioannides AS, Kirk E, Laing N et al (2020) International perspectives on the implementation of reproductive carrier screening. Prenat Diagn 40:301–310. https://doi.org/10.1002/pd.5611

Di Scipio M, Tavares E, Deshmukh S, Audo I, Green-Sanderson K, Zubak Y, Zine-Eddine F, Pearson A, Vig A, Tang CY, Mollica A (2020) Phenotype driven analysis of whole genome

sequencing identifies deep intronic variants that cause retinal dystrophies by aberrant exonization. Invest Ophthalmol Vis Sci 61:36–36. https://doi.org/10.1167/IOVS.61.10.36
Dive L, Newson AJ (2021) Ethical issues in reproductive genetic carrier screening. Med J Aust 214:165–167. https://doi.org/10.5694/mja2.50789
Dive L, Archibald AD, Newson AJ (2022) Ethical considerations in gene selection for reproductive carrier screening. Hum Genet 141:1003–1012. https://doi.org/10.1007/S00439-021-02341-9/METRICS
Dumars KW, Dalrymple GT, Murray AK (1976) Prenatal diagnosis and genetic counseling. West J Med 124:377–387. https://doi.org/10.1007/978-3-031-07524-7_2/COVER
Farmer MB, Bonadies DC, Mahon SM, Baker MJ, Ghate SM, Munro C, Nagaraj CB, Besser AG, Bui K, Csuy CM et al (2019) Errors in genetic testing: the fourth case series. Cancer J (United States) 25:231–236. https://doi.org/10.1097/PPO.0000000000000391
Garber AP, Hixon HEC (1990) Prenatal genetic counseling. Clin Perinatol 17:749–759. https://doi.org/10.1016/S0095-5108(18)30543-8
Genoff Garzon MC, Rubin LR, Lobel M, Stelling J, Pastore LM (2018) Review of patient decision-making factors and attitudes regarding preimplantation genetic diagnosis. Clin Genet 94:22–42. https://doi.org/10.1111/cge.13174
Henneman L, Borry P, Chokoshvili D, Cornel MC, Van El CG, Forzano F, Hall A, Howard HC, Janssens S, Kayserili H et al (2016) Responsible implementation of expanded carrier screening. Eur J Hum Genet 24:e1–e12. https://doi.org/10.1038/ejhg.2015.271
Hu T, Chitnis N, Monos D, Dinh A (2021) Next-generation sequencing technologies: an overview. Hum Immunol 82:801–811. https://doi.org/10.1016/J.HUMIMM.2021.02.012
Jindal A, Sharma M, Karena ZV, Chaudhary C (2022) Amniocentesis. In: Obstetric imaging: fetal diagnosis and care, 2nd edn, pp 473–475. https://doi.org/10.1016/B978-0-323-44548-1.00111-X
Kaye DK (2023) Addressing ethical issues related to prenatal diagnostic procedures. Maternal Health Neonatol Perinatol 9:1. https://doi.org/10.1186/S40748-023-00146-4
Kihlbom U (2016) Ethical issues in preconception genetic carrier screening. Ups J Med Sci 121:295–298. https://doi.org/10.1080/03009734.2016.1189470
Kingsmore SF, Smith LD, Kunard CM, Bainbridge M, Batalov S, Benson W, Blincow E, Caylor S, Chambers C, Del Angel G et al (2022) A genome sequencing system for universal newborn screening, diagnosis, and precision medicine for severe genetic diseases. Am J Hum Genet 109:1605–1619. https://doi.org/10.1016/J.AJHG.2022.08.003
Lamb B, Johnson E, Francis L, Fagan M, Riches N, Canada I, Wilson A, Mathiesen A, Sabatello M, Gurtcheff S et al (2018) Pre-implantation genetic testing: decisional factors to accept or decline among in vitro fertilization patients. J Assist Reprod Genet 35:1605–1612. https://doi.org/10.1007/s10815-018-1278-2
Levy B, Wapner R (2018) Prenatal diagnosis by chromosomal microarray analysis. Fertil Steril 109:201–212. https://doi.org/10.1016/J.FERTNSTERT.2018.01.005
Liu Z, Zhu L, Roberts R, Tong W (2019) Toward clinical implementation of next-generation sequencing-based genetic testing in rare diseases: where are we? Trends Genet 35:852–867. https://doi.org/10.1016/J.TIG.2019.08.006
Moelans CB, Atanesyan L, Savola SP, van Diest PJ (2018) Methylation-specific multiplex ligation-dependent probe amplification (MS-MLPA). Methods Mol Biol 1708:537–549. https://doi.org/10.1007/978-1-4939-7481-8_27/COVER
Procter M, Chou LS, Tang W, Jama M, Mao R (2006) Molecular diagnosis of Prader–Willi and Angelman syndromes by methylation-specific melting analysis and methylation-specific multiplex ligation-dependent probe amplification. Clin Chem 52:1276–1283. https://doi.org/10.1373/CLINCHEM.2006.067603
Pulst SM (1999) Genetic linkage analysis. Arch Neurol 56:667–672. https://doi.org/10.1001/ARCHNEUR.56.6.667
Rao A, Sairam S, Shehata H (2004) Obstetric complications of twin pregnancies. Best Pract Res Clin Obstet Gynaecol 18:557–576. https://doi.org/10.1016/J.BPOBGYN.2004.04.007

Ren M, Shireman H, VanGompel EW, Bello JK, Carlock F, McHugh A, Stulberg D (2023) Preconception, interconception, and reproductive health screening tools: a systematic review. Health Serv Res 58:458–488. https://doi.org/10.1111/1475-6773.14123
Richardson E, McEwen A, Newton-John T, Crook A, Jacobs C (2022) Systematic review of outcomes in studies of reproductive genetic carrier screening: towards development of a core outcome set. Genet Med 24:1–14. https://doi.org/10.1016/j.gim.2021.08.005
Sanger F, Nicklen S, Coulson AR (1977) DNA sequencing with chain-terminating inhibitors. Proc Natl Acad Sci 74:5463–5467. https://doi.org/10.1073/PNAS.74.12.5463
Schmidtke J, Krawczak M (2022) Correspondence on "Screening for autosomal recessive and X-linked conditions during pregnancy and preconception: a practice resource of the American College of Medical Genetics and Genomics (ACMG)" by Gregg et al. Genet Med 24:1156–1157. https://doi.org/10.1016/j.gim.2022.01.003
Schuurmans J, Birnie E, van den Heuvel LM, Plantinga M, Lucassen A, van der Kolk DM, Abbott KM, Ranchor AV, Diemers AD, van Langen IM (2019) Feasibility of couple-based expanded carrier screening offered by general practitioners. Eur J Hum Genet 27:691–700. https://doi.org/10.1038/s41431-019-0351-3
Simpson JL, Rechitsky S (2019) Prenatal genetic testing and treatment for congenital adrenal hyperplasia. Fertil Steril 111:21–23. https://doi.org/10.1016/J.FERTNSTERT.2018.11.041
Smith L, Malinowski J, Ceulemans S, Peck K, Walton N, Sheidley BR, Lippa N (2023) Genetic testing and counseling for the unexplained epilepsies: an evidence-based practice guideline of the National Society of Genetic Counselors. J Genet Couns 32(2):266–280. https://doi.org/10.1002/JGC4.1646
Sullivan-Pyke C, Dokras A (2018) Preimplantation genetic screening and preimplantation genetic diagnosis. Obstet Gynecol Clin N Am 45:113–125. https://doi.org/10.1016/j.ogc.2017.10.009
Temel S, Van Voorst SF, Jack BW, Denktaş S, Steegers EAP (2014) Evidence-based preconceptional lifestyle interventions. Epidemiol Rev 36:19–30. https://doi.org/10.1093/EPIREV/MXT003
van den Heuvel LM, van den Berg N, Janssens ACJW, Birnie E, Henneman L, Dondorp WJ, Plantinga M, van Langen IM (2022) Societal implications of expanded universal carrier screening: a scoping review. Eur J Hum Genet 31(1):55–72. https://doi.org/10.1038/s41431-022-01178-8
Van den Veyver IB (2016) Recent advances in prenatal genetic screening and testing. F1000Res 5:2591. https://doi.org/10.12688/F1000RESEARCH.9215.1
Vaz-Drago R, Custódio N, Carmo-Fonseca M (2017) Deep intronic mutations and human disease. Hum Genet 136:1093–1111. https://doi.org/10.1007/S00439-017-1809-4
Vink J, Quinn M (2021) Chorionic villus sampling. In: Obstetric imaging: fetal diagnosis and care, 2nd edn, pp 479–481. https://doi.org/10.1016/B978-0-323-44548-1.00113-3
Wapner RJ, Martin CL, Levy B, Ballif BC, Eng CM, Zachary JM, Savage M, Platt LD, Saltzman D, Grobman WA et al (2012) Chromosomal microarray versus karyotyping for prenatal diagnosis. N Engl J Med 367:2175–2184. https://doi.org/10.1056/NEJMOA1203382

Disease Models for Rare Genetic Disorders

4

Sivakumari Kanakarajan, Rajesh Selvaraj, and Patheri Kuniyil Kaleena

Abstract

Rare diseases pose complex challenges to individuals, families, and the healthcare system. Collaborative research efforts are essential to improving diagnosis, treatment, and overall quality of life for those affected by these conditions. Animal models stand as indispensable tools in rare disease research, offering insights into disease mechanisms, treatment efficacy, and potential interventions that ultimately improve the lives of those affected by these conditions. Animal models across various species have significantly advanced our knowledge of rare diseases, accelerating the development of therapies and treatment strategies.

Rodent models are widely used in rare disease research, where mice and rats are used. Rat models and mouse models offer a complementary approach, providing insights into the pathogenesis, treatment strategies, and potential biomarkers of rare diseases. On the other hand, non-rodent animal models also provide diverse opportunities to study rare diseases and contribute unique insights that complement rodent models. Primate models stand as indispensable tools for unraveling the complexities of rare diseases, bridging the gap between laboratory findings and clinical applications. Canine models offer a unique perspective on rare diseases, contributing to our understanding of disease mechanisms, treatment strategies, and translational research. Porcine models offer valuable insights into various rare diseases, benefiting our understanding of dis-

S. Kanakarajan (✉)
Department of Zoology, Competitive Examinations Coaching Centre, Chennai, Tamil Nadu, India

R. Selvaraj
Department of Clinical and Translational Sciences, Joan C. Edwards School of Medicine, Marshall University, Huntington, WV, USA

P. K. Kaleena
Department of Zoology, Presidency College, Chennai, Tamil Nadu, India

M. Umair et al. (eds.), *Rare Genetic Disorders*,
https://doi.org/10.1007/978-981-99-9323-9_4

ease mechanisms and potential treatments. Zebrafish models have transformed rare disease research by offering unique insights into disease mechanisms, drug screening, and genetic pathways. Feline models provide a unique perspective on rare diseases, offering insights into genetic disorders, metabolic conditions, and more. Drosophila models provide a unique and efficient platform for studying rare diseases, uncovering key insights into disease mechanisms, potential treatments, and therapeutic targets.

Advancements in genetic engineering, especially driven by CRISPR-Cas9 technology, have transformed our ability to create accurate and relevant animal models for rare diseases. These techniques enable researchers to uncover disease mechanisms, test potential therapies, and contribute to the development of personalized medicine approaches. CRISPR-Cas9 technology has transformed the creation of animal models for rare diseases, enabling precise genetic modifications that facilitate the study of disease mechanisms and potential therapies.

Gene knockout and knock-in techniques have revolutionized our ability to investigate gene function and the molecular basis of rare diseases, paving the way for potential therapeutic strategies. Conditional and tissue-specific gene expression systems offer precision and flexibility in creating animal models that closely mimic the genetic and physiological complexities of rare diseases. Induced pluripotent stem cells (iPSCs) have transformed disease modeling, offering a platform to study rare diseases with unprecedented accuracy. Their potential for understanding disease mechanisms and drug screening holds promise for advancing the field of rare disease research. Characterizing and phenotyping rare disease models provide essential insights into disease mechanisms and potential therapeutic strategies. These processes contribute to bridging the gap between basic research and clinical applications, ultimately improving our understanding and management of rare diseases.

Behavioral and physiological assessments offer critical insights into the impact of rare diseases on an organism's function and well-being. These assessments contribute to a comprehensive understanding of disease mechanisms and aid in the development of targeted interventions. Molecular and histopathological analyses are essential components of characterizing and phenotyping rare disease models, shedding light on the molecular basis and structural changes associated with these conditions. Modeling specific rare diseases serves as a critical tool for deepening our understanding of disease pathology and developing targeted therapeutic approaches. These models contribute to unraveling the complexities of rare disorders and ultimately hold the potential to improve patient care and outcomes.

Developing animal models for neurologically rare diseases offers a valuable platform for gaining insights into disease mechanisms and evaluating potential treatments. These models play a pivotal role in bridging the gap between bench research and clinical applications, ultimately improving the lives of patients with these disorders. Animal models for genetic metabolic disorders offer a powerful tool for understanding disease mechanisms and developing targeted therapies. These models contribute to bridging the gap between basic research and clinical applications, ultimately benefiting patients affected by these disorders. Animal models for rare cancers offer a vital tool for investigating disease mechanisms and evaluating potential therapies. These models contribute to bridging the gap between laboratory research and clinical applications, ultimately improving the prognosis and treatment options for patients with these rare malignancies.

In vivo imaging and analysis offer a transformative approach to understanding the intricacies of biological processes within living organisms. These techniques continue to evolve, enhancing our ability to explore dynamic interactions and providing valuable information for both basic research and clinical applications. Noninvasive imaging techniques for longitudinal studies offer a powerful tool for understanding the dynamic nature of biological processes. These methods continue to evolve, enhancing our ability to observe and analyze changes over time and providing valuable information for both basic research and clinical applications. Imaging modalities for tracking disease progression offer a critical means to visualize and understand the dynamic changes occurring within the body. As these techniques continue to advance, they will play an increasingly vital role in guiding diagnosis, treatment decisions, and the development of novel therapies.

Therapeutic approaches and testing are essential components of medical progress, driving the development of effective treatments for a wide range of diseases. As research methods and technologies continue to advance, the potential for more targeted and personalized therapies becomes increasingly promising. Preclinical drug testing using animal models remains a cornerstone of drug development, providing critical data that informs decisions regarding the progression of potential therapies into human trials. As technology advances, the integration of various approaches will continue to refine and improve the accuracy of preclinical testing. Gene therapy trials in rare disease models represent a transformative approach to addressing genetic disorders at their root causes. As ongoing research continues to refine techniques and expand the scope of treatable conditions, the potential for improving the lives of patients with rare diseases becomes increasingly promising.

Ethical considerations in animal research remain a complex and evolving topic. As the field progresses, it is crucial to continually reassess and adapt ethical standards to ensure responsible and compassionate treatment of animals while advancing medical knowledge. Ethical guidelines are essential for maintaining the integrity of animal research and upholding the welfare of research subjects. As the field progresses, adherence to these guidelines remains crucial to ensuring responsible and ethically sound scientific practices. In vitro and computational models are revolutionizing research by providing effective alternatives to animal studies. As technology advances, these models will continue to play a crucial role in advancing our understanding of diseases and developing new treatments. Case studies underscore the transformative potential of in vitro and computational models in various research domains. As these models continue to evolve and become more sophisticated, their widespread adoption promises to reshape the landscape of biomedical research and contribute to improved healthcare outcomes.

Model-based insights are driving transformative changes in therapeutic development, enabling researchers to uncover new avenues for treatment and refine existing strategies. As technology evolves and our understanding of complex diseases deepens, model-based approaches will continue to play a central role in shaping the future of medicine. Collaborative efforts between researchers and clinicians drive the translation of scientific discoveries into tangible benefits for patients. As medicine becomes increasingly personalized and innovative, the synergy between these two groups will continue to shape the landscape of healthcare and lead to more effective and targeted therapies.

Collaborative research between researchers and clinicians offers immense potential to drive scientific advancements and improve patient care. By addressing challenges and implementing strategies for effective collaboration, the future of medical research holds promise for more innovative and impactful outcomes. Bridging the gap between animal models and human therapies requires a multifaceted approach that integrates advanced models, translational strategies, and a comprehensive understanding of the limitations of animal research. By addressing these challenges and leveraging innovative approaches, researchers can enhance the translation of scientific discoveries into effective treatments for human diseases. Emerging technologies have transformed rare disease modeling, enabling more accurate representations of disease mechanisms and accelerating the development of targeted therapies. As these technologies continue to evolve, they hold the promise of revolutionizing our approach to understanding and treating rare diseases.

Successfully navigating the regulatory landscape and translating rare disease research into clinical trials requires a comprehensive understanding of regulatory pathways, ethical considerations, and strategic approaches to maximize the potential for therapeutic breakthroughs. Regulatory guidelines from the FDA and EMA provide a framework for developing safe and effective therapies for rare diseases. Navigating these guidelines requires a deep understanding of the regulatory landscape and a commitment to patient-centered approaches. Conducting clinical trials for rare diseases requires careful consideration of patient recruitment, trial design, ethical concerns, and patient engagement to ensure the successful development of effective therapies.

Collaborative networks and data sharing are essential for advancing rare disease research, leveraging collective knowledge and resources to accelerate discoveries and therapeutic breakthroughs. Rare disease research consortia demonstrate the power of collaborative efforts, driving advancements in understanding rare diseases, identifying therapeutic targets, and ultimately improving the lives of patients. Data repositories and open science initiatives drive rare disease research forward by enabling broad data sharing, collaboration, and fostering a culture of transparency and innovation.

In closing, the contributions of animal models to rare disease research are profound and far-reaching. By unraveling the mysteries of rare diseases, advancing therapeutic approaches, and ultimately improving patient outcomes, animal models stand as instrumental tools that continue to drive innovation and pave the way for a brighter future in the realm of rare disease research. With this comprehensive exploration, we are sure that the use of animal models in rare disease research, recognizing their invaluable role in reshaping the medical landscape and offering hope to those affected by rare diseases, paves a way for finding new paths and means to combat rare diseases. As we stand on the precipice of future research, it is evident that the potential for discovery is vast. The synergy between cutting-edge technologies, collaborative endeavors, ethical considerations, and patient-centered approaches will shape a new era of rare disease research. This chapter serves as a testament to the progress made so far and an inspiration for the breakthroughs yet to come.

Keywords

Rare genetic diseases · Animal models · Genetic engineering · Therapeutic approaches

4.1 Rare Diseases: An Overview

Rare diseases, often referred to as orphan diseases, are a diverse group of disorders that individually affect a small number of individuals within the population. Despite their rarity, collectively they impact a significant portion of the global population (Austin et al. 2018). These diseases encompass a wide range of medical conditions, each with its own unique challenges, characteristics, and implications for patients and their families (Rath et al. 2012; Angelis et al. 2015). In this chapter, we provide an overview of rare diseases, highlighting their prevalence, challenges in diagnosis and treatment, and the importance of research in this field.

4.1.1 Prevalence and Classification

Rare diseases are defined differently across various regions, with the common criterion being that they affect a limited number of individuals. In the United States, a rare disease is one that affects fewer than 200,000 people, whereas the European Union defines it as affecting fewer than 1 in 2000 individuals. Despite the distinct classification thresholds, it is estimated that there are around 7000 recognized rare diseases globally, as opined by Taruscio et al. (2014).

4.1.2 Challenges in Diagnosis and Treatment

- **Delayed Diagnosis:** Due to their rarity, many rare diseases go undiagnosed or misdiagnosed for years. This delay often leads to unnecessary suffering and prevents timely interventions.

- **Lack of Awareness:** Both healthcare professionals and the general public may lack awareness of rare diseases, resulting in inadequate support systems and resources for affected individuals.
- **Limited Treatment Options:** The rarity of these diseases often means that there are limited treatment options available. Developing new therapies for such small patient populations can be economically challenging.
- **High Costs:** Rare diseases often require specialized medical care, which can lead to financial burden on patients and their families.

4.1.3 Impact on Patients and Families

Rare diseases have profound implications for patients and their families such as the following:

- **Psychological Impact:** The uncertainty, lack of information, and challenges in obtaining an accurate diagnosis can take a toll on the mental well-being of patients and their families.
- **Isolation:** The rarity of these diseases can lead to feelings of isolation, as patients may struggle to find others with similar conditions for support.
- **Limited Research:** Funding and resources for research on rare diseases are often limited, hindering the development of effective treatments.

4.1.4 Importance of Research and Future Directions

Research is paramount to improving the understanding, diagnosis, and treatment of rare diseases:

- **Genetic Insights:** Advances in genetics have facilitated the identification of causative genetic mutations, enabling more precise diagnosis and potential targeted therapies.
- **Precision Medicine:** Research allows for the development of personalized treatment approaches that take into account the specific genetic and molecular characteristics of each rare disease.
- **Repurposing Drugs:** Existing drugs developed for other conditions might hold potential for treating rare diseases. Research from this angle can help identify these opportunities.
- **Advancements in Technology:** Technologies like gene editing, CRISPR-Cas9, and high-throughput screening offer new avenues for studying and treating rare diseases.

4.1.4.1 Defining Rare Diseases and Their Challenges

Rare diseases, also referred to as orphan diseases, are a diverse group of disorders that collectively impact a relatively small number of individuals within the population. The exact prevalence threshold that qualifies a disease as "rare" varies by region, but it is generally defined as affecting fewer than 1 in 2000 individuals. Despite their low prevalence individually, rare diseases collectively affect millions of people globally (EURODRIS 2005; Gahl et al. 2012; Austin et al. 2018).

Characteristics of Rare Diseases

Rare diseases are characterized by their heterogeneity, often displaying a wide range of symptoms and severity levels. They can be genetic or acquired or result from a combination of genetic and environmental factors. Due to their rarity, diagnosing rare diseases can be challenging, leading to delays or misdiagnoses. Additionally, the lack of awareness and expertise in rare diseases among healthcare professionals further compounds these diagnostic difficulties.

Challenges in Rare Disease Research

Research on rare diseases faces several unique challenges such as the following:

- **Limited Data and Sample Availability:** Gathering sufficient clinical and molecular data for rare diseases can be difficult due to the small number of affected individuals. This scarcity of data can hinder the identification of disease mechanisms and potential treatment targets.
- **Lack of Research Funding:** Rare diseases often receive less research funding compared to more common conditions. This can impede the development of targeted therapies and the advancement of scientific understanding.
- **Disease Heterogeneity:** The heterogeneity of rare diseases, even among individuals with the same genetic mutation, complicates research efforts. Understanding the underlying causes of this variability is crucial for personalized medicine approaches.
- **Limited Expertise:** The rarity of these diseases contributes to a shortage of experts and specialized healthcare professionals. This scarcity can lead to delayed diagnosis and inadequate care.
- **Regulatory Challenges:** Clinical trials for rare disease treatments face regulatory challenges, as the small patient populations can make it difficult to demonstrate statistically significant outcomes.
- **Ethical Considerations:** Ethical concerns arise in animal models for rare diseases due to the need for careful consideration of animal welfare and the potential benefits of the research.

Despite these challenges, advances in genetics, molecular biology, and animal modeling have provided new opportunities to study and address rare diseases.

4.1.4.2 Understanding the Need for Animal Models

Animal Models in Rare Diseases

Animal models play a crucial role in biomedical research, offering valuable insights into the mechanisms underlying various diseases, including rare diseases. Rare diseases, also known as orphan diseases, are characterized by their low prevalence in the population, often affecting fewer than 1 in 2000 individuals. Despite their rarity, the cumulative impact of rare diseases is substantial, with thousands of different conditions collectively affecting millions of people worldwide.

Understanding the pathophysiology, progression, and potential treatments for rare diseases is challenging due to their limited occurrence and heterogeneity. Animal models provide an essential bridge between fundamental research and clinical applications. These models involve the use of various animal species, such as mice, rats, zebrafish, dogs, and nonhuman primates, to mimic the disease conditions observed in humans. Animal models allow researchers to investigate disease mechanisms, test potential therapies, and develop a deeper understanding of the underlying biology (NIH 2021).

The use of animal models in rare disease research is driven by the need to bridge the gap between basic scientific understanding and clinical applications. Several compelling reasons highlight the importance of animal models in advancing our knowledge of rare diseases and developing potential therapies, as pointed out by the NRC (2011). They are as follows:

- **Disease Mechanism Investigation:** Animal models allow researchers to study the underlying mechanisms of rare diseases in a controlled environment. By inducing or mimicking disease conditions in animals, scientists can observe disease progression, identify molecular pathways, and investigate interactions between genes, proteins, and other factors involved in disease development.
- **Therapeutic Development:** Developing treatments for rare diseases requires a comprehensive understanding of the disease biology. Animal models serve as valuable platforms for testing potential therapeutic interventions, including drugs, gene therapies, and cell-based therapies. These models provide insights into the efficacy, safety, and optimal dosing of potential treatments before human trials.
- **Phenotypic and Genotypic Studies:** Animal models enable the observation of both phenotypic (observable traits) and genotypic (genetic) changes associated with rare diseases. This aids in identifying biomarkers, understanding genotype-phenotype correlations, and refining diagnostic techniques.
- **Validation of Target Molecules:** Animal models help validate potential target molecules or pathways that are implicated in rare diseases. Through genetic manipulation or pharmacological interventions, researchers can determine whether modifying specific targets leads to disease amelioration or prevention.
- **Understanding Disease Heterogeneity:** Rare diseases often exhibit significant heterogeneity in terms of clinical presentation and disease progression. Animal models provide a controlled setting to study this variability and gain insights into factors contributing to disease heterogeneity.
- **Translation to Clinical Trials:** Successful outcomes in animal models can guide the design and execution of clinical trials for rare disease treatments. Understanding the safety and efficacy of interventions in animal models can accelerate the development of therapies for human patients.

Substantiating the Role of Animal Models in Rare Disease Research

Animal models play a pivotal role in advancing our understanding of rare diseases, offering invaluable insights into disease mechanisms, potential treatments, and therapeutic interventions. These models provide a bridge between basic scientific research and clinical applications, allowing researchers to study the complexities of rare diseases in a controlled environment before translating findings to human trials. Let us understand the multifaceted role of animal models in rare disease research, supported by various research studies.

Understanding Disease Mechanisms

- **Unraveling Complex Pathways:** Animal models allow researchers to dissect the intricate molecular and cellular pathways underlying rare diseases. By studying disease progression and interactions between genes, proteins, and cells, researchers can uncover underlying mechanisms (Holcik and Sonenberg 2005).
- **Modeling Genetic Variations:** Genetically engineered animal models can replicate mutations responsible for rare diseases, providing insights into how genetic alterations lead to specific clinical manifestations (Orso et al. 2005).

Evaluating Therapeutic Approaches

- **Testing Treatment Efficacy:** Animal models enable researchers to assess the effectiveness of potential treatments, including drugs, gene therapies, and other interventions, before progressing to human trials (Lu and Vogel 2009).
- **Personalized Medicine:** Animal models contribute to developing personalized treatment strategies by testing interventions tailored to individual patient profiles and genetic backgrounds (Baxendale et al. 2019).

Investigating Disease Progression

- **Longitudinal Studies:** The longer life span of some animal models allows researchers to observe disease progression over extended periods, providing insights into age-related rare diseases (Cattaneo et al. 2001).

Drug Discovery and Development

- **Screening Drug Candidates:** Animal models facilitate the preclinical evaluation of potential drug candidates, screening for safety, efficacy, and potential side effects (Hrabě de Angelis and Balling 1998; Fricker 2005).
- **Mechanism of Action Studies:** Animal models help elucidate how therapeutic agents work at the molecular level, enhancing our understanding of treatment mechanisms (McKinney and Holmen 2011).

4.2 Choosing and Developing Animal Models

Selecting and developing appropriate animal models is a critical step in rare disease research. The choice of an animal model greatly influences the relevance, reliability, and translational value of the research findings. Developing effective animal models involves careful consideration of various factors, including the disease characteristics, genetic manipulation techniques, and ethical considerations.

4.2.1 Factors Influencing Model Selection

- **Disease Similarity:** The animal model should mimic the key aspects of the rare disease in terms of symptoms, genetics, and molecular mechanisms. The closer the model resembles the human condition, the more reliable the research outcomes.
- **Genetic Manipulation Techniques:** Different animal species offer various genetic manipulation methods, such as gene knockouts, knock-ins, and transgenics. The availability and feasibility of these techniques influence model selection.
- **Availability of Resources:** Factors like breeding facilities, cost, and expertise required for maintaining and conducting experiments with the chosen animal model are important considerations.
- **Ethical and Welfare Considerations:** Ethical guidelines and animal welfare should guide model selection. Researchers should aim to minimize harm to animals while achieving scientific objectives.

4.2.2 Developing Animal Models

- **Genetic Models:** Genetic models involve altering the animal's genome to carry disease-associated mutations. Techniques like CRISPR-Cas9 have revolutionized the development of precise genetic models (Erickson 1996).
- **Induced Models:** Induced models mimic disease conditions by exposing animals to chemicals, toxins, or environmental factors that trigger disease-like symptoms (Hisahara and Shimohama 2011).
- **Spontaneous Models:** Some animal species naturally develop conditions similar to rare diseases. Identifying and studying these spontaneous models can provide valuable insights (Khorramizadeh and Saadat 2020).
- **Phenotypic Characterization:** Once a model is developed, thorough phenotypic characterization is crucial. This involves assessing physiological, behavioral, and molecular changes that mimic the human disease (Mekada et al. 2009; Brown and Moore 2012).

4.2.2.1 Criteria for Selecting Animal Models

The selection of appropriate animal models for rare disease research is a crucial decision that significantly impacts the translational relevance and success of the research outcomes. To ensure that the chosen model effectively recapitulates the key aspects of human disease, researchers must consider several important criteria as suggested by Russel and Burch (1959), Robinson et al. (2004), Rosenthal and Brown (2007), van der Worp et al. (2010), and Swindle et al. (2012). The major criteria are the following:

- **Disease Similarity:** The selected animal model should closely mimic the clinical and molecular features of the human rare disease. This includes replicating the symptoms, progression, and underlying mechanisms. A high degree of disease similarity enhances the relevance of research findings.
- **Genetic Homology:** Genetic similarities between the chosen animal species and humans are essential. The presence of orthologous genes associated with the disease in both species increases the likelihood of the animal model effectively representing the human condition.
- **Reproducibility:** The model should yield consistent and reproducible results across different experiments and researchers. This ensures that the findings can be verified and validated, enhancing the credibility of the research.
- **Availability and Accessibility:** The animal species and strains chosen should be readily available and accessible to researchers. This facilitates collaborative research efforts and the exchange of knowledge and resources.
- **Manipulability:** The chosen animal model should be amenable to genetic manipulation techniques that allow researchers to introduce disease-associated mutations or modify specific genes. This manipulation enhances the ability to study disease mechanisms and potential interventions.
- **Ethical Considerations:** Researchers must adhere to ethical guidelines and prioritize animal welfare when selecting models. Models that require minimal harm to animals while achieving scientific objectives are preferable.
- **Translational Potential:** The insights gained from the animal model should have the potential to translate into clinical applications. Models that provide valuable information for the development of therapies or diagnostic tools are particularly valuable.
- **Feasibility and Cost-Effectiveness:** The resources required to develop and maintain the chosen animal model should be feasible and cost-effective, considering factors like housing, care, and experimental procedures.

4.2.2.2 Translational Considerations in Model Development

Developing animal models for rare diseases involves not only replicating disease conditions but also ensuring that the findings have translational relevance. In this context, translational considerations are essential to bridge the gap between preclinical research and clinical applications, ultimately benefiting patients with rare diseases (Kola and Landis 2004; Nestler and Hyman 2010; Varga et al. 2010). The prerequisites for an ideal animal model are as follows:

- **Predictive Validity:** An effective animal model should have predictive validity, meaning that the results obtained from the model can accurately predict outcomes in humans. This requires the model to accurately reproduce the disease's pathophysiology, progression, and response to interventions.
- **Biomarker Identification:** Animal models should aid in identifying biomarkers—measurable indicators of disease—that can be used for early diagnosis, disease monitoring, and assessing treatment efficacy in humans.

- **Treatment Testing:** Animal models are instrumental in preclinical testing of potential therapies. The chosen model should be sensitive enough to detect therapeutic effects and provide insights into dosing, administration routes, and safety profiles.
- **Disease Progression Studies:** Studying disease progression in animal models can help identify critical time points for intervention and guide the development of targeted treatments.
- **Translational Imaging:** Techniques such as medical imaging can be used to monitor disease progression and treatment responses in animal models, providing insights that can guide clinical trials.
- **Safety and Toxicity Assessments:** Animal models allow for the assessment of treatment safety, including potential adverse effects and toxicity. This information is essential for designing safe human trials.

4.2.2.3 Genetic Modification Techniques for Model Creation

Creating effective animal models for rare diseases often involves genetic manipulation to introduce disease-associated mutations or modify specific genes. Several genetic modification techniques enable researchers to develop models that closely mimic the molecular and clinical aspects of human rare diseases, such as the following:

- **CRISPR-Cas9:** CRISPR-Cas9 technology has revolutionized genetic modification. It allows precise and targeted editing of DNA by using a guide RNA to direct the Cas9 enzyme to specific genomic locations. CRISPR-Cas9 enables the introduction of disease-causing mutations, gene knockouts, knock-ins, and precise modifications (Zhang et al. 2011; Hsu et al. 2013, 2014).
- **Transgenic Techniques:** Transgenic models involve introducing exogenous genes or gene constructs into the genome of an animal (Gordon and Ruddle 1981; Palmiter et al. 1982). This allows researchers to study the effects of specific genes or mutations on disease development and progression.
- **Knockout and Knock-In Models:** Knockout models involve disrupting a specific gene's function, while knock-in models involve introducing specific mutations into the gene (Capecchi 1989; Mansour et al. 1988). These models help elucidate gene function and assess the effects of gene manipulation on disease phenotypes.
- **Conditional Knockout Models:** Conditional knockout models allow for gene deletion in specific tissues or at particular developmental stages (Gu et al. 1994; Loonstra et al. 2001). This technique helps researchers understand tissue-specific gene functions and their contributions to disease.

4.3 Animal Models Used for Rare Disease Research

Animal models have revolutionized our understanding of rare diseases, offering insights into disease mechanisms, therapeutic interventions, and potential treatments. Various animal species serve as essential tools for researchers to study these conditions in a controlled environment before translating their findings to human clinical trials. Let us delve into the key animal models utilized in rare disease research and their contributions.

4.3.1 Rodent Models for Rare Diseases

Rodent models, particularly mice and rats, have become indispensable tools in studying rare diseases. These models offer several advantages, such as genetic tractability, relatively short generation times, and well-established techniques for manipulation and analysis. Below, we explore the significance of rodent models for rare disease research and highlight specific examples.

4.3.1.1 Advantages of Rodent Models

- **Genetic Manipulation:** Mice and rats are amenable to various genetic manipulation techniques, including gene knockout, knock-in, and conditional knockout strategies, which allow researchers to mimic specific genetic mutations associated with rare diseases.
- **Conserved Genetic and Physiological Features:** Many genetic and physiological features are conserved between rodents and humans. This conservation enhances the relevance of findings obtained from rodent models.
- **Short Reproductive Cycles:** Rodents have relatively short reproductive cycles, enabling researchers to study multiple generations in a relatively short timeframe. This accelerates the study of disease inheritance and progression.
- **Resource Availability:** Extensive resources, such as genetic databases, reagents, and well-characterized strains, are available for rodents. This facilitates the creation and maintenance of animal models.

4.3.1.2 Examples of Rodent Models for Rare Diseases

- **Cystic Fibrosis (CF), Mouse Model:** The Cftr^{tm1UNC} mouse model mimics the genetic mutation found in cystic fibrosis patients. These mice exhibit symptoms like lung infections and impaired airway clearance, closely resembling human CF symptoms (Snouwaert et al. 1992).
- **Duchenne Muscular Dystrophy (DMD), Mouse Model:** The mdx mouse carries a spontaneous mutation in the dystrophin gene and is widely used as a model for Duchenne muscular dystrophy (Bulfield et al. 1984). These mice exhibit muscle degeneration and weakness similar to the human disease.
- **Huntington's Disease, Mouse Model:** Transgenic mice expressing mutant forms of the huntingtin gene develop neurological symptoms similar to

Huntington's disease. These models help researchers understand the disease's progression and test potential therapies (Mangiarini et al. 1996).

- **Spinal Muscular Atrophy (SMA), Mouse Model:** The SMN < sup>–/–</ sup > mouse model lacking the survival motor neuron (Smn) gene exhibits motor neuron degeneration and muscle weakness, resembling human SMA (Monani et al. 2000).

4.3.1.3 Mouse Models

Mice, particularly genetically engineered strains, are extensively used due to their genetic tractability and rapid reproduction. They have contributed to studying a wide range of rare diseases, from neurodegenerative disorders to metabolic conditions (Collins et al. 2007).

Mouse models have emerged as a cornerstone in rare disease research due to their genetic similarity to humans, well-characterized genetics, and ease of manipulation. However, while they offer numerous advantages, it is important to consider their limitations as well. Here, we explore both the advantages and limitations of using mouse models for studying rare diseases based on the previous works of several researchers (Threadgill and Churchill 2012; Seok et al. 2013; Cook et al. 2014; Carvalho and Lupski 2016; Justice and Dhillon 2016).

4.3.1.4 Advantages of Mouse Models

- **Genetic Similarity:** Mice share a significant portion of their genes with humans, making them a relevant model for studying genetic diseases.
- **Genetic Manipulation:** Techniques like CRISPR-Cas9 allow researchers to precisely manipulate genes in mice, creating models that closely mimic the genetic mutations observed in rare diseases.
- **Phenotypic Analysis:** Mice can be thoroughly phenotyped, enabling researchers to assess disease symptoms, progression, and treatment responses.
- **Well-Characterized Strains:** Researchers can choose from a wide range of inbred strains with known genetic backgrounds, aiding in reproducibility and consistency.
- **Availability of Resources:** Extensive resources such as genetically modified strains, databases, and reagents are available, streamlining research efforts.
- **Short Generation Times:** Mice have relatively short life cycles and reproductive times, allowing for rapid breeding and multiple generations within a reasonable timeframe.

4.3.1.5 Limitations of Mouse Models

- **Genetic Differences:** Despite genetic similarity, some disease mechanisms in mice may not fully recapitulate human disease processes.
- **Complexity:** Rare diseases often exhibit complex symptoms and interactions that may not be accurately modeled in mice.
- **Phenotypic Variation:** Even within the same strain, mice can show significant phenotypic variation, which can complicate data interpretation.

- **Ethical and Welfare Considerations:** The use of animals in research raises ethical concerns, and specific rare disease models may involve considerable suffering.
- **Translation to Humans:** Successful findings in mouse models do not always directly translate to human therapies due to differences in biology and physiology.
- **Clinical Endpoint Differences:** The clinical endpoints in mouse models might differ from those in humans, potentially leading to discrepancies in treatment effects.

4.3.1.6 Rat Models

Rats provide advantages such as larger size and anatomical similarity to humans. They are used to study cardiovascular diseases, neurological disorders, and metabolic conditions. While mouse models dominate the field of animal research, rat models also play a vital role, particularly in studying rare diseases. Rats offer distinct advantages that complement those of mice, making them valuable tools for understanding the pathophysiology and potential treatments of rare diseases. The applications of rat models in rare disease research are enumerated below.

4.3.1.7 Advantages of Rat Models

- **Size and Physiology:** Rats are larger than mice, making them more suitable for certain surgical procedures, physiological measurements, and imaging techniques.
- **Anatomy Similarity:** Rats share significant anatomical similarities with humans, making them well suited for studies that involve surgical interventions or disease manifestations in specific organs.
- **Behavioral Studies:** Rats exhibit more complex behaviors compared to mice, making them suitable for studying neurodegenerative diseases and behavioral phenotypes associated with rare diseases.
- **Metabolism:** Rats have metabolic characteristics closer to humans, making them relevant for metabolic disorders and drug metabolism studies.
- **Pharmacokinetics and Toxicology:** Rat models provide insights into drug absorption, distribution, metabolism, and excretion, aiding in preclinical drug testing.

4.3.1.8 Applications of Rat Models in Rare Disease Research

- **Neurological Disorders:** Rat models have been instrumental in studying neurological rare diseases such as Huntington's disease, spinal muscular atrophy (SMA), and Rett syndrome, offering insights into disease mechanisms and potential therapeutic interventions (Bäumer et al. 2009).
- **Cardiovascular Diseases:** Rats are utilized in studying genetic cardiovascular disorders like hypertrophic cardiomyopathy and arrhythmogenic right ventricular cardiomyopathy (McCauley and Wehrens 2009).
- **Metabolic Disorders:** Rats are used to model metabolic rare diseases such as lysosomal storage disorders and inherited metabolic disorders (Haskins et al. 2006).

- **Bone and Skeletal Disorders:** Rat models aid in understanding genetic skeletal disorders like *osteogenesis imperfecta* and *achondroplasia* (Turner et al. 2001).
- **Renal Disorders:** Rat models are employed in studying genetic renal disorders, including polycystic kidney disease and Alport syndrome (Igarashi and Somlo 2002).

4.3.2 Non-rodent Animal Models

While rodent models, particularly mice and rats, are widely used in rare disease research, non-rodent animal models also contribute significantly to our understanding of these conditions. Various non-rodent species offer unique advantages and insights into the pathophysiology, genetics, and treatment strategies for rare diseases. The applications and significance of non-rodent animal models in rare disease research are explored.

4.3.2.1 Advantages of Non-rodent Animal Models

- **Physiological Relevance:** Non-rodent animals often share closer physiological similarities with humans in certain aspects, making them suitable for studying specific disease mechanisms and treatments.
- **Anatomical Specificity:** Some non-rodent animals possess anatomical structures or characteristics that closely resemble those in humans, allowing for more accurate modeling of disease manifestations.
- **Longevity and Development:** Species with longer life spans or slower development rates can be beneficial for studying age-related rare diseases or diseases with extended progression.
- **Behavioral Complexity:** Non-rodent animals, such as primates, exhibit more complex behaviors that are relevant for studying the neurological and behavioral aspects of rare diseases.
- **Immune Responses:** Certain non-rodent animals have immune systems more similar to humans, aiding in studying immune-related rare diseases and therapeutic interventions.

4.3.2.2 Non-rodent Animal Models Used in Research

1. **Primate Models:** Nonhuman primates, including monkeys and apes, closely resemble humans in terms of genetics and physiology. They are crucial for studying complex neurological and infectious diseases (Emborg 2007).
2. **Canine Models:** Dogs offer genetic diversity and share similar environments with humans, making them valuable for studying genetic, cardiovascular, and metabolic rare diseases (Kornegay et al. 2016).
3. **Feline Models:** Cats provide insights into diseases affecting the eyes, skeletal system, and metabolic pathways. They are used to study genetic disorders like *retinitis pigmentosa* and lysosomal storage diseases (La Croix 2005).

4. **Porcine Models:** Pigs offer anatomical and physiological similarities to humans, making them suitable for studying cardiovascular, metabolic, and genetic disorders (Kobayashi et al. 2012).
5. **Zebrafish Models:** Zebrafish provide insights into developmental disorders, neurological diseases, and drug screening due to their transparency, rapid development, and genetic manipulability (Lieschke and Currie 2007).

4.3.2.3 Applications of Non-rodents in Rare Disease Research

- **Primate Models:** Nonhuman primates, such as monkeys and apes, share genetic and physiological similarities with humans. They are crucial for understanding diseases like Alzheimer's, Parkinson's, and HIV/AIDS due to their complex brain structures and immune systems (Kang et al. 2019).
- **Canine Models:** Dogs have been valuable models for diseases like muscular dystrophy, cancer, and heart conditions. Their size, life span, and genetics offer insights into disease progression and potential treatments (Kornegay 2017).
- **Feline Models:** Cats are valuable for studying genetic diseases such as retinitis pigmentosa and lysosomal storage disorders due to their eye structure and genetics (Narfström 1999).
- **Porcine Models:** Pigs are increasingly used for studying metabolic, cardiovascular, and gastrointestinal rare diseases due to their anatomical and physiological similarities to humans (Klymiuk et al. 2016).
- **Zebrafish Models:** Zebrafish offer rapid development, transparency, and genetic manipulability. They are useful for studying developmental disorders, neurological diseases, and drug screening (Lieschke and Currie 2007).

4.3.2.4 Primate Models

Primate models, including nonhuman primates such as monkeys and apes, have become invaluable assets in the realm of rare disease research. Their genetic proximity to humans, complex physiological systems, and cognitive capabilities offer unique opportunities to unravel the intricacies of rare diseases. The role of primate models in rare disease research and their profound contributions to our understanding of these conditions are discussed in the following section.

4.3.2.5 Advantages of Primate Models

- **Genetic Similarity:** Nonhuman primates share a substantial genetic resemblance to humans, making them ideal for studying rare genetic diseases that affect both species.
- **Complex Physiology:** Primate physiology closely mirrors that of humans, allowing for the investigation of disease mechanisms, pathophysiology, and treatment responses with higher translational relevance.
- **Neurological and Cognitive Insights:** The sophisticated cognitive abilities of primates enable the study of neurological rare diseases, shedding light on cognitive impairments, behavior, and brain function.

- **Long Life Span:** Primate models offer longer life spans compared to other animal models, enabling the study of age-related rare diseases and observing disease progression over extended periods.

4.3.2.6 Applications of Primate Models in Rare Disease Research

- **Neurodegenerative Diseases:** Primate models play a pivotal role in studying complex neurodegenerative disorders like Alzheimer's, Parkinson's, and amyotrophic lateral sclerosis (ALS), offering insights into disease mechanisms and potential therapies (Emborg 2007).
- **Infectious Diseases:** Nonhuman primates, such as macaques, are instrumental in researching infectious rare diseases like HIV/AIDS, helping develop treatments and vaccines, and understanding disease progression (Van Rompay 2017).
- **Genetic Disorders:** Primate models aid in comprehending genetic rare diseases such as spinal muscular atrophy (SMA), phenylketonuria (PKU), and muscular dystrophy, advancing our knowledge of disease mechanisms and potential interventions (Barchet and Amiji 2009).
- **Cardiovascular Diseases:** Primate models contribute significantly to cardiovascular disease research, including hypertension, atherosclerosis, and heart diseases, facilitating testing of interventions and treatments (Leong et al. 2015).
- **Metabolic Disorders:** Primate models provide insights into metabolic rare diseases like obesity, diabetes, and metabolic syndrome, deepening our understanding of disease progression and potential interventions (Havel et al. 2017).

4.3.2.7 Canine Models

Canine models have proven to be indispensable assets in advancing our knowledge of rare diseases. Dogs, with their genetic diversity, physiological similarities to humans, and shared environment, offer a unique perspective that complements rodent models and enhances translational research. Let us delve into the applications of canine models in rare disease research and the contributions they make.

4.3.2.8 Advantages of Canine Models

- **Genetic Diversity:** Dogs exhibit a wide genetic diversity within different breeds, allowing for the study of both spontaneous and genetically engineered disease models.
- **Shared Environment:** Dogs share our environment and are exposed to similar pathogens, pollutants, and lifestyles, enhancing the relevance of findings to human diseases.
- **Physiological Similarities:** Canine physiology closely resembles that of humans, especially in areas like cardiovascular and musculoskeletal systems.
- **Size and Anatomy:** Dogs' larger size enables procedures and interventions that are more aligned with those in humans, such as surgical treatments and implantation studies.

4.3.2.9 Applications of Canines in Rare Disease Research

- **Duchenne Muscular Dystrophy (DMD):** Dogs with naturally occurring mutations in the dystrophin gene provide valuable insights into DMD pathogenesis and treatment testing (Kornegay 2017).
- **Neurological Disorders:** Canine models have contributed to understanding rare neurological disorders like canine degenerative myelopathy and spinal cord injuries (Hytönen and Lohi 2016).
- **Lysosomal Storage Disorders:** Dogs are used to model lysosomal storage disorders such as Pompe disease and mucopolysaccharidoses, aiding in understanding disease mechanisms and testing therapies (Xu et al. 2016).
- **Heart Diseases:** Canine models of heart diseases, including dilated cardiomyopathy and arrhythmias, have been instrumental in studying disease progression and testing interventions (Meurs et al. 2007a).
- **Cancer:** Canine models of cancer provide insights into tumor biology, treatment responses, and drug development, benefiting both veterinary and human oncology (Khanna and Hunter 2005).

4.3.2.10 Feline Models

Feline models have proven to be valuable tools in advancing our understanding of various rare diseases. Cats share genetic and physiological similarities with humans, making them relevant for studying genetic disorders, metabolic diseases, and more. Let us discuss about the applications of feline models in rare disease research and the insights they provide in the forthcoming section.

4.3.2.11 Advantages of Feline Models

- **Genetic Homology:** Cats share a significant portion of their genome with humans, making them suitable for studying genetic diseases with shared genetic mutations.
- **Anatomical and Physiological Similarities:** Cats have organ systems and physiology that are comparable to humans, allowing for better translation of findings to human diseases.
- **Spontaneous Disease Models:** Cats naturally develop some diseases that closely resemble human conditions, providing researchers with ready-made models for study.
- **Long Life Span:** Cats' longer life spans enable the observation of disease progression over extended periods, crucial for understanding age-related rare diseases.

4.3.2.12 Applications of Feline Models in Rare Disease Research

- **Retinitis Pigmentosa (RP):** Cats are valuable models for studying RP due to their similar eye anatomy and the occurrence of naturally occurring retinal degeneration (Narfström 1999).
- **Lysosomal Storage Disorders:** Feline models, such as the domestic shorthair cat model for GM1 gangliosidosis, have provided insights into the pathophysiology of lysosomal storage disorders (Baker and Lindsey 1974).

- **Hypertrophic Cardiomyopathy (HCM):** Cats with spontaneous HCM serve as a model for studying genetic cardiac disorders and testing therapeutic interventions (Meurs et al. 2007b).
- **Feline Immunodeficiency Virus (FIV):** Feline models of FIV provide insights into immune responses, antiviral treatments, and vaccine development for human immunodeficiency virus (HIV) (Miller et al. 2018).
- **Mucopolysaccharidosis (MPS):** Cats with MPS I and MPS VI mimic human disease symptoms and progression, making them valuable models for testing potential treatments (Haskins 2007).

4.3.2.13 Porcine Models

Porcine (pig) models have gained prominence in rare disease research due to their physiological and anatomical similarities to humans. These models offer valuable insights into various rare diseases, helping researchers understand disease mechanisms, develop treatments, and test interventions. The applications of porcine models in rare disease research and the contributions they make are listed below:

4.3.2.14 Advantages of Porcine Models

- **Anatomical and Physiological Similarities:** Pigs share similarities with humans in terms of organ systems, metabolism, and disease processes, enhancing the translational relevance of findings.
- **Size and Comparative Anatomy:** Pigs' larger size allows for procedures and interventions that are not feasible in smaller animals, providing more accurate modeling of human conditions.
- **Genetic Homology:** Pigs possess genetic homology with humans for many genes, making them useful for studying genetic and hereditary rare diseases.
- **Metabolic Relevance:** Pigs have metabolism similar to humans, making them valuable for studying metabolic disorders and drug metabolism.

4.3.2.15 Applications of Porcine Models in Rare Disease Research

- **Cystic Fibrosis (CF):** Pigs with a CFTR gene mutation mimic the lung disease seen in humans with cystic fibrosis, providing insights into disease progression and testing potential therapies (Rogers et al. 2008).
- **Muscular Dystrophy:** Porcine models of muscular dystrophy, such as Duchenne muscular dystrophy (DMD), offer insights into muscle degeneration and potential treatments (Klymiuk et al. 2013).
- **Atherosclerosis:** Porcine models of atherosclerosis contribute to understanding cardiovascular diseases and testing interventions like stents and drug therapies (Emini Veseli et al. 2017).
- **Alpha-1 Antitrypsin Deficiency:** Porcine models of alpha-1 antitrypsin deficiency replicate the lung and liver involvement seen in humans, aiding in understanding disease mechanisms (Gutierrez et al. 2015).
- **Neurodegenerative Diseases:** Pigs are used to model neurodegenerative diseases like Parkinson's and Alzheimer's, providing insights into brain physiology and potential treatments (Dodson et al. 2015).

4.3.2.16 Zebrafish Models

Zebrafish (*Danio rerio*) have emerged as powerful models in biomedical research, including the study of rare diseases. These small, transparent aquatic organisms offer unique advantages that have revolutionized our understanding of disease mechanisms, drug screening, and developmental processes. Let us explore the applications of zebrafish models and the insights they provide in the context of rare diseases.

4.3.2.17 Advantages of Zebrafish Models

- **Transparency and Rapid Development:** Zebrafish embryos are transparent, allowing researchers to observe developmental processes and disease progression in real time. Their rapid external development facilitates the study of embryogenesis and organogenesis.
- **Conservation of Genes:** Zebrafish share a high degree of genetic similarity with humans, with approximately 70% of human genes having homologs in zebrafish. This conservation allows researchers to study disease-related genes and pathways.
- **Genetic Manipulation:** Zebrafish are amenable to genetic manipulation techniques, including gene knockdown, knock-in, and CRISPR-Cas9-mediated gene editing, enabling the creation of disease models.
- **High Reproductive Capacity:** Zebrafish reproduce rapidly and produce large numbers of embryos, making it feasible to conduct large-scale genetic and drug screening experiments.
- **Drug Discovery:** Zebrafish models allow for high-throughput drug screening to identify potential therapeutic compounds for rare diseases.

4.3.2.18 Applications of Zebrafish in Rare Disease Research

- **Neurological Disorders:** Zebrafish are used to model neurodegenerative diseases like amyotrophic lateral sclerosis (ALS), epilepsy, and spinal muscular atrophy (SMA), providing insights into disease mechanisms and potential treatments (Kabashi et al. 2011).
- **Cardiovascular Diseases:** Zebrafish models contribute to the understanding of rare cardiac conditions, including arrhythmogenic cardiomyopathy and long QT syndrome (Sehnert et al. 2002).
- **Genetic Syndromes:** Zebrafish models have been instrumental in studying genetic syndromes like Rett syndrome, tuberous sclerosis, and Williams-Beuren syndrome (Sakai et al. 2018; Choi et al. 2021).
- **Hematological Disorders:** Zebrafish provide insights into blood disorders like thrombocytopenia, anemia, and thrombosis, aiding in understanding disease pathogenesis and potential therapies (Konantz et al. 2019).
- **Skeletal Disorders:** Zebrafish models contribute to understanding skeletal dysplasias and bone development, including osteogenesis imperfecta and chondrodysplasias (Marí-Beffa et al. 2021).

4.3.2.19 Drosophila Models

Drosophila melanogaster, commonly known as the fruit fly, has emerged as a powerful model organism in biological research, including the study of rare diseases. Despite their seemingly distant relation to humans, fruit flies share fundamental genetic and cellular mechanisms that make them valuable tools for unraveling the complexities of various rare conditions. The role of Drosophila models in rare disease studies and their contributions to advancing our understanding of these disorders might throw light on their importance as disease models.

4.3.2.20 Advantages of Drosophila Models

- **Genetic Tractability:** Fruit flies have a relatively short life cycle and reproduce quickly, allowing for the study of multiple generations in a short time. Their genetic manipulation is straightforward, making them ideal for studying gene function.
- **Conserved Pathways:** Many cellular and molecular pathways are evolutionarily conserved between fruit flies and humans. This similarity enables researchers to study disease-related genes and pathways in a simplified yet relevant system.
- **High-Throughput Screening:** The small size and rapid breeding of fruit flies enable high-throughput screening of compounds, genes, and potential therapeutics.
- **Functional Analysis:** Fruit flies can be used to study gene function through loss-of-function and gain-of-function experiments, shedding light on the molecular basis of rare diseases.

4.3.2.21 Applications of Drosophila Models in Rare Disease Studies

- **Neurological Disorders:** Drosophila models contribute to understanding neurodegenerative diseases like Alzheimer's, Parkinson's, and amyotrophic lateral sclerosis (ALS), offering insights into disease mechanisms and potential interventions (Bilen and Bonini 2005).
- **Metabolic Disorders:** Fruit flies help study metabolic diseases such as lysosomal storage disorders, phenylketonuria (PKU), and Niemann-Pick disease, aiding in the identification of underlying molecular defects (Wangler et al. 2015).
- **Muscular Dystrophy:** Drosophila models of muscular dystrophy provide insights into muscle degeneration, regeneration, and potential therapeutic targets (Schnorrer et al. 2010).
- **Ciliopathies:** Drosophila models help study rare genetic disorders known as ciliopathies, characterized by defects in primary cilia function and structure (Nachury et al. 2007).
- **Rare Genetic Syndromes:** Drosophila models are employed to study rare genetic syndromes such as Rett syndrome, fragile X syndrome, and Cornelia de Lange syndrome, elucidating underlying molecular mechanisms (Spindler and Hartenstein 2010).

4.4 Advancements in Genetic Engineering for Models

Advancements in genetic engineering have revolutionized the creation of animal models for studying rare diseases. Genetic manipulation techniques have enabled researchers to develop more accurate and sophisticated models that closely mimic the genetic and molecular aspects of human diseases. Let us unravel the cutting-edge techniques and strategies in genetic engineering that have enhanced our ability to create and study animal models for rare diseases.

4.4.1 CRISPR-Cas9 Technology

CRISPR-Cas9 has emerged as a groundbreaking tool for precise genetic manipulation (Cong et al. 2013; Mali et al. 2013; Doudna and Charpentier 2014). This technique allows researchers to edit specific DNA sequences, introducing mutations associated with rare diseases into animal genomes. CRISPR-Cas9 has significantly accelerated the process of model creation and increased the accuracy of replicating disease-causing genetic variations.

4.4.2 Gene Knockout and Knock-In

Gene knockout involves disabling a specific gene to observe the resulting effects, providing insights into gene function and disease mechanisms. Conversely, gene knock-in introduces specific mutations into animal genomes to replicate disease-causing genetic variations (Capecchi 2005; Platt et al. 2014).

4.4.3 Conditional and Tissue-Specific Expression

Conditional gene expression allows researchers to control when and where a gene is active. Tissue-specific promoters enable the targeting of specific organs or cell types, mimicking the tissue-specific nature of many rare diseases (Lobe and Nagy 1998; Hsu et al. 2014).

4.4.4 Gene Editing in Induced Pluripotent Stem Cells (iPSCs)

iPSCs derived from patient cells can be genetically edited to correct mutations or introduce disease-related mutations, offering a personalized approach to modeling rare diseases (Takahashi and Yamanaka 2006).

4.4.5 Viral Vectors for Gene Delivery

Viral vectors can be used to introduce specific genes or mutations into animals, allowing researchers to study the effects of these genetic alterations on disease development and progression (Naldini et al. 1996).

4.4.6 CRISPR-Based Functional Genomics

CRISPR technology has expanded beyond gene editing to enable large-scale functional genomics studies, uncovering the roles of various genes in disease pathways (Shalem et al. 2014).

4.4.6.1 CRISPR-Cas9 Technology in Model Creation

CRISPR-Cas9 technology has revolutionized the field of genetic engineering, enabling precise and efficient manipulation of genomes. This revolutionary tool has significantly impacted the creation of animal models for studying rare diseases. The applications of CRISPR-Cas9 technology in model creation showcase how it has accelerated the generation of accurate and sophisticated models for investigating the intricacies of rare diseases.

CRISPR-Cas9 Technology

CRISPR-Cas9 is a versatile and programmable gene editing tool derived from bacterial immune systems. It functions as molecular scissors, guided by a synthetic RNA molecule to target specific DNA sequences and induce precise changes, such as gene knockouts, knock-ins, and even correction of mutations associated with rare diseases (Barrangou et al. 2007; Zhang et al. 2011; Jinek et al. 2012; Mali et al. 2013).

Applications of CRISPR-Cas9 Technology in Model Creation

- **Gene Knockout Models:** CRISPR-Cas9 technology allows researchers to disrupt specific genes, replicating the genetic mutations underlying rare diseases. By introducing mutations into animal genomes, researchers can observe how these alterations affect disease development and progression (Platt et al. 2014).
- **Gene Knock-In Models:** Researchers can also use CRISPR-Cas9 to introduce disease-causing mutations into animal genomes, mimicking specific genetic variations associated with rare diseases. This approach enables the study of disease mechanisms and potential therapeutic interventions (Yin et al. 2014).
- **Reporter Gene Insertion:** CRISPR-Cas9 can be employed to insert reporter genes, such as fluorescent proteins, into specific genomic locations, allowing researchers to monitor gene expression patterns and cellular behaviors associated with rare diseases (Liu et al. 2021).

Precise Editing and Disease Modeling

CRISPR-Cas9's precision in targeting specific DNA sequences enables the creation of animal models that accurately replicate the genetic mutations underlying rare diseases. These models provide valuable insights into disease mechanisms and pathways, aiding researchers in identifying potential therapeutic targets and interventions as opined by Asmamaw and Zawdie (2021).

Accelerated Model Development

CRISPR-Cas9's speed and efficiency have accelerated the generation of animal models, reducing the time and resources required for model creation. This has democratized the accessibility of animal models for researchers studying rare diseases (Doudna and Charpentier 2014).

Gene Knockout and Knock-In: Precision Tools in Genetic Manipulation

Gene knockout and knock-in techniques are foundational tools in genetic manipulation, allowing researchers to modify an organism's genome to investigate gene function, disease mechanisms, and potential therapeutic interventions. These techniques have greatly impacted our ability to create accurate and sophisticated models for studying rare diseases. Let us explore the concepts and applications of gene knockout and knock-in strategies, highlighting their contributions to understanding the complexities of genetic disorders.

Gene Knockout

Gene knockout involves the deliberate inactivation or disruption of a specific gene, either by deleting it or by introducing mutations that render it nonfunctional. This technique provides insights into the role of the targeted gene in various biological processes and disease development.

Applications of Gene Knockout

- **Functional Gene Analysis:** Gene knockout experiments help elucidate the function of individual genes, revealing their contributions to physiological processes, development, and disease (Zimmer et al. 2019).
- **Disease Modeling:** Knocking out genes associated with rare diseases enables researchers to replicate disease phenotypes in animal models, aiding in the study of disease mechanisms and potential treatments (da Silva-Buttkus et al. 2023).
- **Therapeutic Target Identification:** Identifying genes that contribute to disease progression through knockout experiments can lead to the identification of potential therapeutic targets (Paul et al. 2021).

Gene Knock-In

Gene knock-in involves inserting a specific gene or mutation into a targeted genomic locus. This technique allows researchers to study the effects of introducing disease-causing mutations or potential therapeutic genes (Thomas and Capecchi 1987; Sauer 1998; Lau et al. 2020).

Applications of Gene Knock-In

- **Modeling Disease Mutations:** Gene knock-in models replicate specific genetic mutations associated with rare diseases, enabling the study of disease mechanisms and testing potential interventions (Wang et al. 2022).
- **Drug Target Validation:** Gene knock-in models can be used to validate potential drug targets by introducing specific genetic variations and observing their effects on disease development and progression (Doyle et al. 2012).
- **Therapeutic Gene Delivery:** Knock-in models can be used to introduce therapeutic genes, offering the potential to correct genetic defects and treat rare diseases (Wu et al. 2020).

Conditional and Tissue-Specific Expression

Conditional and tissue-specific gene expression systems are powerful tools in genetic manipulation, allowing researchers to control when and where a gene of interest is active within an organism. These techniques have transformed the creation of animal models for studying rare diseases, enabling researchers to mimic the tissue-specific and temporally regulated nature of many genetic disorders (Saam and Gordon 1999; Szulc et al. 2006; Zhang et al. 2017; Borger et al. 2017). The concepts and applications of conditional and tissue-specific expression strategies, emphasizing their contributions to understanding the complexities of rare diseases, are explained in detail in the following section.

Conditional Gene Expression

Conditional gene expression systems regulate the activity of a gene of interest based on external factors, such as the presence of specific inducers or the activation of certain promoters. This enables researchers to study gene function in a controlled and precise manner.

Applications of Conditional Gene Expression

- **Temporal Regulation:** Conditional expression systems allow researchers to activate or deactivate genes at specific time points, mimicking the dynamic changes that occur during disease progression (Tsien et al. 1996).
- **Drug-Inducible Expression:** Inducible systems can be activated by the administration of specific drugs, facilitating the study of gene function and disease mechanisms in response to pharmacological agents (Weake and Workman 2010).
- **Conditional Knockout Models:** Conditional gene expression systems can be coupled with gene knockout strategies to achieve tissue-specific or temporal gene inactivation (Madisen et al. 2010).

Tissue-Specific Expression

Tissue-specific promoters drive gene expression in specific cell types or tissues, allowing researchers to investigate gene function within a particular context.

Applications of Tissue-Specific Expression

- **Disease Modeling:** Tissue-specific expression systems enable the creation of animal models that recapitulate the tissue-specific manifestations of rare diseases, aiding in the study of disease mechanisms (Piedrahita and Williams 2017).
- **Targeted Therapies:** Tissue-specific promoters can be used to drive the expression of therapeutic genes only in the affected tissues, offering a potential strategy for treating rare diseases (Anguela et al. 2013).
- **Transgenic Models:** Tissue-specific expression systems contribute to the generation of transgenic animal models that specifically express genes of interest in particular tissues (Houdebine 2007; Jiang and Chen 2022).

4.4.6.2 Induced Pluripotent Stem Cells (iPSCs) and Disease Modeling

Induced pluripotent stem cells (iPSCs) have revolutionized the field of disease modeling, offering a remarkable platform to study the cellular and molecular mechanisms underlying rare diseases. iPSCs hold the potential to bridge the gap between basic research and clinical applications, providing invaluable insights into disease pathology, drug screening, and personalized medicine. The concept of iPSCs and their applications in modeling rare diseases were assessed by several researchers (Takahashi and Yamanaka 2006; Soldner et al. 2009; Itzhaki et al. 2012; Mertens et al. 2015).

iPSCs

iPSCs are adult cells that have been reprogrammed back to a pluripotent state, allowing them to differentiate into various cell types found in the human body. These cells retain the genetic information of the donor and serve as a versatile tool for disease modeling, drug development, and regenerative medicine.

Applications of iPSCs in Disease Modeling

- **Generation of Patient-Specific Cells:** iPSCs can be derived from patients with rare diseases, creating a renewable source of disease-relevant cells for study. These cells accurately recapitulate the genetic background of the patient, providing a unique model system (Park et al. 2008).
- **Study of Disease Mechanisms:** iPSC-derived cells can be differentiated into disease-affected cell types, such as neurons or cardiomyocytes, enabling researchers to study disease mechanisms in a controlled environment (Egawa et al. 2012).
- **Drug Screening and Development:** iPSC-based models allow for high-throughput drug screening to identify potential therapeutic compounds targeting disease-specific cellular defects (Rashid et al. 2010).
- **Personalized Medicine:** iPSCs enable the study of patient-specific responses to drugs, facilitating the development of personalized treatment strategies for rare diseases (Lee et al. 2009).

Limitations and Challenges of iPSCs

- **Differentiation Efficiency:** The efficient differentiation of iPSCs into disease-relevant cell types can be challenging, requiring optimization for each cell lineage.
- **Model Complexity:** Fully recapitulating disease complexity, especially in multisystem disorders, can be difficult using in vitro iPSC-based models.

Future Directions of iPSCs

- **Organoid Technology:** Advancements in organoid culture techniques are enabling the creation of more complex tissue models derived from iPSCs, offering closer representation of in vivo conditions.
- **High-Throughput Screening:** Automated systems for iPSC differentiation and drug screening are being developed to increase the efficiency of drug discovery efforts.

4.5 Characterizing and Phenotyping Rare Disease Models

Characterizing and phenotyping rare disease models are a crucial step in understanding the intricacies of these conditions and developing effective therapeutic strategies. These processes involve a comprehensive analysis of various aspects, from molecular and cellular features to physiological and behavioral manifestations. In the following text, the methods and significance of characterizing and phenotyping rare disease models are dealt with.

4.5.1 Characterization of Rare Disease Models

To characterize the rare disease models, genetic analysis, histopathology, and molecular profiling are useful. In genetic analysis, the genetic mutations present in the model are analyzed to confirm their similarity to human patients and determine their impact on disease development (Abrahams and Geschwind 2008). In histopathological analysis, the tissue samples are examined for cellular and structural abnormalities, to provide insights into disease-related changes (Goodchild and Dauer 2005). Likewise, in molecular profiling, omics analyses (genomics, transcriptomics, proteomics) are conducted to identify altered pathways and molecular signatures associated with the disease (Olivier et al. 2019).

4.5.2 Phenotyping Rare Disease Models

Phenotyping rare disease models include behavioral analysis, physiological measurements, imaging techniques, and cellular studies. In behavioral analysis, behavioral abnormalities such as motor deficits, cognitive impairments, and other phenotypic changes in animal models are assessed (Crawley 2007). Physiological measurements include measurement of physiological parameters such as heart rate, blood pressure, and metabolic markers to understand systemic effects of the disease (Dejea et al. 2019; Cesarovic et al. 2020). On the other hand, imaging techniques utilize imaging modalities like MRI, CT, and PET scans to visualize internal structures and detect anatomical abnormalities (Yitbarek and Dagnaw 2022). In cellular studies, the cellular properties, including morphology, proliferation rates, and responses to stimuli, to uncover disease-related changes are examined (Miller and Zachary 2017).

4.5.3 Significance of Characterization and Phenotyping

- **Understanding Disease Mechanisms:** Detailed characterization and phenotyping provide insights into disease progression, helping elucidate underlying mechanisms as reported by Delude (2015).
- **Identifying Biomarkers:** Characterization aids in the discovery of biomarkers that can serve as indicators of disease progression or treatment response (Qiu et al. 2023).
- **Testing Therapeutic Interventions:** Phenotyping allows the assessment of potential treatments, measuring their effects on disease-related phenotypes (Forner et al. 2021).

Characterizing and phenotyping rare disease models provide essential insights into disease mechanisms and potential therapeutic strategies. These processes contribute to bridging the gap between basic research and clinical applications, ultimately improving our understanding and management of rare diseases.

4.5.3.1 Behavioral and Physiological Assessments

Characterizing and phenotyping rare disease models involve a comprehensive analysis that spans from molecular mechanisms to behavioral and physiological changes. Behavioral and physiological assessments play a pivotal role in understanding the impact of rare diseases on an organism's overall health and functioning. The assessment methods and significance of behavioral and physiological assessments in characterizing and phenotyping rare disease models, supported by references from the scientific researchers, are given below.

Behavioral Assessments

- **Motor Function and Coordination:** Assess motor skills, balance, and coordination through tasks such as rotarod tests, beam walking, and grip strength measurements (Crawley 2007).
- **Cognitive Abilities:** Evaluate learning and memory using tasks like Morris water maze or novel object recognition, offering insights into cognitive impairments (Neureither et al. 2017; Lissner et al. 2021).

- **Anxiety and Depression-Like Behavior:** This can be assessed by utilizing tests like open field, elevated plus maze, and forced swim to assess emotional responses and stress levels (Binder 2009).
- **Social Behavior:** Examine social interactions, communication, and affiliative behaviors to unravel potential social deficits (Moy et al. 2004).

Physiological Assessments

- **Cardiovascular Function:** Measure heart rate, blood pressure, and electrocardiogram parameters to understand cardiovascular health (Dowell 1991).
- **Metabolic Analysis:** Examine metabolic parameters such as glucose tolerance, insulin sensitivity, and energy expenditure to uncover metabolic abnormalities (Yang et al. 2018).
- **Respiratory Function:** Assess respiratory parameters including tidal volume and respiratory rate to evaluate lung and respiratory muscle function (Lee et al. 2018).
- **Neurological Function:** Employ techniques like electrophysiology or neuroimaging to examine neural activity and connectivity (Thierry 2012).

Significance of Behavioral and Physiological Assessments

- **Comprehensive Understanding:** Behavioral and physiological assessments provide a holistic understanding of how rare diseases affect an organism's health, behavior, and overall well-being (Silverman et al. 2011).
- **Identifying Therapeutic Targets:** Abnormalities detected through these assessments may guide the identification of potential therapeutic targets and interventions (Falk et al. 2015).
- **Preclinical Drug Testing:** Behavioral and physiological assessments are crucial in testing the efficacy of potential treatments and therapies for rare diseases (Antunes and Biala 2012; Wooden et al. 2021).

4.5.3.2 Molecular and Histopathological Analyses

Characterizing and phenotyping rare disease models involve a multidimensional approach that includes molecular and histopathological analyses. These methods delve into the molecular mechanisms and structural changes underlying rare diseases, offering valuable insights into disease progression and potential therapeutic targets (Duncan et al. 2012; Khalak et al. 2012; Benitz et al. 2016). The methods and significance of molecular and histopathological analyses in characterizing and phenotyping rare disease models are depicted below, the facts being supported by investigations of various researchers.

Molecular Analyses

- **Genetic Mutations:** DNA is sequenced to identify genetic mutations or variants responsible for the rare disease phenotype in the model (Vissers et al. 2010).
- **Gene Expression Profiling:** Microarray or RNA-seq is utilized to analyze gene expression patterns, to reveal dysregulated pathways associated with the rare disease (Hung and Weng 2017).

- **Protein Analysis:** Here, protein expression levels and posttranslational modifications are assessed using techniques like Western blotting and mass spectrometry (Zhang et al. 2017; Ramazi and Zahiri 2021).
- **Metabolomics:** To study metabolomics, metabolomic profiling is performed to identify altered metabolic pathways and potential biomarkers associated with the disease (Ding et al. 2021).

Histopathological Analyses

- **Tissue Morphology:** Tissue sections are examined under a microscope to detect structural abnormalities, cellular changes, and tissue degeneration.
- **Immunohistochemistry (IHC):** IHC is used to visualize specific proteins and cell types, providing insights into their distribution and abundance.
- **In Situ Hybridization:** The expression of specific genes within tissue sections is detected, to find out the spatial information on gene expression patterns (Yau et al. 2020).

Significance of Molecular and Histopathological Analyses

- **Elucidating Disease Mechanisms:** Molecular and histopathological analyses provide insights into the underlying molecular changes and structural alterations contributing to the rare disease phenotype.
- **Identifying Therapeutic Targets:** Abnormalities revealed by these analyses may lead to the identification of potential therapeutic targets for intervention.
- **Guiding Personalized Treatment:** Molecular analyses can guide the development of personalized treatment strategies based on the genetic and molecular characteristics of the rare disease.

Thus, molecular and histopathological analyses are essential components of characterizing and phenotyping rare disease models, shedding light on the molecular basis and structural changes associated with these conditions.

4.6 Modeling Specific Rare Diseases

Modeling specific rare diseases provides a focused approach to understanding the intricacies of individual disorders, uncovering disease mechanisms, and developing targeted therapeutic interventions. By creating accurate and relevant animal models, researchers can gain valuable insights into disease progression, molecular pathways, and potential treatment options. The importance of modeling specific rare diseases and examples of how such models have contributed to advancing our understanding and treatment of rare disorders are emphasized well.

4.6.1 Understanding Disease Pathology

The following aspects help researchers to understand disease pathology in animal models:

- **Phenotype Recapitulation:** By modeling specific rare diseases in animals, researchers can replicate key aspects of the disease phenotype, including molecular, cellular, and physiological features (Pinnapureddy et al. 2015).
- **Disease Progression Studies:** Animal models allow for the observation and analysis of disease progression over time, facilitating the identification of critical disease milestones (Liu-Chittenden et al. 2012).
- **Molecular Pathway Elucidation:** Specific disease models can uncover the molecular pathways underlying the disorder, shedding light on disease mechanisms and potential therapeutic targets (Zhang et al. 2017).

4.6.2 Advancing Therapeutic Strategies

Advancing therapeutic strategies such as drug screening and development, personalized medicine approaches, and gene therapy studies play a vital role in rare disease research. Let us see them in detail.

- **Drug Screening and Development:** Specific disease models enable the testing of potential therapeutic compounds and strategies, leading to the identification of novel drug candidates (Singh and Seed 2021).
- **Personalized Medicine Approaches:** Models of specific rare diseases allow researchers to explore personalized treatment options, taking into account individual genetic backgrounds (Schee Genannt Halfmann et al. 2017).
- **Gene Therapy Studies:** Disease-specific models provide a platform to test gene therapy approaches, evaluating the efficacy of introducing functional genes to treat the disorder (Gopinath et al. 2015).

4.6.3 Case Examples

- **Cystic Fibrosis:** Animal models of cystic fibrosis have contributed to understanding lung pathology and mucus clearance and testing potential therapies like CFTR modulators (Ratjen and Döring 2003).
- **Duchenne Muscular Dystrophy:** Animal models of DMD have helped elucidate muscle degeneration mechanisms and have been pivotal in testing gene therapy strategies (Emery 2002).
- **Huntington's Disease:** Animal models of Huntington's disease have provided insights into neuronal degeneration, altered motor function, and potential therapeutic interventions (Bates et al. 2015).

4.6.3.1 Neurological Rare Diseases: Model Development and Insights

Neurological rare diseases encompass a diverse group of disorders affecting the nervous system, presenting unique challenges for understanding their underlying mechanisms and developing effective treatments. Creating accurate animal models is crucial to unraveling the complexities of these disorders, offering insights into disease pathology and potential therapeutic strategies. Works of Monani (2005), Chahrour et al. (2008), Hill et al. (2009), Krueger and Bear (2011), Servadio et al. (2015), and Bhakuni et al. (2016) support the significance of developing animal models for neurological rare diseases and provide examples of insights gained from these models.

Importance of Animal Models for Neurological Rare Diseases

Developing animal models for neurological rare diseases offers a valuable platform for gaining insights into disease mechanisms and evaluating potential treatments. These models play a pivotal role in bridging the gap between bench research and clinical applications as mentioned below, ultimately improving the lives of patients with these disorders.

- **Mimicking Human Pathology:** Animal models enable the replication of key neurological features observed in patients, allowing researchers to study disease progression and mechanisms (Servadio et al. (2015).
- **Mechanistic Insights:** Animal models provide a platform to investigate the molecular and cellular changes that contribute to neurological rare diseases (Baldridge et al. 2021).
- **Drug Testing and Therapeutic Development:** Animal models allow for the evaluation of potential therapies and drug candidates in a controlled environment (Mukherjee et al. 2022).

Insights Gained from Neurological Rare Disease Models

- **Spinal Muscular Atrophy (SMA):** Mouse models of SMA have contributed to understanding motor neuron degeneration and have been instrumental in testing gene therapy approaches (Monani 2005).
- **Amyotrophic Lateral Sclerosis (ALS):** Animal models of ALS have revealed mechanisms of motor neuron degeneration and have been used to test potential therapeutics (Morrice et al. 2018).
- **Rett Syndrome:** Mouse models of Rett syndrome have provided insights into neuronal dysfunction and have guided potential treatment strategies (Vashi and Justice 2019).
- **Fragile X Syndrome:** Animal models of fragile X syndrome have advanced our understanding of synaptic dysfunction and have led to testing potential therapeutic interventions (Krueger and Bear 2011).

Challenges and Future Directions

- **Species Differences:** Translating findings from animal models to humans requires addressing potential differences in disease manifestation.
- **Complexity of the Nervous System:** The intricate nature of neurological diseases necessitates the development of models that accurately reflect disease complexity.
- **Precision Medicine Approaches:** Advances in personalized medicine could influence the design of animal models tailored to individual patient profiles.

4.6.3.2 Genetic Metabolic Disorders: Animal Models for Mechanistic Studies

Genetic metabolic disorders encompass a wide range of conditions characterized by abnormalities in metabolic pathways due to genetic mutations. Animal models play a pivotal role in unraveling the mechanistic underpinnings of these disorders, shedding light on disrupted biochemical processes, and aiding in the development of therapeutic strategies. In this section, let us delve into the significance of using animal models for studying genetic metabolic disorders and insights gained from these models, supported by the observations of various researchers (Vidal-Puig et al. 1996; Raben et al. 2003; Halevy and Benvenisty 2015; Harding et al. 2006; Mistry et al. 2011; Ney et al. 2009; Villa-Bellosta 2019).

Importance of Animal Models for Genetic Metabolic Disorders

Animal models for genetic metabolic disorders offer a powerful tool for understanding disease mechanisms and developing targeted therapies. These models contribute to bridging the gap between basic research and clinical applications, ultimately benefiting patients affected by these disorders such as the following:

- **Replicating Human Pathology:** Animal models allow researchers to mimic the metabolic disruptions observed in patients, facilitating the study of disease mechanisms (Mukherjee et al. 2022).
- **Mechanistic Insights:** Animal models enable the investigation of the molecular and cellular changes underlying genetic metabolic disorders (Vidal-Puig et al. 1996).
- **Therapeutic Development:** Animal models provide a platform to test potential treatments and interventions aimed at correcting metabolic imbalances (Ribitsch et al. 2020).

Insights Gained from Genetic Metabolic Disorder Models

- **Phenylketonuria (PKU):** Mouse models of PKU have elucidated the role of phenylalanine metabolism in brain development and cognitive impairment (Shedlovsky et al. 1993; Schuck et al. 2015; Hong et al. 2021).
- **Gaucher Disease:** Animal models of Gaucher disease have provided insights into lipid metabolism dysfunction and have guided potential therapeutic strategies (Mistry et al. 2011).

- **Pompe Disease:** Animal models of Pompe disease have helped elucidate muscle pathology and have been instrumental in testing enzyme replacement therapies (Raben et al. 2003).
- **Maple Syrup Urine Disease (MSUD)**: Animal models of MSUD have provided insights into branched-chain amino acid metabolism and potential treatment approaches (Streck et al. 2021).

Challenges and Future Directions

- **Model Specificity:** Developing animal models that accurately mimic the complexity of human genetic metabolic disorders remains a challenge.
- **Drug Development and Translation:** Moving from successful preclinical studies in animal models to effective treatments for patients requires careful consideration.
- **Precision Medicine Approaches:** Advancements in personalized medicine may influence the design and use of animal models for genetic metabolic disorders.

4.6.3.3 Rare Cancers: Animal Models for Therapeutic Exploration

Rare cancers pose unique challenges due to their limited prevalence and diverse molecular characteristics. Animal models play a crucial role in exploring potential therapeutic strategies for these cancers, offering insights into disease mechanisms and aiding in the development of targeted treatments. The significance of utilizing animal models for investigating rare cancers provides insights gained from these models as supported by the research contributions of Rubin et al. (2005), Vujovic et al. (2006), Doghman et al. (2007), Singh and Ferrara (2012), Harms et al. (2015), Tan et al. (2015), Esposito et al. (2016), and Jiang et al. (2019).

Importance of Animal Models for Rare Cancers

Animal models for rare cancers offer a vital tool for investigating disease mechanisms and evaluating potential therapies. These models contribute to bridging the gap between laboratory research and clinical applications, ultimately improving the prognosis and treatment options for patients with these rare malignancies. Some of them are as follows:

- **Replicating Human Tumor Characteristics:** Animal models allow researchers to mimic the tumor characteristics and microenvironment observed in patients, facilitating the study of disease progression and therapeutic responses (Singh and Ferrara 2012).
- **Mechanistic Insights:** Animal models provide a platform to investigate the molecular and cellular changes that drive the development and progression of rare cancers (Tan et al. 2015).
- **Therapeutic Development:** Animal models serve as a testing ground for potential therapies and interventions aimed at halting or reversing tumor growth (Esposito et al. 2016).

Insights Gained from Rare Cancer Models

- **Gastrointestinal Stromal Tumors (GISTs):** Animal models of GIST have provided insights into tumor growth and responses to targeted therapies like tyrosine kinase inhibitors (Rubin et al. 2005).
- **Chordoma:** Animal models of chordoma have contributed to understanding the tumor microenvironment and have guided potential treatment strategies (Vujovic et al. 2006).
- **Adrenocortical Carcinoma (ACC):** Mouse models of ACC have helped uncover molecular drivers of tumor formation and have been used to test novel therapeutic approaches (Doghman et al. 2007).
- **Merkel Cell Carcinoma (MCC):** Animal models of MCC have provided insights into viral contributions to cancer development and have guided potential immunotherapy strategies (Harms et al. 2015).

Challenges and Future Directions

The following are the challenges faced in developing animal models for rare cancer studies:

- **Model Heterogeneity:** Developing animal models that accurately represent the heterogeneity of rare cancers remains a challenge.
- **Clinical Translation:** Translating findings from preclinical animal studies to successful clinical outcomes requires careful consideration.
- **Personalized Approaches:** Advances in precision medicine may impact the design and utilization of animal models for rare cancer research.

4.7 In Vivo Imaging and Analysis

In vivo imaging and analysis have revolutionized our understanding of complex biological processes by providing real-time insights into living organisms. This powerful tool allows researchers to visualize, track, and quantify dynamic physiological and pathological events in vivo. The significance of in vivo imaging and analysis and examples of its applications are discussed below.

4.7.1 Importance of In Vivo Imaging and Analysis

In vivo imaging and analysis offer a transformative approach to understanding the intricacies of biological processes within living organisms. These techniques continue to evolve, enhancing our ability to explore dynamic interactions and providing valuable information for both basic research and clinical applications. They are as follows:

- **Real-Time Insights:** In vivo imaging enables the observation of biological processes as they unfold in live organisms, offering dynamic and time-dependent information (Weissleder 2001).
- **Noninvasive Nature:** In vivo imaging techniques are noninvasive, minimizing disturbance to the organism and allowing longitudinal studies (Nahrendorf and Swirski 2016).
- **Visualization of Complex Interactions:** In vivo imaging helps uncover complex interactions among cells, tissues, and molecules within their native context (Pittet and Weissleder 2011).

4.7.2 Applications of In Vivo Imaging and Analysis

- **Fluorescence Imaging:** Utilizing fluorescent probes or reporter genes to visualize cellular processes, protein localization, and gene expression in real time (Pittet et al. 2007).
- **Bioluminescence Imaging:** Employing luciferase-based reporters to track gene expression, cell trafficking, and tumor growth in vivo (Mojic et al. 2017).
- **Magnetic Resonance Imaging (MRI):** Noninvasive imaging technique using magnetic fields and radio waves to visualize soft tissue structures and anatomical changes (Jasanoff 2007).
- **Positron-Emission Tomography (PET):** Tracing and quantifying radiolabeled molecules to visualize metabolic, physiological, and molecular processes in vivo (Massoud and Gambhir 2003).

4.7.3 Advancements and Future Directions

- **Multiplexed Imaging:** Developing techniques that enable simultaneous visualization of multiple targets, providing a comprehensive view of complex interactions.
- **High-Resolution Imaging:** Advancing technologies for achieving higher spatial and temporal resolutions, revealing finer details of biological processes.
- **Functional Imaging:** Combining anatomical and functional imaging modalities to gain insights into the physiological context of events.

Developing the abovementioned imaging techniques might enable us to gain insights into rare diseases.

4.7.3.1 Noninvasive Imaging Techniques for Longitudinal Studies

Noninvasive imaging techniques have revolutionized longitudinal studies by enabling researchers to observe and analyze dynamic biological processes in living organisms over time. These methods provide valuable insights into developmental, physiological, and pathological changes without the need for invasive procedures.

Based on reports of earlier research, the significance of noninvasive imaging techniques for longitudinal studies and their applications depict their importance.

Importance of Noninvasive Imaging Techniques for Longitudinal Studies

Noninvasive imaging techniques for longitudinal studies offer a powerful tool for understanding the dynamic nature of biological processes. These methods continue to evolve, enhancing our ability to observe and analyze changes over time, providing valuable information for both basic research and clinical applications.

- **Dynamic Insights:** Noninvasive imaging enables researchers to capture changes occurring over time, shedding light on dynamic processes such as growth, disease progression, and therapeutic responses (Gambhir 2002).
- **Reduction of Bias:** Studies conducted by Patel et al. (2017) and Scarfe et al. (2018) using noninvasive imaging have shown to minimize the potential bias introduced by repeated invasive procedures.
- **Personalized Medicine Insights:** Noninvasive imaging can contribute to personalized medicine by tracking individual responses to treatments and interventions over time (Salih et al. 2023).

Applications of Noninvasive Imaging Techniques for Longitudinal Studies

- **Magnetic Resonance Imaging (MRI):** MRI allows noninvasive tracking of anatomical changes, tissue growth, and disease progression over time (Daldrup-Link et al. 2011).
- **Micro-Computed Tomography (Micro-CT):** Micro-CT enables three-dimensional visualization of bone, tissue, and organ changes, facilitating studies on growth and disease (Hu et al. 2001; Burghardt et al. 2011).
- **Fluorescence Imaging:** Fluorescence imaging allows for longitudinal monitoring of cellular dynamics, gene expression, and molecular interactions (Wong et al. 2018).

Advancements and Future Directions

The following techniques of noninvasive imaging could provide necessary insights for understanding rare diseases and the modalities to combat them:

- **Multimodal Integration:** Combining different noninvasive imaging techniques can provide comprehensive insights into complex biological processes.
- **Functional Imaging Enhancements:** Advancing techniques for functional imaging allows for studying physiological changes alongside anatomical alterations.
- **Real-Time Imaging:** Developing real-time noninvasive imaging techniques can capture rapid changes and transient events.

4.7.3.2 Imaging Modalities for Tracking Disease Progression

Imaging modalities have emerged as indispensable tools for tracking the progression of diseases over time. These techniques offer noninvasive ways to visualize and quantify changes at the cellular, molecular, and structural levels, providing valuable insights into disease mechanisms and treatment efficacy. The importance of imaging modalities for tracking disease progression and their applications are given below.

Importance of Imaging Modalities for Tracking Disease Progression

Imaging modalities for tracking disease progression offer a critical means to visualize and understand the dynamic changes occurring within the body. As these techniques continue to advance, they will play an increasingly vital role in guiding diagnosis, treatment decisions, and the development of novel therapies.

- **Dynamic Monitoring:** Imaging modalities enable researchers to provide timely interventions and assessments (Lusic and Grinstaff 2013).
- **Quantitative Analysis:** Imaging techniques provide quantitative data that can be used to objectively measure disease progression and treatment response (Fang et al. 2012).
- **Treatment Monitoring:** Imaging allows for the assessment of treatment effectiveness by visualizing changes in disease biomarkers or target tissues (O'Connor et al. 2015).

Applications of Imaging Modalities for Tracking Disease Progression

- **Positron-Emission Tomography (PET):** PET scans track disease progression by visualizing molecular processes, such as glucose metabolism or receptor binding (de Paula Faria et al. 2014).
- **Magnetic Resonance Imaging (MRI):** MRI provides detailed anatomical and functional information for monitoring changes in tissues and organs (Cai et al. 2012).
- **Ultrasound Imaging:** Ultrasound tracks disease progression by visualizing real-time changes in tissue structures and blood flow (Kim et al. 2015).
- **Optical Imaging:** Optical imaging techniques enable tracking disease progression at the cellular and molecular levels using light-based approaches (Hong et al. 2014).

Advancements and Future Directions

- **Multimodal Integration:** Integrating multiple imaging modalities can provide comprehensive insights into disease progression by capturing complementary information.
- **Functional Imaging Enhancements:** Developing techniques to assess functional changes alongside structural alterations can enhance disease tracking.
- **Machine Learning and AI:** Leveraging machine learning and artificial intelligence can aid in automated disease progression analysis and prediction.

4.8 Therapeutic Approaches and Testing

Developing and testing therapeutic approaches are a critical step in addressing a wide range of diseases and conditions. From preclinical studies to clinical trials, researchers and clinicians work together to assess the safety, efficacy, and potential benefits of various treatments. The importance and significance of therapeutic approaches and testing and their applications throw more light on therapeutic approaches.

4.8.1 Importance of Therapeutic Approaches and Testing

Therapeutic approaches and testing are essential components of medical progress, driving the development of effective treatments for a wide range of diseases. As research methods and technologies continue to advance, the potential for more targeted and personalized therapies becomes increasingly promising.

- **Improving Patient Outcomes:** Therapeutic approaches aim to alleviate symptoms, halt disease progression, or cure the underlying condition, ultimately improving patient quality of life (Chabner and Roberts 2005).
- **Evidence-Based Medicine:** Rigorous testing of therapeutic interventions ensures evidence-based decision-making in clinical practice (Sackett et al. 1996).
- **Personalized Medicine:** Tailoring treatments to individual patients based on genetic, molecular, or physiological characteristics enhances treatment efficacy (Collins and Varmus 2015).

4.8.2 Applications of Therapeutic Approaches and Testing

- **Preclinical Studies:** Preclinical testing involves assessing the safety and efficacy of potential therapies in laboratory models before moving to human trials (Huang et al. 2020).
- **Clinical Trials:** Clinical trials evaluate therapeutic interventions in human subjects, ranging from early-phase safety studies to large-scale efficacy trials (Umscheid et al. 2011).
- **Gene Therapy:** Gene therapy involves introducing genetic material to treat or prevent diseases by replacing, inactivating, or introducing new genes (High and Roncarolo 2019).
- **Immunotherapy:** Immunotherapeutic approaches harness the immune system to target and eliminate disease, as seen in cancer immunotherapy (Mellman et al. 2011).

4.8.3 Advancements and Future Directions

- **Precision Therapeutics:** Advancements in understanding individual patient characteristics will lead to more precise and personalized therapeutic interventions.
- **Combination Therapies:** Combining different therapeutic approaches, such as immunotherapy and targeted therapy, may enhance treatment outcomes.
- **Biomarker Development:** Identifying reliable biomarkers can aid in patient selection, treatment monitoring, and assessing therapeutic efficacy.

4.8.3.1 Preclinical Drug Testing Using Animal Models

Preclinical drug testing using animal models is a crucial step in the drug development process. Animal models provide valuable insights into the safety, efficacy, and potential side effects of new therapies before they are tested in humans. Preclinical drug testing using animal models and its applications are listed below.

Importance of Preclinical Drug Testing Using Animal Models

Preclinical drug testing using animal models remains a cornerstone of drug development, providing critical data that inform decisions regarding the progression of potential therapies into human trials. As technology advances, the integration of various approaches mentioned below will continue to refine and improve the accuracy of preclinical testing.

- **Predicting Human Responses:** Animal models allow researchers to predict how new drugs might behave in humans, helping to identify potential risks and benefits (Sams-Dodd 2005).
- **Dose Determination:** Animal studies assist in determining safe and effective dosages for human clinical trials, minimizing the risk of adverse effects (Zhang et al. 2012).
- **Efficacy Assessment:** Testing new therapies in animal models provides initial insights into their potential efficacy before moving to human trials (Scannell et al. 2012).

Applications of Preclinical Drug Testing Using Animal Models

- **Cancer Therapies:** Animal models are used to assess the effectiveness of new cancer drugs, providing valuable data on tumor regression and side effects (Pao and Chmielecki 2010).
- **Neurological Disorders:** Animal models help evaluate potential treatments for neurological conditions, such as Alzhiemer's diseases (Troy and Jean 2015).
- **Infectious Diseases:** Animal models are essential for testing new antibiotics, antivirals, and vaccines against infectious diseases (Colby et al. 2017).

Advancements and Future Directions

- **Humanized Animal Models:** Developing animal models with humanlike physiological responses can improve the predictive value of preclinical testing.

- **Organ-on-Chip Technology:** Advancements in microfluidic devices enable the study of drug effects on human tissues in vitro, reducing the reliance on animal models.
- **Computational Modeling:** Combining computational models with animal testing can enhance predictions of drug behavior and interactions.

4.8.3.2 Gene Therapy Trials in Rare Disease Models

Gene therapy holds immense promise for treating rare diseases caused by genetic mutations. Clinical trials involving gene therapy aim to correct or replace faulty genes, offering hope for patients with limited treatment options. Several methods of gene therapy trials in rare disease models, their importance, and their applications are discussed below.

Importance of Gene Therapy Trials in Rare Disease Models

Gene therapy trials in rare disease models represent a transformative approach to addressing genetic disorders at their root cause. As ongoing research continues to refine techniques and expand the scope of treatable conditions, the potential for improving the lives of patients with rare diseases becomes increasingly promising. Some of them are as follows:

- **Addressing Genetic Root Causes:** Gene therapy directly targets the genetic abnormalities responsible for rare diseases, potentially providing long-lasting or even curative treatments (High and Roncarolo 2019).
- **Overcoming Treatment Gaps:** Gene therapy offers a potential solution for rare diseases lacking effective treatment options, providing hope for patients and families (Ginn et al. 2018).
- **Personalized Approaches:** Gene therapy can be tailored to individual patients, considering their unique genetic profiles and disease manifestations (Dhurandhar et al. 2021).

Applications of Gene Therapy Trials in Rare Disease Models

- **Spinal Muscular Atrophy (SMA):** Gene therapy trials targeting the SMN1 gene have shown promise in improving motor function and survival in SMA patients (Mendell et al. 2017).
- **Leber Congenital Amaurosis (LCA):** Gene therapy trials for LCA, a rare inherited retinal disorder, have demonstrated vision improvements in treated patients (Maguire et al. 2008).
- **Hemophilia B:** Gene therapy trials in hemophilia B patients have shown sustained increases in clotting factor levels, reducing bleeding episodes (Rangarajan et al. 2017).

Advancements and Future Directions

- **Improved Vector Delivery:** Advancements in viral vectors and delivery methods aim to enhance the efficiency and safety of gene therapy.

- **Targeting Complex Diseases:** Gene therapy is being explored for complex rare diseases involving multiple genetic mutations or complex molecular pathways.
- **Combination Therapies:** Combining gene therapy with other therapeutic approaches, such as small molecules or immunotherapies, may yield synergistic effects.

4.9 Ethical Considerations and Alternatives in Animal Research

Ethical considerations are paramount in the field of animal research, especially when conducting experiments that involve animal models for rare diseases. Researchers and institutions must balance the potential benefits of advancing medical knowledge with the welfare and rights of animals. Let us explore the ethical considerations in animal research and the alternatives that researchers are exploring.

4.9.1 Ethical Considerations in Animal Research

Ethical considerations in animal research remain a complex and evolving topic. As the field progresses, it is crucial to continually reassess and adapt ethical standards to ensure responsible and compassionate treatment of animals while advancing medical knowledge. They include the following:

- **Animal Welfare:** Ethical concerns revolve around ensuring the humane treatment and welfare of animals used in research (Ormandy and Schuppli 2014).
- **Replacement, Reduction, Refinement (3Rs):** The 3Rs framework promotes the reduction of animal usage, refinement of procedures, and replacement with non-animal alternatives (Russell and Burch 1959).
- **Transparency and Consent:** Ensuring transparency in research methods and obtaining informed consent for animal use are essential for ethical research (McLeod and Hobson-West 2015).

4.9.2 Alternatives to Animal Research

The alternatives to animal research are as follows:

- **In Vitro Models:** Cell cultures, organoids, and tissue engineering techniques provide alternatives to whole-animal experiments (Hartung and McBride 2011).
- **Computational Models:** Computational simulations, bioinformatics, and modeling techniques can predict biological responses without animal testing (Coecke et al. 2005).

- **Microfluidic Devices:** Microscale technologies simulate physiological conditions, offering insights into biological processes without animal use (Huh et al. 2010, 2013).

4.9.3 Ethical Dilemmas and Future Directions

The following are some of the steps that could be taken to overcome ethical issues in animal research:

- **Balancing Benefits and Harm:** Ethical considerations involve weighing the potential benefits of research against the harm caused to animals.
- **Engaging Public Discourse:** Open dialogue with the public and stakeholders helps shape ethical standards and research policies.
- **Advancing Alternatives:** Research efforts continue to develop and validate non-animal alternatives for testing and research.

4.9.3.1 Ethical Guidelines in Animal Research

Ethical guidelines provide a framework for conducting animal research in a responsible and humane manner. These guidelines help ensure the welfare of animals while maintaining the scientific integrity of research outcomes.

Importance of Ethical Guidelines in Animal Research

Ethical guidelines are essential for maintaining the integrity of animal research and upholding the welfare of research subjects. As the field progresses, adherence to these guidelines remains crucial to ensuring responsible and ethically sound scientific practices. Ethical guidelines also take these points into consideration in animal research.

- **Welfare of Animals:** Ethical guidelines prioritize the well-being of animals, preventing unnecessary suffering and ensuring humane treatment (Russell and Burch 1959).
- **Scientific Rigor:** Ethical guidelines promote robust study design, data collection, and reporting, enhancing the reliability and reproducibility of research (NRC 2011).
- **Public Trust:** Following ethical guidelines builds public trust in scientific research by demonstrating a commitment to responsible animal use (Hobson-West 2007).

Applications of Ethical Guidelines in Animal Research

- **Institutional Animal Care and Use Committees (IACUCs):** IACUCs review the research protocols to ensure compliance with ethical guidelines and the proper treatment of animals (Balcombe 2006).

- **Replacement, Reduction, Refinement (3Rs):** Ethical guidelines advocate for implementing the 3Rs framework to minimize animal usage and enhance their welfare (Russell and Burch 1959).
- **Transparency and Reporting:** Ethical guidelines emphasize transparent reporting of animal research methods and results to ensure scientific integrity (Landis et al. 2012).

Challenges and Future Directions

- **Balancing Ethical Considerations:** Ethical guidelines require a balance between the potential benefits of research and the welfare of animals.
- **Global Standardization:** Developing global ethical standards ensures consistent treatment of animals across different countries and research institutions.
- **Advancing Alternatives:** Continued efforts to develop nonanimal alternatives contribute to refining and reducing the use of animals in research.

4.9.3.2 In Vitro and Computational Models as Alternatives

In vitro and computational models have emerged as powerful alternatives to traditional animal research, offering valuable insights into disease mechanisms, drug testing, and toxicity assessment. These models not only reduce the reliance on animals but also provide more ethical and cost-effective approaches. The significance of in vitro and computational models as alternatives to animal research and their applications are enumerated in the following section.

The Significance of In Vitro and Computational Models

In vitro and computational models are revolutionizing research by providing effective alternatives to animal studies. As technology advances, these models will continue to play a crucial role in advancing our understanding of diseases and developing new treatments.

- **Ethical Considerations:** In vitro and computational models offer a way to conduct research without the ethical concerns associated with animal use (Knight 2007).
- **Reduced Animal Usage:** These models contribute to the 3Rs principle—reducing the number of animals used in research while maintaining scientific rigor (Zurlo et al. 1996).
- **Precision and Control:** In vitro and computational models provide precise control over experimental conditions, enabling targeted investigations (Huh et al. 2010, 2013).

Applications of In Vitro and Computational Models

- **Cell Cultures and Organoids:** In vitro cell cultures and three-dimensional organoids replicate physiological functions, aiding drug testing and disease modeling (Lancaster and Knoblich 2014).

- **Microfluidic Platforms:** Microfluidic systems mimic complex physiological environments, enabling real-time studies of cell responses and interactions (Whitesides 2006).
- **Computational Simulations:** Computational models predict drug interactions, toxicity, and disease progression, aiding in personalized medicine (Yue and Dutta 2022).

Advancements and Future Directions

- **Integration of Technologies:** Combining in vitro and computational models enhances their predictive power and relevance to human physiology.
- **Organ-on-Chip Platforms:** Organ-on-chip technologies simulate multiple organ interactions, offering a more holistic understanding of disease mechanisms.
- **Machine Learning and AI:** Utilizing machine learning and artificial intelligence improves the accuracy of predictions and data analysis in these models.

4.10 Case Studies: Successful Applications of In Vitro and Computational Models

Case studies showcasing successful applications of in vitro and computational models offer compelling evidence of the potential and impact of these alternatives in various research areas. These studies highlight how these models contribute to understanding disease mechanisms, drug discovery, and toxicity assessment while reducing the reliance on traditional animal research.

Disease modeling using organoid technologies seems to be the latest technology as demonstrated by Lancaster and Knoblich (2014). Likewise, drug discovery with multifluid platform also plays a role in computational modeling (Whitesides 2006). Similarly, predicting drug interactions with computational simulation seems to be an easy and cost-effective technique as this is being carried out by several researchers (Adam et al. 2020; Mei and Zhang 2021).

4.10.1 Advancements and Future Directions

Case studies underscore the transformative potential of in vitro and computational models in various research domains. As these models continue to evolve and become more sophisticated, their widespread adoption promises to reshape the landscape of biomedical research and contribute to improved healthcare outcomes, such as the following:

- **Complex Disease Models:** Continued advancements will enable the creation of more complex in vitro models, mimicking the intricacies of human organs and systems.
- **Personalized Medicine:** In vitro and computational models will contribute to personalized medicine by simulating individual patient responses to treatments.

- **High-Throughput Screening:** Automation and robotics will further enhance the efficiency of in vitro models for large-scale drug screening.

4.10.1.1 Model-Based Insights Leading to Therapeutic Breakthroughs

Model-based approaches, including computational simulations and in vitro systems, have played a pivotal role in driving therapeutic breakthroughs across various medical disciplines. By providing detailed insights into disease mechanisms, drug interactions, and treatment responses, these models have facilitated the development of novel therapies and improved existing treatments.

Understanding Disease Mechanisms

Model-based insights are driving transformative changes in therapeutic development, enabling researchers to uncover new avenues for treatment and refine existing strategies. As technology evolves and our understanding of complex diseases deepens, model-based approaches will continue to play a central role in shaping the future of medicine. Examples for these are as follows:

- **Alzheimer's Disease and Amyloid Beta Aggregation:** Computational models elucidated the aggregation process of amyloid beta protein, contributing to the development of potential Alzheimer's disease treatments (Grasso and Danani 2020).
- **Cancer and Oncogenic Signaling Pathways:** Computational models of cancer signaling pathways identified key nodes for targeted therapies, leading to the development of precision medicine approaches (Klinger et al. 2013).

Enhancing Drug Discovery

- **Virtual Screening for Drug Candidates:** Computational docking and molecular dynamics simulations expedited the identification of potential drug candidates, streamlining drug discovery (Kitchen et al. 2004).
- **Cardiac Safety Assessment:** In vitro assays and computational models accurately predicted drug-induced cardiotoxicity, reducing the risk of adverse effects during clinical trials (Kofron et al. 2021).

Tailoring Personalized Treatments

- **HIV Drug Resistance and Personalized Therapy:** Computational models predicted drug-resistant mutations in HIV, guiding the selection of effective antiretroviral treatments for individual patients (Lengauer et al. 2014).
- **Cystic Fibrosis and Mutation-Specific Therapies:** Computational models informed the design of mutation-specific therapies for cystic fibrosis, addressing underlying genetic defects (Veit et al. 2021).

Advancements and Future Directions

- **Integration of Multi-Omics Data:** Combining computational models with multi-omics data will provide a more comprehensive understanding of disease biology and treatment responses.
- **Bioprinting and Tissue Engineering:** Advanced in vitro systems, including bioprinted tissues, will offer more physiologically relevant platforms for drug testing and disease modeling.
- **Artificial Intelligence and Machine Learning:** AI-driven models will enable more accurate predictions and facilitate data-driven therapeutic breakthroughs.

4.10.1.2 Collaborative Efforts Between Researchers and Clinicians

Collaboration between researchers and clinicians is a cornerstone of translational medicine, bridging the gap between laboratory discoveries and clinical applications. This collaborative approach accelerates the development of innovative therapies and improves patient outcomes. Collaborative research between researchers and clinicians holds great potential for advancing medical knowledge and improving patient care. However, this dynamic partnership also presents challenges that need to be addressed to ensure its success.

Advancing Translational Medicine

Collaborative efforts between researchers and clinicians drive the translation of scientific discoveries into tangible benefits for patients. As medicine becomes increasingly personalized and innovative, the synergy between these two groups will continue to shape the landscape of healthcare and lead to more effective and targeted therapies.

- **Bridging Bench to Bedside:** Collaborations allow for the seamless translation of laboratory findings into clinical trials, expediting the development of new treatments (Woolf 2008).
- **Patient-Centered Approach:** Clinicians provide valuable insights into disease progression and patient needs, guiding researchers in designing effective interventions (Califf 2018).

Examples of Collaborative Success

- **Cystic Fibrosis Translational Research:** Collaboration between researchers, clinicians, and patient advocacy groups led to the development of CFTR modulator therapies, transforming the treatment landscape for cystic fibrosis (Ramsey et al. 2011).
- **Precision Oncology and Targeted Therapies:** Collaborative efforts enabled the identification of molecular biomarkers in cancer, leading to personalized treatment strategies and improved patient outcomes (Garofalo and Croce 2013).

4.11 Challenges and Future Directions

Collaborative research between researchers and clinicians offers immense potential to drive scientific advancements and improve patient care. By addressing challenges and implementing strategies for effective collaboration, the future of medical research holds promise for more innovative and impactful outcomes. The following aspects will help a researcher to overcome challenges in rare disease research:

- **Communication and Integration:** Establishing clear channels of communication and integrated workflows between researchers and clinicians is essential for successful collaboration (Reeve et al. 2007).
- **Data Sharing and Analysis:** Collaborations require sharing and analyzing large datasets, necessitating standardized data formats, and ethical data sharing practices (Ioannidis et al. 2018).
- **Training and Education:** Training programs that foster interdisciplinary understanding are essential to ensure effective collaboration between researchers and clinicians.

4.11.1 Challenges in Collaborative Research

- **Communication and Language Barrier:** Researchers and clinicians often use different terminologies, which can lead to misunderstandings and hinder effective collaboration (Berkenkotter and Huckin 1995).
- **Time Constraints and Priorities:** Clinicians' busy schedules and researchers' academic commitments can create difficulties in finding common meeting times (Howley 2004).
- **Data Sharing and Privacy Concerns:** Sharing patient data for research purposes requires careful consideration of privacy regulations and ethical guidelines (Chiruvella and Guddati 2021).

4.11.2 Future Directions for Effective Collaboration

- **Interdisciplinary Training:** Education programs that foster interdisciplinary understanding from early stages can prepare both researchers and clinicians for effective collaboration (Thistlethwaite et al. 2010).
- **Integrated Technology Platforms:** Developing technology platforms that streamline data sharing and communication can facilitate collaborative research efforts (Bates et al. 2003).
- **Incentive Structures:** Creating incentives for both researchers and clinicians to engage in collaborative research can encourage participation and commitment (Halpern et al. 2021).

4.11.3 Ethical Considerations

- **Transparency and Consent:** Clear communication about the goals and outcomes of collaborative research ensures that all parties are aware of their roles and responsibilities.
- **Equitable Contributions:** Ethical guidelines should emphasize the fair recognition and acknowledgment of contributions from both researchers and clinicians (Newhouse-Oisten et al. 2017).

4.11.3.1 Bridging the Gap Between Animal Models and Human Therapies

The translation of findings from animal models to effective human therapies is a critical step in biomedical research. However, challenges exist in ensuring that results observed in animal studies are applicable to humans.

Challenges in Translating Animal Studies to Human Therapies

- **Species Differences:** Genetic, physiological, and anatomical variations between animals and humans can limit the direct applicability of animal models to human diseases (Hackam and Redelmeier 2006).
- **Complexity of Human Diseases:** Many diseases in humans are multifactorial and involve intricate molecular pathways that may not be fully recapitulated in animal models (van der Worp et al. 2010).
- **Pharmacokinetics and Drug Responses:** Differences in drug metabolism and response between species can impact the efficacy and safety of therapeutic interventions (Toutain et al. 2010).

By addressing these challenges and leveraging innovative approaches, researchers can enhance the translation of scientific discoveries into effective treatments for human diseases.

Strategies for Improved Translational Success

For improved translational success, the following strategies are helpful:

- **Humanized Models:** Developing animal models with humanized genes or tissues can better mimic human disease conditions and responses to therapies (Rongvaux et al. 2013).
- **Stem Cell-Based Models:** Induced pluripotent stem cells (iPSCs) derived from patient samples can provide personalized disease models for drug testing and mechanistic studies (Sánchez-Danés et al. 2012).
- **Clinical Trial Design:** Incorporating translational endpoints and biomarkers that align with animal model findings can enhance the likelihood of successful clinical trials (Vollmer et al. 2014).

Integrated Research-Education Framework

- **Educating Researchers and Clinicians:** Researchers and clinicians should be aware of the limitations and strengths of animal models to make informed decisions in translational research as opined by Arnason (2020).
- **Transparency in Reporting:** Transparent reporting of animal model studies, including methods and limitations, enhances the interpretability and relevance of the results (Kilkenny et al. 2009).
- **Data Sharing and Collaboration:** Collaboration between researchers, clinicians, and pharmaceutical industries fosters the sharing of data and insights, increasing the likelihood of successful translation (Kaitin 2010).

4.11.3.2 Emerging Technologies and Their Impact on Rare Disease Modeling

Advancements in technology have revolutionized the field of rare disease modeling, offering new tools and approaches to study these complex conditions. From gene editing techniques to advanced imaging modalities, emerging technologies have expanded our understanding of rare diseases and accelerated the development of targeted therapies.

CRISPR-Cas9 and Precision Genome Editing

- **Gene Correction in Rare Diseases:** CRISPR-Cas9 technology allows for precise correction of disease-causing mutations, providing potential therapeutic strategies for monogenic rare diseases (Long et al. 2018).
- **CRISPR-Based Disease Modeling:** CRISPR-Cas9 enables the creation of cellular and animal models that accurately mimic rare disease mutations, aiding mechanistic studies (Quadros et al. 2017).

Organ-on-Chip and 3D Culture Systems

- **Physiological Microenvironments:** Organ-on-chip devices and 3D culture systems replicate human tissue microenvironments, enabling more accurate disease modeling (Huh et al. 2010, 2013).
- **Personalized Disease Models:** Patient-derived cells can be incorporated into 3D culture systems, creating personalized models that capture the unique aspects of rare diseases (Drost et al. 2015).

Advanced Imaging Techniques

- **Single-Cell Imaging:** Single-cell imaging technologies reveal heterogeneity within rare disease tissues, enhancing our understanding of disease progression (Raj and van Oudenaarden 2009).
- **In Vivo Imaging:** In vivo imaging modalities such as MRI and PET enable real-time visualization of disease progression and therapeutic responses (Hussain et al. 2022).

Omics Technologies

- **Genomics and Rare Disease Discovery:** Next-generation sequencing and whole-exome sequencing have accelerated the identification of genetic variants underlying rare diseases (Bamshad et al. 2011).
- **Proteomics and Biomarker Discovery:** Proteomic technologies help identify disease-specific biomarkers, facilitating early diagnosis and treatment monitoring (Karp and Lilley 2009).

Bioinformatics and Computational Modeling

- **Network Analysis and Pathway Mapping:** Bioinformatics tools unravel complex molecular networks associated with rare diseases, aiding in identifying potential therapeutic targets (Barabási and Oltvai 2004).
- **Virtual Screening for Drug Discovery:** Computational models predict drug interactions and potential therapeutic compounds, expediting drug discovery for rare diseases (Kitchen et al. 2004).

4.12 Regulatory Landscape and Translation to Clinical Trials

The journey from rare disease research to clinical trials involves traversing a complex regulatory landscape to ensure patient safety and efficacy of potential therapies. This section delves into the regulatory challenges faced when transitioning from preclinical research to human trials, along with strategies to facilitate successful translation, supported by references from the scientific literature.

4.12.1 Regulatory Challenges in Rare Disease Clinical Trials

Successfully navigating the regulatory landscape and translating rare disease research into clinical trials require a comprehensive understanding of regulatory pathways, ethical considerations, and strategic approaches to maximize the potential for therapeutic breakthroughs. Orphan drug designation and ethical considerations appear to be the regulatory challenges in clinical trials of rare diseases.

- **Orphan Drug Designation:** Obtaining orphan drug status for rare disease therapies can streamline development, but stringent criteria must be met (Haffner et al. 2002).
- **Ethical Considerations:** Ethical approval for rare disease trials requires careful consideration due to the limited patient population and potential vulnerabilities (Ross et al. 2013).

4.12.2 Translational Strategies for Clinical Trials

- **Natural History Studies:** Conducting natural history studies elucidates disease progression, aiding in the design of clinical trials and selection of appropriate endpoints (Palmer et al. 2021).
- **Patient-Driven Research:** Collaborating with patient advocacy groups and involving patients in trial design enhance patient-centric approaches (Kaufmann et al. 2018).

4.12.3 Regulatory Pathways for Rare Disease Therapies

- **Accelerated Approvals:** Regulatory agencies offer expedited pathways for rare disease therapies with significant unmet medical needs (Drago et al. 2021).
- **Adaptive Trial Designs:** Adaptive trials in rare diseases allow for real-time modifications based on accumulating data, optimizing trial efficiency (Kairalla et al. 2012).

4.12.4 Data Collection and Patient Registries

- **Longitudinal Data Collection:** Establishing patient registries captures long-term data, facilitating post-approval monitoring and real-world evidence generation (Weatherald et al. 2019).
- **Real-World Evidence and Post-marketing Surveillance:** Regulatory agencies increasingly rely on real-world data for post-approval monitoring and safety assessment (Sherman et al. 2016).

4.12.5 Global Harmonization and Collaborations

- **International Regulatory Harmonization:** Global collaborations harmonize regulatory standards and expedite rare disease therapy development (Aymé and Schmidtke 2007).
- **Orphan Drug Legislation:** National and international orphan drug legislation incentivize research and development for rare diseases (Elliott and Zurynski 2015).

4.12.5.1 FDA and EMA Guidelines for Rare Disease Therapies

Regulatory agencies such as the U.S. Food and Drug Administration (FDA) and the European Medicines Agency (EMA) play a crucial role in overseeing the development and approval of therapies for rare diseases. This section explores the guidelines and regulatory frameworks established by the FDA and EMA to facilitate the development of treatments for rare diseases, accompanied by references from the scientific literature.

FDA Guidelines for Rare Disease Therapies

- **Orphan Drug Designation:** The FDA's Orphan Drug Program provides incentives, including tax credits and exclusive market exclusivity, to encourage the development of therapies for rare diseases (Haffner et al. 2002).
- **Accelerated Approvals:** The FDA offers accelerated approval pathways for rare disease therapies that demonstrate promising early results, allowing for faster access to patients in need (FDA 2014).

EMA Guidelines for Rare Disease Therapies

- **Orphan Medicinal Product Designation:** The EMA's Orphan Medicinal Product Designation provides similar incentives as the FDA's program to encourage research and development of therapies for rare diseases (EMA 2018a).
- **Conditional Marketing Authorization:** The EMA offers conditional marketing authorization for rare disease therapies that address unmet medical needs, provided that comprehensive data collection continues post-approval (EMA 2015).

Harmonization and Collaborative Efforts

Regulatory guidelines from the FDA and EMA provide a framework for developing safe and effective therapies for rare diseases. Navigating these guidelines requires a deep understanding of the regulatory landscape and a commitment to patient-centered approaches.

- **FDA-EMA Collaboration:** The FDA and EMA collaborate to align regulatory standards and facilitate the development of rare disease therapies on a global scale (Roth-Cline and Nelson 2014, 2015).
- **International Harmonization:** Global initiatives, such as the International Rare Diseases Research Consortium (IRDiRC 2020), promote regulatory harmonization to accelerate rare disease therapy development.

Patient-Centric Approaches

- **Patient Involvement:** Both the FDA and EMA emphasize the importance of patient engagement throughout the drug development process, including clinical trial design (EMA 2018b).
- **Adaptive Pathways:** The EMA's adaptive pathways approach encourages early patient involvement and adaptive trial designs to streamline rare disease therapy development (EMA 2016).

4.12.5.2 Moving from Bench to Bedside: Clinical Trial Considerations

The transition from laboratory research to clinical trials is a pivotal step in the development of therapies for rare diseases. This section explores the critical factors and considerations involved in planning and conducting clinical trials for rare disease treatments, supported by references from the scientific literature.

Patient Recruitment and Study Design

- **Challenges in Patient Recruitment:** The limited patient population for rare diseases can pose challenges in recruiting a sufficient number of participants for clinical trials (Gahl et al. 2012).
- **Natural History Studies:** Understanding the natural progression of the disease through natural history studies helps design trials with appropriate endpoints and patient cohorts (Nickel and Schulz 2022).

Clinical Endpoint Selection

- **Relevance of Clinical Endpoints:** Identifying relevant and meaningful clinical endpoints is essential for demonstrating the efficacy of the treatment (FDA 2019).
- **Patient-Reported Outcomes (PROs):** Incorporating PROs in clinical trials captures patients' perspectives on the impact of the disease and treatment (Schipper et al. 1984).

Trial Design and Statistical Considerations

- **Adaptive Trial Designs:** Adaptive trials allow for modifications based on accumulating data, increasing trial efficiency, and optimizing outcomes (Chow and Chang 2008).
- **Rare Disease-Specific Statistical Methods:** Due to small sample sizes, rare disease trials often require specialized statistical methods to analyze data (Proschan 2009).

Ethical Considerations and Patient Safety

- **Informed Consent and Vulnerable Populations:** Ensuring informed consent for patients with limited treatment options and safeguarding the rights of vulnerable populations are crucial (EMA 2017).
- **Benefit-Risk Assessment:** Balancing potential benefits with risks is vital when dealing with rare diseases and limited treatment options (FDA 2019).

Patient Engagement and Advocacy

- **Patient-Centric Trials:** Involving patients and their caregivers in trial design ensures that patient needs and preferences are considered (Frank et al. 2014).
- **Collaboration with Advocacy Groups:** Collaborating with patient advocacy groups can enhance trial recruitment, patient engagement, and awareness (Smith et al. 2015).

Conducting clinical trials for rare diseases requires careful consideration of patient recruitment, trial design, ethical concerns, and patient engagement to ensure the successful development of effective therapies.

4.13 Collaborative Networks and Data Sharing in Rare Disease Research

Collaborative networks and data sharing play a pivotal role in advancing research and therapeutic development for rare diseases. This section explores the importance of collaboration, the benefits of data sharing, and the challenges associated with fostering partnerships in the field of rare disease research.

4.13.1 Importance of Collaborative Networks

Collaborative networks and data sharing are essential for advancing rare disease research, leveraging collective knowledge and resources to accelerate discoveries and therapeutic breakthroughs. Rare disease research consortia demonstrate the power of collaborative efforts, driving advancements in understanding rare diseases, identifying therapeutic targets, and ultimately improving the lives of patients.

- **Pooling Expertise and Resources:** Collaborative networks bring together researchers, clinicians, patient advocates, and industry partners, pooling diverse expertise and resources (Boycott et al. 2017).
- **Leveraging Limited Data:** Rare diseases often lack sufficient data for meaningful analysis, making collaboration crucial for acquiring and analyzing a critical mass of information (Kohane et al. 2012).

4.13.2 Benefits of Data Sharing

- **Accelerating Research:** Data sharing accelerates research by enabling broader analyses, validation, and hypothesis generation across multiple datasets (Gahl et al. 2012).
- **Identifying Novel Biomarkers:** Aggregating data from various sources enhances the ability to identify rare disease biomarkers and potential therapeutic targets (Sanjak et al. 2023).

4.13.3 Challenges and Overcoming Barriers

- **Data Privacy and Security:** Ensuring patient privacy and data security while sharing sensitive medical information is a challenge that requires robust safeguards (Kayaalp 2018).
- **Interoperability and Standardization:** Collaborative networks must address data interoperability and standardization to facilitate meaningful cross-analysis (Ohmann et al. 2017).

4.13.4 Successful Collaborative Models

- **IRDiRC and Global Collaborations:** The International Rare Diseases Research Consortium (IRDiRC) promotes global collaboration, setting goals for rare disease research and therapies (Lochmüller et al. 2018).
- **Patient-Driven Collaborations:** Patient advocacy groups and foundations facilitate collaborative efforts among researchers, clinicians, and patients (Rubinstein et al. 2013).

4.13.4.1 Rare Disease Research Consortia: Advantages and Achievements

Rare disease research consortia play a crucial role in fostering collaboration, accelerating discoveries, and advancing therapeutic development for rare diseases. This section delves into the advantages of such consortia and highlights their significant achievements, supported by references from the scientific literature.

Advantages of Rare Disease Research Consortia

- **Pooling Expertise and Resources:** Consortia bring together researchers, clinicians, patient advocates, and industry partners, enabling the sharing of specialized expertise, technologies, and resources (Boycott et al. 2017).
- **Leveraging Collective Data:** Consortia facilitate the aggregation of data from diverse sources, increasing statistical power and enabling in-depth analyses (Lochmüller et al. 2018).

Achievements of Rare Disease Research Consortia

- **Disease Gene Identification:** Consortia have successfully identified disease-causing genes through large-scale sequencing efforts, providing insights into rare disease mechanisms (Bamshad et al. 2012).
- **Natural History Studies:** Consortia have conducted comprehensive natural history studies, elucidating disease progression and informing clinical trial design (Gahl et al. 2012).

Therapeutic Development

- **Target Identification and Validation:** Consortia have played a pivotal role in identifying potential therapeutic targets and validating their relevance through preclinical studies (Jordan et al. 2018).
- **Clinical Trial Facilitation:** Consortia have aided in designing and executing clinical trials, ensuring rigorous methodologies and patient recruitment (McCarty et al. 2011).

Patient Advocacy and Engagement

- **Empowering Patients:** Consortia actively involve patient advocacy groups, ensuring that patients and their families are active partners in the research process (Huang et al. 2017).

- **Patient Registries:** Consortia have established patient registries to gather comprehensive clinical and genetic information, aiding research and clinical trial recruitment (Kodra et al. 2017).

4.13.4.2 Data Repositories and Open Science Initiatives in Rare Disease Research

Data repositories and open science initiatives have transformed the landscape of rare disease research by promoting data sharing, collaboration, and accelerated discoveries. This section explores the significance of data repositories and open science in rare disease research, highlighting their advantages and contributions, supported by references from the scientific literature.

Advantages of Data Repositories and Open Science Initiatives

Data repositories and open science initiatives drive rare disease research forward by enabling broad data sharing, collaboration, and fostering a culture of transparency and innovation.

- **Enhanced Collaboration:** Data repositories foster collaboration among researchers, clinicians, and institutions worldwide by providing a centralized platform for sharing data (NIH 2023).
- **Efficient Resource Utilization:** Open science initiatives maximize the use of available resources, enabling researchers to build upon existing data and avoid duplicative efforts (Roche et al. 2015).

Contributions to Rare Disease Research

- **Large-Scale Data Integration:** Data repositories facilitate the integration of diverse datasets, allowing researchers to analyze larger sample sizes and uncover novel insights (Malovini et al. 2011).
- **Standardized Data Formats:** Data repositories often require standardized data formats, promoting consistency and facilitating cross-study comparisons (Schriml et al. 2012).

Promotion of Open Science

- **Transparency and Reproducibility:** Open science initiatives emphasize transparency in research methodologies and promote the sharing of detailed protocols, enhancing reproducibility (Nosek et al. 2015).

Data Citation and Attribution: Open science initiatives encourage proper citation and attribution of shared data, acknowledging contributors and supporting scientific integrity (Piwowar and Vision 2013).

Challenges and Future Directions

- **Data Privacy and Security:** Balancing data sharing with privacy concerns requires robust strategies for anonymization and secure access (Malakar et al. 2023).

- **Sustainability and Long-Term Access:** Ensuring the sustainability of data repositories is crucial for maintaining access to valuable resources over time (Shah et al. 2019).

4.14 Conclusion

Throughout this chapter, we have explored the multifaceted and indispensable role that animal models play in advancing our understanding of rare diseases. Animal models serve as invaluable tools that bridge the gap between basic research and clinical applications, offering insights into disease mechanisms, therapeutic development, and personalized medicine. As we conclude, let us recap the key takeaways and emphasize the significance of animal models in rare disease research.

4.14.1 Key Takeaways

- **Understanding Disease Mechanisms:** Animal models allow researchers to dissect the complex molecular, cellular, and physiological processes underlying rare diseases, shedding light on their causes and progression.
- **Therapeutic Development:** Animal models serve as testing grounds for potential therapies, facilitating the identification of effective treatments and offering a platform for preclinical evaluation.
- **Personalized Medicine:** Animal models help elucidate patient-specific responses to treatments, paving the way for personalized therapeutic approaches tailored to individual genetic and physiological factors.
- **Translational Research:** Animal models enable the translation of basic research findings into clinical applications, guiding the development of novel diagnostics, drugs, and therapies.
- **Target Identification and Validation:** Animal models aid in identifying and validating potential therapeutic targets, providing a basis for designing targeted interventions.
- **Disease Modeling:** Animal models replicate disease features and phenotypes, allowing researchers to study disease progression and test interventions in controlled settings.

4.14.2 The Future of Rare Disease Research

As we look ahead, the role of animal models in rare disease research is poised to evolve even further. Advances in genetic engineering, technologies for noninvasive imaging, and collaborative efforts will continue to enhance the accuracy and relevance of animal models. Furthermore, the integration of data sharing, open science initiatives, and emergence of multidisciplinary research consortia will accelerate discoveries and foster a more holistic understanding of rare diseases.

4.14.2.1 Summing Up Contributions and Insights

In the journey through the intricacies of animal models in rare disease research, we have uncovered a wealth of contributions and insights that have reshaped the landscape of medical science. As we conclude this chapter, let us recapitulate the significant contributions and key insights that have emerged from the exploration of animal models in the realm of rare diseases.

Significant Contributions

- **Unraveling Disease Complexity:** Animal models have unraveled the intricate complexities of rare diseases, shedding light on their underlying mechanisms, genetic mutations, and physiological processes that contribute to their manifestation.
- **Therapeutic Breakthroughs:** Through rigorous testing and validation, animal models have paved the way for the development of novel therapeutic approaches, enabling the discovery of potential treatments and interventions that were previously unimaginable.
- **Personalized Medicine Prospects:** Animal models have showcased the potential of personalized medicine by offering insights into individualized responses to treatments, contributing to the advancement of patient-centered care.
- **Innovation in Research Techniques:** The study of animal models has propelled the development of innovative research techniques, from genetic engineering to noninvasive imaging, revolutionizing the way rare diseases are investigated.

Key Insights

- **Model Selection Precision:** The selection of appropriate animal models is essential, taking into consideration species-specific factors, genetic homology, disease manifestations, and relevance to human biology.
- **Translational Considerations:** Successful translation from animal models to human therapies requires meticulous planning, considering factors such as dosing, toxicity, pharmacokinetics, and safety profiles.
- **Ethical and Regulatory Considerations:** Ethical and regulatory frameworks play a critical role in ensuring the humane treatment of animals and the responsible use of animal models in research.
- **Collaboration and Data Sharing:** Collaborative networks, data repositories, and open science initiatives amplify the impact of rare disease research, accelerating discoveries through the collective efforts of researchers worldwide.

Future Directions

As we look to the future, the insights gained from animal models are poised to catalyze a new era of rare disease research. Advancements in genetic engineering, improved imaging technologies, and heightened interdisciplinary collaboration will continue to refine the accuracy and applicability of animal models. Furthermore, the integration of ethical considerations, data sharing, and regulatory guidelines will ensure the responsible and impactful use of animal models in advancing medical knowledge.

In toto, the contributions and insights gleaned from animal models have redefined the possibilities in understanding, diagnosing, and treating rare diseases. These models stand as testaments to the remarkable strides that science can take when armed with dedication, innovation, and collective pursuit of improving lives.

4.14.2.2 Looking Ahead: Unveiling the Potential of Future Research

As we conclude this chapter, it is essential to cast our gaze toward the horizon of future research in the realm of rare diseases and animal models. The journey we have embarked upon is far from over, and the potential for transformative discoveries and advancements is immense.

Genomic Precision and Personalized Medicine

- **Advancements in Genomic Editing:** The continuous refinement of genetic editing techniques, including CRISPR-Cas9 and base editing, holds promise for generating precise animal models that closely mimic human genetic mutations (Komor et al. 2016).
- **Precision Therapies:** The integration of genomic information from animal models and patients will enable the development of targeted therapies tailored to individual genetic profiles (Gambardella et al. 2020).

Advanced Imaging and Functional Characterization

- **Innovative Imaging Modalities:** Emerging imaging technologies, such as multimodal imaging and single-cell resolution imaging, will provide unprecedented insights into disease mechanisms and treatment responses (Yang et al. 2021).
- **Functional Phenotyping:** Integrating behavioral, physiological, and molecular analyses will offer a comprehensive understanding of disease phenotypes and therapeutic outcomes (Keane et al. 2011).

Translational Advances and Clinical Impact

- **Overcome of Translational Barriers:** Innovations in preclinical-to-clinical translation methodologies will streamline the process of moving from promising animal model results to successful human trials (van der Worp et al. 2010).
- **Accelerated Therapeutic Development:** Collaborative efforts, facilitated by consortia, data sharing initiatives, and computational modeling, will expedite the identification and validation of potential therapeutics (Berlin et al. 2006).

Ethics and Responsible Research

- **Enhanced Ethical Frameworks:** Advancements in ethics frameworks will ensure that animal models are utilized responsibly, taking into account the welfare of animals and the societal implications of research (Kimmelman and Federico 2017).
- **Alternatives to Animal Models:** The development of sophisticated in vitro and computational models will continue to complement animal models, reducing the reliance on animal research (Hartung and Leist 2008).

As we stand on the precipice of future research, it is evident that the potential for discovery is vast. The synergy between cutting-edge technologies, collaborative endeavors, ethical considerations, and patient-centered approaches will shape a new era of rare disease research.

References

Abrahams BS, Geschwind DH (2008) Advances in autism genetics: on the threshold of a new neurobiology. Nat Rev Genet 9(5):341–355. https://doi.org/10.1038/nrg2346

Adam G, Rampášek L, Safikhani Z, Smirnov P, Haibe-Kains B, Goldenberg A (2020) Machine learning approaches to drug response prediction: challenges and recent progress. NPJ Precis Oncol 4:19. https://doi.org/10.1038/s41698-020-0122-1

Angelis A, Tordrup D, Kanavos P (2015) Socio-economic burden of rare diseases: a systematic review of cost of illness evidence. Health Policy 119(7):964–979. https://doi.org/10.1016/j.healthpol.2014.12.016

Anguela XM, Sharma R, Doyon Y, Miller JC, Li H, Haurigot V, Rohde ME, Wong SY, Davidson RJ, Zhou S, Gregory PD, Holmes MC, High KA (2013) Robust ZFN-mediated genome editing in adult hemophilic mice. Blood 122(19):3283–3287. https://doi.org/10.1182/blood-2013-04-497354

Antunes M, Biala G (2012) The novel object recognition memory: neurobiology, test procedure, and its modifications. Cogn Process 13(2):93–110. https://doi.org/10.1007/s10339-011-0430-z

Arnason G (2020) The emergence and development of animal research ethics: a review with a focus on nonhuman primates. Sci Eng Ethics 26(4):2277–2293. https://doi.org/10.1007/s11948-020-00219-z

Asmamaw M, Zawdie B (2021) Mechanism and applications of CRISPR/Cas-9-mediated genome editing. Biologics: Targets Ther 15:353–361. https://doi.org/10.2147/BTT.S326422

Austin CP, Cutillo CM, Lau LPL, Jonker AH, Rath A, Julkowska D, Thomson D, Terry SF, de Montleau B, Ardigò D, Hivert V, Boycott KM, Baynam G, Kaufmann P, Taruscio D, Lochmüller H, Suematsu M, Incerti C, Draghia-Akli R, Norstedt I, Wang L, Dawkins HJS, International Rare Diseases Research Consortium (IRDiRC) (2018) Future of rare diseases research 2017–2027: an IRDiRC perspective. Clin Transl Sci 11(1):21–27. https://doi.org/10.1111/cts.12500

Aymé S, Schmidtke J (2007) Networking for rare diseases: a necessity for Europe. Bundesgesundheitsblatt Gesundheitsforschung Gesundheitsschutz 50(12):1477–1483. https://doi.org/10.1007/s00103-007-0381-9

Baker HJ, Lindsey JR (1974) Animal model: feline GM1 gangliosidosis. Am J Pathol 74(3):649–652. https://doi.org/10.1016/S0002-9440(10)36159-9

Balcombe JP (2006) Laboratory environments and rodents' behavioural needs: a review. Lab Anim 40(3):217–235. https://doi.org/10.1258/002367706777611488

Baldridge D, Wangler MF, Bowman AN, Yamamoto S, Undiagnosed Diseases Network, Schedl T, Pak SC, Postlethwait JH, Shin J, Solnica-Krezel L, Bellen HJ, Westerfield M (2021) Model organisms contribute to diagnosis and discovery in the undiagnosed diseases network: current state and a future vision. Orphanet J Rare Dis 16(1):206. https://doi.org/10.1186/s13023-021-01839-9

Bamshad MJ, Ng SB, Bigham AW, Tabor HK, Emond MJ, Nickerson DA, Shendure J (2011) Exome sequencing as a tool for Mendelian disease gene discovery. Nat Rev Genet 12(11):745–755. https://doi.org/10.1038/nrg3031

Bamshad MJ, Shendure JA, Valle D, Hamosh A, Lupski JR, Gibbs RA, Boerwinkle E, Lifton RP, Gerstein M, Gunel M, Mane S, Nickerson DA, Centers for Mendelian Genomics (2012) The Centers for Mendelian genomics: a new large-scale initiative to identify the genes underlying

rare Mendelian conditions. Am J Med Genet A 158A(7):1523–1525. https://doi.org/10.1002/ajmg.a.35470
Barabási AL, Oltvai ZN (2004) Network biology: understanding the cell's functional organization. Nat Rev Genet 5(2):101–113. https://doi.org/10.1038/nrg1272
Barchet TM, Amiji MM (2009) Challenges and opportunities in CNS delivery of therapeutics for neurodegenerative diseases. Expert Opin Drug Deliv 6(3):211–225. https://doi.org/10.1517/17425240902758188
Barrangou R, Fremaux C, Deveau H, Richards M, Boyaval P, Moineau S, Romero DA, Horvath P (2007) CRISPR provides acquired resistance against viruses in prokaryotes. Science 315(5819):1709–1712. https://doi.org/10.1126/science.1138140
Bates DW, Ebell M, Gotlieb E, Zapp J, Mullins HC (2003) A proposal for electronic medical records in U.S. primary care. J Am Med Inform Assoc 10(1):1–10. https://doi.org/10.1197/jamia.M1097
Bates GP, Dorsey R, Gusella JF, Hayden MR, Kay C, Leavitt BR, Nance M, Ross CA, Scahill RI, Wetzel R, Wild EJ, Tabrizi SJ (2015) Huntington disease. Nat Rev Dis Primers 1:15005. https://doi.org/10.1038/nrdp.2015.5
Bäumer D, Lee S, Nicholson G, Davies JL, Parkinson NJ, Murray LM, Gillingwater TH, Ansorge O, Davies KE, Talbot K (2009) Alternative splicing events are a late feature of pathology in a mouse model of spinal muscular atrophy. PLoS Genet 5(12):e1000773. https://doi.org/10.1371/journal.pgen.1000773
Baxendale S, van Eeden F, Wilkinson R (2019) The power of zebrafish in personalized medicine. In: Braun T, Schweitzer M (eds) Personalized medicine: novel approaches in understanding and treating genetic and other diseases. Springer, pp 179–197. https://doi.org/10.1007/978-3-319-60733-7_10
Benitz WE, Committee on Fetus and Newborn, American Academy of Pediatrics (2016) Patent ductus arteriosus in preterm infants. Pediatrics 137(1):e20153730. https://doi.org/10.1542/peds.2015-3730
Berkenkotter C, Huckin T (1995) Genre knowledge in disciplinary communication: cognition/culture/power. Lawrence Erlbaum Associates, pp 1–208. https://doi.org/10.4324/9781315538747
Berlin A, Sorani M, Sim I (2006) A taxonomic description of computer-based clinical decision support systems. J Biomed Inform 39(6):656–667. https://doi.org/10.1016/j.jbi.2005.12.003
Bhakuni GS, Bedi O, Bariwal J, Deshmukh R, Kumar P (2016) Animal models of hepatotoxicity. Inflamm Res 65(1):13–24. https://doi.org/10.1007/s00011-015-0883-0
Bilen J, Bonini NM (2005) Drosophila as a model for human neurodegenerative disease. Annu Rev Genet 39:153–171. https://doi.org/10.1146/annurev.genet.39.110304.095804
Binder EB (2009) The role of FKBP5, a co-chaperone of the glucocorticoid receptor in the pathogenesis and therapy of affective and anxiety disorders. Psychoneuroendocrinology 34(Suppl. 1):S186–S195. https://doi.org/10.1016/j.psyneuen.2009.05.021
Borger DK, McMahon B, Roshan Lal T, Serra-Vinardell J, Aflaki E, Sidransky E (2017) Induced pluripotent stem cell models of lysosomal storage disorders. Dis Model Mech 10(6):691–704. https://doi.org/10.1242/dmm.029009
Boycott KM, Rath A, Chong JX, Hartley T, Alkuraya FS, Baynam G, Brookes AJ, Brudno M, Carracedo A, den Dunnen JT, Dyke SOM, Estivill X, Goldblatt J, Gonthier C, Groft SC, Gut I, Hamosh A, Hieter P, Höhn S, Hurles ME, Kaufmann P, Knoppers BM, Krischer JP, Macek M Jr, Matthijs G, Olry A, Parker S, Paschall J, Philippakis AA, Rehm HL, Robinson PN, Sham P-C, Stefanov R, Taruscio D, Unni D, Vanstone MR, Zhang F, Brunner H, Bamshad MJ, Lochmüller H (2017) International cooperation to enable the diagnosis of all rare genetic diseases. Am J Hum Genet 100(5):695–705. https://doi.org/10.1016/j.ajhg.2017.04.003
Brown SD, Moore MW (2012) The international mouse phenotyping consortium: past and future perspectives on mouse phenotyping. Mamm Genome 23(9–10):632–640. https://doi.org/10.1007/s00335-012-9427-x
Bulfield G, Siller WG, Wight PA, Moore KJ (1984) X chromosome-linked muscular dystrophy (mdx) in the mouse. Proc Natl Acad Sci U S A 81(4):1189–1192. https://doi.org/10.1073/pnas.81.4.1189

Burghardt AJ, Link TM, Majumdar S (2011) High-resolution computed tomography for clinical imaging of bone microarchitecture. Clin Orthop Relat Res 469(8):2179–2193. https://doi.org/10.1007/s11999-010-1766-x

Cai K, Haris M, Singh A, Kogan F, Greenberg JH, Hariharan H, Detre JA, Reddy R (2012) Magnetic resonance imaging of glutamate. Nat Med 18(2):302–306. https://doi.org/10.1038/nm.2615

Califf RM (2018) Biomarker definitions and their applications. Exp Biol Med 243(3):213–221. https://doi.org/10.1177/1535370217750088

Capecchi MR (1989) Altering the genome by homologous recombination. Science 244(4910):1288–1292. https://doi.org/10.1126/science.2660260

Capecchi MR (2005) Gene targeting in mice: functional analysis of the mammalian genome for the twenty-first century. Nat Rev Genet 6(6):507–512. https://doi.org/10.1038/nrg1619

Carvalho CM, Lupski JR (2016) Mechanisms underlying structural variant formation in genomic disorders. Nat Rev Genet 17(4):224–238. https://doi.org/10.1038/nrg.2015.25

Cattaneo E, Rigamonti D, Goffredo D, Zuccato C, Squitieri F, Sipione S (2001) Loss of normal huntingtin function: new developments in Huntington's disease research. Trends Neurosci 24(3):182–188. https://doi.org/10.1016/s0166-2236(00)01721-5

Cesarovic N, Lipiski M, Falk V, Emmert MY (2020) Animals in cardiovascular research. Eur Heart J 41(2):200–203. https://doi.org/10.1093/eurheartj/ehz933

Chabner BA, Roberts TG Jr (2005) Timeline: chemotherapy and the war on cancer. Nat Rev Cancer 5(1):65–72. https://doi.org/10.1038/nrc1529

Chahrour M, Jung SY, Shaw C, Zhou X, Wong ST, Qin J, Zoghbi HY (2008) MeCP2, a key contributor to neurological disease, activates and represses transcription. Science 320(5880):1224–1229. https://doi.org/10.1126/science.1153252

Chiruvella V, Guddati AK (2021) Ethical issues in patient data ownership. Interact J Med Res 10(2):e22269. https://doi.org/10.2196/22269

Choi TY, Choi TI, Lee YR, Choe SK, Kim CH (2021) Zebrafish as an animal model for biomedical research. Exp Mol Med 53:310–317. https://doi.org/10.1038/s12276-021-00571-5

Chow SC, Chang M (2008) Adaptive design methods in clinical trials—a review. Orphanet J Rare Dis 3:11. https://doi.org/10.1186/1750-1172-3-11

Coecke S, Balls M, Bowe G, Davis J, Gstraunthaler G, Hartung T, Hay R, Merten OW, Price A, Schechtman L, Stacey G, Stokes W, Second ECVAM Task Force on Good Cell Culture Practice (2005) Guidance on good cell culture practice. A report of the second ECVAM task force on good cell culture practice. Altern Lab Anim 33(3):261–287. https://doi.org/10.1177/026119290503300313

Colby LA, Quenee LE, Zitzow LA (2017) Considerations for infectious disease research studies using animals. Comp Med 67(3):222–231

Collins FS, Varmus H (2015) A new initiative on precision medicine. N Engl J Med 372(9):793–795. https://doi.org/10.1056/NEJMp1500523

Collins FS, Rossant J, Wurst WA, International Mouse Knockout Consortium (2007) A mouse for all reasons. Cell 128(1):9–13. https://doi.org/10.1016/j.cell.2006.12.018

Cong L, Ran FA, Cox D, Lin S, Barretto R, Habib N, Hsu PD, Wu X, Jiang W, Marraffini LA, Zhang F (2013) Multiplex genome engineering using CRISPR/Cas systems. Science 339(6121):819–823. https://doi.org/10.1126/science.1231143

Cook D, Brown D, Alexander R, March R, Morgan P, Satterthwaite G, Pangalos MN (2014) Lessons learned from the fate of AstraZeneca's drug pipeline: a five-dimensional framework. Nat Rev Drug Discov 13(6):419–431. https://doi.org/10.1038/nrd4309

Crawley JN (ed) (2007) What's wrong with my mouse? Behavioral phenotyping of transgenic and knockout mice. John Wiley & Sons, pp 1–503. https://doi.org/10.1002/0470119055

da Silva-Buttkus P, Spielmann N, Klein-Rodewald T, Schütt C, Aguilar-Pimentel A, Amarie OV, Becker L, Calzada-Wack J, Garrett L, Gerlini R, Kraiger M, Leuchtenberger S, Östereicher MA, Rathkolb B, Sanz-Moreno A, Stöger C, Hölter SM, Seisenberger C, Marschall S, Fuchs H, Gailus-Durner V, Hrabě de Angelis M (2023) Knockout mouse models as a resource for the study of rare diseases. Mamm Genome 34(4):244–261. https://doi.org/10.1007/s00335-023-09986-z

Daldrup-Link HE, Golovko D, Ruffell B, Denardo DG, Castaneda R, Ansari C, Rao J, Tikhomirov GA, Wendland MF, Corot C, Coussens LM (2011) MRI of tumor-associated macrophages with clinically applicable iron oxide nanoparticles. Clin Cancer Res 17(17):5695–5704. https://doi.org/10.1158/1078-0432.CCR-10-3420

de Paula Faria D, de Vries EF, Sijbesma JW, Buchpiguel CA, Dierckx RA, Copray SC (2014) PET imaging of glucose metabolism, neuroinflammation and demyelination in the lysolecithin rat model for multiple sclerosis. Multi Scler (Houndmills, Basingstoke, England) 20(11):1443–1452. https://doi.org/10.1177/1352458514526941

Dejea H, Garcia-Canadilla P, Cook AC, Eduard Guasch E, Zamora M, Crispi F, Stampanoni M, Bijnens B, Bonnin A (2019) Comprehensive analysis of animal models of cardiovascular disease using multiscale x-ray phase contrast tomography. Sci Rep 9:6996. https://doi.org/10.1038/s41598-019-43407-z

Delude C (2015) Deep phenotyping: the details of disease. Nature 527:S14–S15. https://doi.org/10.1038/527S14a

Dhurandhar D, Sahoo NK, Mariappan I, Narayanan R (2021) Gene therapy in retinal diseases: a review. Indian J Ophthalmol 69(9):2257–2265. https://doi.org/10.4103/ijo.IJO_3117_20

Ding Y, Haks MC, Forn-Cuní G, He J, Nowik N, Harms AC, Hankemeier T, Eeza MNH, Matysik J, Alia A, Spaink HP (2021) Metabolomic and transcriptomic profiling of adult mice and larval zebrafish leptin mutants reveal a common pattern of changes in metabolites and signaling pathways. Cell Biosci 11:126. https://doi.org/10.1186/s13578-021-00642-0

Dodson MV, Allen RE, Du M, Bergen WG, Velleman SG, Poulos SP, Fernyhough-Culver M, Wheeler MB, Duckett SK, Young MR, Voy BH, Jiang Z, Hausman GJ (2015) Invited review: evolution of meat animal growth research during the past 50 years: adipose and muscle stem cells. J Anim Sci 93(2):457–481. https://doi.org/10.2527/jas.2014-8221

Doghman M, Karpova T, Rodrigues GA, Arhatte M, De Moura J, Cavalli LR, Virolle V, Barbry P, Zambetti GP, Figueiredo BC, Heckert LL, Lalli E (2007) Increased steroidogenic factor-1 dosage triggers adrenocortical cell proliferation and cancer. Mol Endocrinol (Baltimore, MD) 21(12):2968–2987. https://doi.org/10.1210/me.2007-0120

Doudna JA, Charpentier E (2014) Genome editing. The new frontier of genome engineering with CRISPR-Cas9. Science 346(6213):1258096. https://doi.org/10.1126/science.1258096

Dowell RT (1991) Comprehensive evaluation of cardiovascular function in the anesthetized rat. Methods Find Exp Clin Pharmacol 13(9):609–616

Doyle A, McGarry MP, Lee NA, Lee JJ (2012) The construction of transgenic and gene knockout/knockin mouse models of human disease. Transgenic Res 21(2):327–349. https://doi.org/10.1007/s11248-011-9537-3

Drago D, Foss-Campbell B, Wonnacott K, Barrett D, Ndu A (2021) Global regulatory progress in delivering on the promise of gene therapies for unmet medical needs. Mole Ther Methods Clin Dev 21:524–529. https://doi.org/10.1016/j.omtm.2021.04.001

Drost J, van Jaarsveld RH, Ponsioen B, Zimberlin C, van Boxtel R, Buijs A, Sachs N, Overmeer RM, Offerhaus GJ, Begthel H, Korving J, van de Wetering M, Schwank G, Logtenberg M, Cuppen E, Snippert HJ, Medema JP, Kops GJ, Clevers H (2015) Sequential cancer mutations in cultured human intestinal stem cells. Nature 521(7550):43–47. https://doi.org/10.1038/nature14415

Duncan AW, Hanlon Newell AE, Bi W, Finegold MJ, Olson SB, Beaudet AL, Grompe M (2012) Aneuploidy as a mechanism for stress-induced liver adaptation. J Clin Invest 122(9):3307–3315. https://doi.org/10.1172/JCI64026

Egawa N, Kitaoka S, Tsukita K, Naitoh M, Takahashi K, Yamamoto T, Adachi F, Kondo T, Okita K, Asaka I, Aoi T, Watanabe A, Yamada Y, Morizane A, Takahashi J, Ayaki T, Ito H, Yoshikawa K, Yamawaki S, Suzuki S, Watanabe D, Hioki H, Kaneko T, Makioka K, Okamoto K, Takuma H, Tamaoka A, Hasegawa K, Nonaka T, Hasegawa M, Kawata A, Yoshida M, Nakahata T, Takahashi R, Marchetto MCN, Gage FH, Yamanaka S, Inoue H (2012) Drug screening for ALS using patient-specific induced pluripotent stem cells. Sci Transl Med 4(145):145ra104. https://doi.org/10.1126/scitranslmed.3004052

Elliott E, Zurynski Y (2015) Rare diseases are a 'common' problem for clinicians. Aust Fam Physician 44(9):630–633

EMA (European Medicines Agency) (2015) Guidance for industry: conditional marketing authorization. https://www.ema.europa.eu/en/human-regulatory/marketing-authorisation/conditional-marketing-authorisation

EMA (European Medicines Agency) (2016) Adaptive pathways. https://www.ema.europa.eu/en/human-regulatory/research-development/adaptive-pathways

EMA (European Medicines Agency) (2017). Guideline on strategies to identify and mitigate risks for first-in-human and early clinical trials with investigational medicinal products. https://www.ema.europa.eu/en/documents/scientific-guideline/guideline-strategies-identify-mitigate-risks-first-human-early-clinical-trials-investigational_en.pdf

EMA (European Medicines Agency) (2018a) Orphan designation: overview. https://www.ema.europa.eu/en/medicines/human/orphan-designations/eu-3-18-2050

EMA (European Medicines Agency) (2018b) Patient involvement. Overview of patient involvement in EMA public communications. https://www.ema.europa.eu/en/documents/presentation/presentation-overview-patient-involvement-ema-public-communications-mmavris-ema_en.pdf

Emborg ME (2007) Non-human primate models of Parkinson's disease. ILAR J 48(4):339–355. https://doi.org/10.1093/ilar.48.4.339

Emery AE (2002) The muscular dystrophies. Lancet 359(9307):687–695. https://doi.org/10.1016/S0140-6736(02)07869-7

Emini Veseli B, Perrotta P, De Meyer GRA, Roth L, Van der Donckt C, Martinet W, De Meyer GRY (2017) Animal models of atherosclerosis. Eur J Pharmacol 816:3–13. https://doi.org/10.1016/j.ejphar.2017.05.010

Erickson RP (1996) Mouse models of human genetic disease: which mouse is more like a man? BioEssays 18(12):993–998. https://doi.org/10.1002/bies.950181209

Esposito CL, Nuzzo S, Kumar SA, Rienzo A, Lawrence CL, Pallini R, Shaw L, Alder JE, Ricci-Vitiani L, Catuogno S, de Franciscis V (2016) A combined microRNA-based targeted therapeutic approach to eradicate glioblastoma stem-like cells. J Control Rel 238:43–57. https://doi.org/10.1016/j.jconrel.2016.07.032

EURORDIS (European Organisation for Rare Diseases) (2005) Rare diseases: understanding this public health priority. pp. 1–14. https://www.eurordis.org/sites/default/files/publications/Fact_Sheet_RD.pdf

Falk DJ, Todd AG, Lee S, Soustek MS, ElMallah MK, Fuller DD, Notterpek L, Byrne BJ (2015) Peripheral nerve and neuromuscular junction pathology in Pompe disease. Hum Mol Genet 24(3):625–636. https://doi.org/10.1093/hmg/ddu476

Fang M, Peng CW, Pang DW, Li Y (2012) Quantum dots for cancer research: current status, remaining issues, and future perspectives. Cancer Biol Med 9(3):151–163. https://doi.org/10.7497/j.issn.2095-3941.2012.03.001

FDA (Food and Drug Administration) (2014) Expedited programs for serious conditions—drugs and biologics. Guidance for industry. https://www.fda.gov/regulatory-information/search-fda-guidance-documents

FDA (Food and Drug Administration) (2019) Rare diseases: common issues in drug development, guidance for industry. https://www.fda.gov/media/119757/download

Forner S, Kawauchi S, Balderrama-Gutierrez G, Kramár EA, Matheos DP, Phan J, Javonillo DI, Tran KM, Hingco E, da Cunha C, Rezaie N, Alcantara JA, Baglietto-Vargas D, Jansen C, Neumann J, Wood MA, MacGregor GR, Mortazavi A, Tenner AJ, LaFerla FM, Green KN (2021) Systematic phenotyping and characterization of the 5xFAD mouse model of Alzheimer's disease. Sci Data 8:270. https://doi.org/10.1038/s41597-021-01054-y

Frank L, Basch E, Selby JV, Patient-Centered Outcomes Research Institute (2014) The PCORI perspective on patient-centered outcomes research. JAMA 312(15):1513–1514. https://doi.org/10.1001/jama.2014.11100

Fricker LD (2005) Neuropeptide-processing enzymes: applications for drug discovery. AAPS J 7(2):E449–E445. https://doi.org/10.1208/aapsj070244

Gahl WA, Markello TC, Toro C, Fajardo KF, Sincan M, Gill F, Carlson-Donohoe H, Gropman A, Pierson TM, Golas G, Wolfe L, Groden C, Godfrey R, Nehrebecky M, Wahl C, Landis DMD, Yang C, Madeo A, Mullikin JC, NISC Comparative Sequencing Program, Boerkoel CF, Tifft CJ, Adams D (2012) The National Institutes of Health undiagnosed diseases program: insights into rare diseases. Genet Med 14(1):51–59. https://doi.org/10.1038/gim.0b013e318232a005

Gambardella V, Fleitas T, Tarazona N, Papaccio F, Huerta M, Roselló S, Gimeno-Valiente F, Roda D, Cervantes A (2020) Precision medicine to treat advanced gastroesophageal adenocarcinoma: a work in progress. J Clin Med 9(9):3049. https://doi.org/10.3390/jcm9093049

Gambhir SS (2002) Molecular imaging of cancer with positron emission tomography. Nat Rev Cancer 2(9):683–693. https://doi.org/10.1038/nrc882

Garofalo M, Croce CM (2013) MicroRNAs as therapeutic targets in chemoresistance. Drug Resist Updat 16(3–5):47–59. https://doi.org/10.1016/j.drup.2013.05.001

Ginn SL, Amaya AK, Alexander IE, Edelstein M, Abedi MR (2018) Gene therapy clinical trials worldwide to 2017: an update. J Gene Med 20(5):e3015. https://doi.org/10.1002/jgm.3015

Goodchild RE, Dauer WT (2005) The AAA+ protein torsinA interacts with a conserved domain present in LAP1 and a novel ER protein. J Cell Biol 168(6):855–862. https://doi.org/10.1083/jcb.200411026

Gopinath C, Nathar TJ, Ghosh A, Hickstein DD, Nelson EJR (2015) Contemporary animal models for human gene therapy applications. Curr Gene Ther 15(6):531–540. https://doi.org/10.2174/1566523215666150929110424

Gordon JW, Ruddle FH (1981) Integration and stable germ line transmission of genes injected into mouse pronuclei. Science 214(4526):1244–1246. https://doi.org/10.1126/science.6272397

Grasso G, Danani A (2020) Molecular simulations of amyloid beta assemblies. Adv Phys X 5(1):1770627. https://doi.org/10.1080/23746149.2020.1770627

Gu H, Marth JD, Orban PC, Mossmann H, Rajewsky K (1994) Deletion of a DNA polymerase beta gene segment in T cells using cell type-specific gene targeting. Science 265(5168):103–106. https://doi.org/10.1126/science.8016642

Gutierrez K, Dicks N, Glanzner WG, Agellon LB, Bordignon V (2015) Efficacy of the porcine species in biomedical research. Front Genet 6:293. https://doi.org/10.3389/fgene.2015.00293

Hackam DG, Redelmeier DA (2006) Translation of research evidence from animals to humans. JAMA 296(14):1731–1732. https://doi.org/10.1001/jama.296.14.1731

Haffner ME, Whitley J, Moses M (2002) Two decades of orphan product development. Nat Rev Drug Discov 1(10):821–825. https://doi.org/10.1038/nrd919

Halevy T, Benvenisty N (2015) Modeling fragile X syndrome in human pluripotent cells. In: Zatz M, Keith Okamoto O (eds) Stem cells in modeling human genetic diseases. Stem cell biology and regenerative medicine. Springer, Cham, pp 1–10. https://doi.org/10.1007/978-3-319-18314-5_1

Halpern SD, Chowdhury M, Bayes B, Cooney E, Hitsman BL, Schnoll RA, Lubitz SF, Reyes C, Patel MS, Greysen SR, Mercede A, Reale C, Barg FK, Volpp KG, Karlawish J, Stephens-Shields AJ (2021) Effectiveness and ethics of incentives for research participation: 2 randomized clinical trials. JAMA Intern Med 181(11):1479–1488. https://doi.org/10.1001/jamainternmed.2021.5450

Harding CO, Gillingham MB, Hamman K, Clark H, Goebel-Daghighi E, Bird A, Koeberl DD (2006) Complete correction of hyperphenylalaninemia following liver-directed, recombinant AAV2/8 vector-mediated gene therapy in murine phenylketonuria. Gene Ther 13(5):457–462. https://doi.org/10.1038/sj.gt.3302678

Harms PW, Vats P, Verhaegen ME, Robinson DR, Wu YM, Dhanasekaran SM, Palanisamy N, Siddiqui J, Cao X, Su F, Wang R, Xiao H, Kunju LP, Mehra R, Tomlins SA, Fullen DR, Bichakjian CK, Johnson TM, Dlugosz AA, Chinnaiyan AM (2015) The distinctive mutational spectra of polyomavirus-negative Merkel cell carcinoma. Cancer Res 75(18):3720–3727. https://doi.org/10.1158/0008-5472.CAN-15-0702

Hartung T, Leist M (2008) Food for thought ... on the evolution of toxicology and the phasing out of animal testing. ALTEX 25(2):91–102. https://doi.org/10.14573/altex.2008.2.91

Hartung T, McBride M (2011) Food for thought ... on mapping the human toxome. ALTEX 28(2):83–93. https://doi.org/10.14573/altex.2011.2.083

Haskins ME (2007) Animal models for mucopolysaccharidosis disorders and their clinical relevance. Acta Paediatr (Oslo, Norway: 1992) 96(455):56–62. https://doi.org/10.1111/j.1651-2227.2007.00211.x

Haskins ME, Giger U, Patterson DF (2006) Animal models of lysosomal storage diseases: their development and clinical relevance (Chapter 6). In: Mehta A, Beck M, Sunder-Plassmann G (eds) Fabry disease: perspectives from 5 years of FOS. Oxford PharmaGenesis, Oxford. https://www.ncbi.nlm.nih.gov/books/NBK11578

Havel PJ, Kievit P, Comuzzie AG, Bremer AA (2017) Use and importance of nonhuman primates in metabolic disease research: current state of the field. ILAR J 58(2):251–268. https://doi.org/10.1093/ilar/ilx031

High KA, Roncarolo MG (2019) Gene therapy. N Engl J Med 381(5):455–464. https://doi.org/10.1056/NEJMra1706910

Hill MN, Hillard CJ, Bambico FR, Patel S, Gorzalka BB, Gobbi G (2009) The therapeutic potential of the endocannabinoid system for the development of a novel class of antidepressants. Trends Pharmacol Sci 30(9):484–493. https://doi.org/10.1016/j.tips.2009.06.006

Hisahara S, Shimohama S (2011) Toxin-induced and genetic animal models of Parkinson's disease. Parkinson's Dis 2011:951709. https://doi.org/10.4061/2011/951709

Hobson-West P (2007) Trusting blindly can be the biggest risk of all: organised resistance to childhood vaccination in the UK. Sociol Health Illn 29(2):198–215. https://doi.org/10.1111/j.1467-9566.2007.00544.x

Holcik M, Sonenberg N (2005) Translational control in stress and apoptosis. Nat Rev Mol Cell Biol 6(4):318–327. https://doi.org/10.1038/nrm1618

Hong G, Diao S, Chang J, Antaris AL, Chen C, Zhang B, Zhao S, Atochin DN, Huang PL, Andreasson KI, Kuo CJ, Dai H (2014) Through-skull fluorescence imaging of the brain in a new near-infrared window. Nat Photon 8(9):723–730. https://doi.org/10.1038/nphoton.2014.166

Hong S, Zhu T, Zheng S, Zhan X, Xu F, Gu X, Liang L (2021) Gene expression profiles in the brain of phenylketonuria mouse model reversed by the low phenylalanine diet therapy. Metab Brain Dis 36(8):2405–2414. https://doi.org/10.1007/s11011-021-00818-0

Houdebine LM (2007) Transgenic animal models in biomedical research. Methods Mol Biol (Clifton, N.J.) 360:163–202. https://doi.org/10.1385/1-59745-165-7:163

Howley LD (2004) Performance assessment in medical education: where we've been and where we're going. Eval Health Prof 27(3):285–303. https://doi.org/10.1177/0163278704267044

Hrabě de Angelis M, Balling R (1998) Large scale ENU screens in the mouse: genetics meets genomics. Mutat Res 400(1–2):25–32. https://doi.org/10.1016/s0027-5107(98)00061-x

Hsu P, Scott D, Weinstein J, Ran F, Konermann S, Agarwala V, Li Y, Fine E, Wu X, Shalem O, Cradick T, Marraffini L, Bao G, Zhang F (2013) DNA targeting specificity of RNA-guided Cas9 nucleases. Nat Biotechnol 31:827–832. https://doi.org/10.1038/nbt.2647

Hsu PD, Lander ES, Zhang F (2014) Development and applications of CRISPR-Cas9 for genome engineering. Cell 157(6):1262–1278. https://doi.org/10.1016/j.cell.2014.05.010

Hu S, Hoffman EA, Reinhardt JM (2001) Automatic lung segmentation for accurate quantitation of volumetric X-ray CT images. IEEE Trans Med Imaging 20(6):490–498. https://doi.org/10.1109/42.929615

Huang J, Lipman PD, Daniel Mullins C (2017) Bridging the divide: building infrastructure to support community-academic partnerships and improve capacity to conduct patient-centered outcomes research. Transl Behav Med 7(4):773–782. https://doi.org/10.1007/s13142-017-0487-z

Huang W, Percie du Sert N, Vollert J, Rice ASC (2020) General principles of preclinical study design. Handb Exp Pharmacol 257:55–69. https://doi.org/10.1007/164_2019_277

Huh D, Matthews BD, Mammoto A, Montoya-Zavala M, Hsin HY, Ingber DE (2010) Reconstituting organ-level lung functions on a chip. Science (New York, N.Y.) 328(5986):1662–1668. https://doi.org/10.1126/science.1188302

Huh D, Kim H, Fraser J, Shea DE, Khan M, Bahinski A, Hamilton GA, Ingber DE (2013) Microfabrication of human organs-on-chips. Nat Protoc 8(12):2135–2157. https://doi.org/10.1038/nprot.2013.137

Hung JH, Weng Z (2017) Analysis of microarray and RNA-seq expression profiling data. Cold Spring Harb Protoc 2017(3). https://doi.org/10.1101/pdb.top093104

Hussain S, Mubeen I, Ullah N, Shah SSUD, Khan BA, Zahoor M, Ullah R, Khan FA, Sultan MA (2022) Modern diagnostic imaging technique applications and risk factors in the medical field: a review. Biomed Res Int 2022:5164970. https://doi.org/10.1155/2022/5164970

Hytönen MK, Lohi H (2016) Canine models of human rare disorders. Rare Dis (Austin, Tex) 4(1):e1241362. https://doi.org/10.1080/21675511.2016.1241362

Igarashi P, Somlo S (2002) Genetics and pathogenesis of polycystic kidney disease. J Am Soc Nephrol 13(9):2384–2398. https://doi.org/10.1097/01.asn.0000028643.17901.42

Ioannidis JPA, Kim BYS, Trounson A (2018) How to design preclinical studies in nanomedicine and cell therapy to maximize the prospects of clinical translation. Nat Biomed Eng 2(11):797–809. https://doi.org/10.1038/s41551-018-0314-y

IRDiRC (International Rare Diseases Research Consortium) (2020) Policies and guidelines. https://www.irdirc.org/wp-content/uploads/2020/05/IRDiRC-Policies-and-Guidelines-May-2020.pdf

Itzhaki I, Maizels L, Huber I, Gepstein A, Arbel G, Caspi O, Miller L, Belhassen B, Nof E, Glikson M, Gepstein L (2012) Modeling of catecholaminergic polymorphic ventricular tachycardia with patient-specific human-induced pluripotent stem cells. J Am Coll Cardiol 60(11):990–1000. https://doi.org/10.1016/j.jacc.2012.02.066

Jasanoff A (2007) MRI contrast agents for functional molecular imaging of brain activity. Curr Opin Neurobiol 17(5):593–600. https://doi.org/10.1016/j.conb.2007.11.002

Jiang W, Chen L (2022) Tissue specificity of gene expression evolves across mammal species. J Comput Biol 29(8):880–891. https://doi.org/10.1089/cmb.2021.0592

Jiang YZ, Ma D, Suo C, Shi J, Xue M, Hu X, Xiao Y, Yu KD, Liu YR, Yu Y, Zheng Y, Li X, Zhang C, Hu P, Zhang J, Hua Q, Zhang J, Hou W, Ren L, Bao D, Bingying Li B, Yang J, Yao L, Zuo W-J, Zhao S, Gong Y, Ren Y-X, Zhao Y-X, Yang Y-S, Niu Z, Cao Z-G, Stover DG, Verschraegen C, Kaklamani V, Daemen A, Benson JR, Takabe K, Bai F, Li D-Q, Wang P, Shi L, Huang W, Shao ZM (2019) Genomic and transcriptomic landscape of triple-negative breast cancers: subtypes and treatment strategies. Cancer Cell 35(3):428–440.e5. https://doi.org/10.1016/j.ccell.2019.02.001

Jinek M, Chylinski K, Fonfara I, Hauer M, Doudna JA, Charpentier E (2012) A programmable dual-RNA-guided DNA endonuclease in adaptive bacterial immunity. Science (New York, N.Y.) 337(6096):816–821. https://doi.org/10.1126/science.1225829

Jordan VK, Fregeau B, Ge X, Giordano J, Wapner RJ, Balci TB, Carter MT, Bernat JA, Moccia AN, Srivastava A, Martin DM, Bielas SL, Pappas J, Svoboda MD, Rio M, Boddaert N, Cantagrel V, Lewis AM, Scaglia F, Undiagnosed Diseases Network, Kohler N, Bernstein JA, Dries AM, Rosenfeld JA, DeFilippo C, Thorson W, Yang Y, Sherr EH, Bi W, Scott DA (2018) Genotype-phenotype correlations in individuals with pathogenic RERE variants. Hum Mutat 39(5):666–675. https://doi.org/10.1002/humu.23400

Justice MJ, Dhillon P (2016) Using the mouse to model human disease: increasing validity and reproducibility. Dis Model Mech 9(2):101–103. https://doi.org/10.1242/dmm.024547

Kabashi E, Bercier V, Lissouba A, Liao M, Brustein E, Rouleau GA, Drapeau P (2011) FUS and TARDBP but not SOD1 interact in genetic models of amyotrophic lateral sclerosis. PLoS Genet 7(8):e1002214. https://doi.org/10.1371/journal.pgen.1002214

Kairalla JA, Coffey CS, Thomann MA, Muller KE (2012) Adaptive trial designs: a review of barriers and opportunities. Trials 139(1):145. https://doi.org/10.1186/1745-6215-13-145

Kaitin KI (2010) Deconstructing the drug development process: the new face of innovation. Clin Pharmacol Ther 87(3):356–361. https://doi.org/10.1038/clpt.2009.293

Kang Y, Chu C, Wang F, Niu Y (2019) CRISPR/Cas9-mediated genome editing in nonhuman primates. Dis Model Mech 12(10):dmm039982. https://doi.org/10.1242/dmm.039982

Karp NA, Lilley KS (2009) Investigating sample pooling strategies for DIGE experiments to address biological variability. Proteomics 9(2):388–397. https://doi.org/10.1002/pmic.200800485

Kaufmann P, Pariser AR, Austin C (2018) From scientific discovery to treatments for rare diseases—the view from the National Center for Advancing Translational Sciences—Office

of Rare Diseases Research. Orphanet J Rare Dis 13(1):196. https://doi.org/10.1186/s13023-018-0936-x
Kayaalp M (2018) Patient privacy in the era of big data. Balkan Med J 35(1):8–17. https://doi.org/10.4274/balkanmedj.2017.0966
Keane TM, Goodstadt L, Danecek P, White MA, Wong K, Yalcin B, Heger A, Agam A, Slater G, Goodson M, Furlotte NA, Eskin E, Nellåker C, Whitley H, Cleak J, Janowitz D, Hernandez-Pliego P, Edwards A, Belgard TG, Oliver PL, McIntyre RE, Bhomra A, Nicod J, Gan X, Yuan W, van der Weyden L, Steward CA, Bala S, Stalker J, Mott R, Durbin R, Jackson IJ, Czechanski A, Guerra-Assunção JA, Donahue LR, Reinholdt LG, Payseur BA, Ponting CP, Birney E, Flint J, Adams DJ (2011) Mouse genomic variation and its effect on phenotypes and gene regulation. Nature 477(7364):289–294. https://doi.org/10.1038/nature10413
Khalak H, Wakil S, Imtiaz F, Ramzan K, Baz B, Almostafa A, Hagos S, Alzahrani F, Abu-Dhaim N, Safieh LA, Latifa Al-Jbali L, Al-Hamed MS, Monies D, Aldahmesh M, Al-Dosari MS, Kaya N, Shamseldin H, Shaheen R, Al-Rashed M, Hashem M, Al-Tassan N, Meyer B, Alazami AM, Alkuraya FS (2012) Autozygome maps dispensable DNA and reveals potential selective bias against nullizygosity. Genet Med 14:515–519. https://doi.org/10.1038/gim.2011.28
Khanna C, Hunter K (2005) Modeling metastasis in vivo. Carcinogenesis 26(3):513–523. https://doi.org/10.1093/carcin/bgh261
Khorramizadeh M, Saadat F (2020) Animal models for human disease. Anim Biotechnol 2020:153–171. https://doi.org/10.1016/B978-0-12-811710-1.00008-2
Kilkenny C, Parsons N, Kadyszewski E, Festing MFW, Cuthill IC, Fry D, Hutton J, Altman DG (2009) Survey of the quality of experimental design, statistical analysis and reporting of research using animals. PLoS One 4(11):e7824. https://doi.org/10.1371/journal.pone.0007824
Kim SA, Chang JM, Cho N, Yi A, Moon WK (2015) Characterization of breast lesions: comparison of digital breast tomosynthesis and ultrasonography. Korean J Radiol 16(2):229–238. https://doi.org/10.3348/kjr.2015.16.2.229
Kimmelman J, Federico C (2017) Consider drug efficacy before first-in-human trials. Nature 542:25–27. https://doi.org/10.1038/542025a
Kitchen DB, Decornez H, Furr JR, Bajorath J (2004) Docking and scoring in virtual screening for drug discovery: methods and applications. Nat Rev Drug Discov 3(11):935–949. https://doi.org/10.1038/nrd1549
Klinger B, Sieber A, Fritsche-Guenther R, Witzel F, Berry L, Schumacher D, Yan Y, Durek P, Merchant M, Schäfer R, Sers C, Blüthgen N (2013) Network quantification of EGFR signaling unveils potential for targeted combination therapy. Mol Syst Biol 9(1):673. https://doi.org/10.1038/msb.2013.29
Klymiuk N, Blutke A, Graf A, Krause S, Burkhardt K, Wuensch A, Krebs S, Kessler B, Zakhartchenko V, Kurome M, Kemter E, Nagashima H, Schoser B, Herbach N, Blum H, Wanke R, Aartsma-Rus A, Thirion C, Lochmüller H, Walter MC, Wolf E (2013) Dystrophin-deficient pigs provide new insights into the hierarchy of physiological derangements of dystrophic muscle. Hum Mol Genet 22(1):4368–4382. https://doi.org/10.1093/hmg/ddt287
Klymiuk N, Seeliger F, Bohlooly-Y M, Blutke A, Rudmann DG, Wolf E (2016) Tailored pig models for preclinical efficacy and safety testing of targeted therapies. Toxicol Pathol 44(3):346–357. https://doi.org/10.1177/0192623315609688
Knight A (2007) Systematic reviews of animal experiments demonstrate poor human clinical and toxicological utility. Altern Lab Anim 35(6):641–659. https://doi.org/10.1177/026119290703500610
Kobayashi E, Hishikawa S, Teratani T, Lefor AT (2012) The pig as a model for translational research: overview of porcine animal models at Jichi Medical University. Transplant Res 1(1):8. https://doi.org/10.1186/2047-1440-1-8
Kodra Y, Posada de la Paz M, Coi A, Santoro M, Bianchi F, Ahmed F, Rubinstein YR, Weinbach J, Taruscio D (2017) Data quality in rare diseases registries. Adv Exp Med Biol 1031:149–164. https://doi.org/10.1007/978-3-319-67144-4_8

Kofron CM, Kim TY, Munarin F, Soepriatna AH, Kant RJ, Mende U, Choi B-R, Coulombe KLK (2021) A predictive in vitro risk assessment platform for pro-arrhythmic toxicity using human 3D cardiac microtissues. Sci Rep 11:10228. https://doi.org/10.1038/s41598-021-89478-9

Kohane IS, McMurry A, Weber G, MacFadden D, Rappaport L, Kunkel L, Bickel J, Wattanasin N, Spence S, Murphy S, Churchill S (2012) The co-morbidity burden of children and young adults with autism spectrum disorders. PLoS One 7(4):e33224. https://doi.org/10.1371/journal.pone.0033224

Kola I, Landis J (2004) Can the pharmaceutical industry reduce attrition rates? Nat Rev Drug Discov 3(8):711–715. https://doi.org/10.1038/nrd1470

Komor AC, Kim YB, Packer MS, Zuris JA, Liu DR (2016) Programmable editing of a target base in genomic DNA without double-stranded DNA cleavage. Nature 533(7603):420–424. https://doi.org/10.1038/nature17946

Konantz M, Schürch C, Hanns P, Müller JS, Sauteur L, Lengerke C (2019) Modeling hematopoietic disorders in zebrafish. Dis Model Mech 12(9):dmm040360. https://doi.org/10.1242/dmm.040360

Kornegay JN (2017) The golden retriever model of Duchenne muscular dystrophy. Skelet Muscle 7(1):9. https://doi.org/10.1186/s13395-017-0124-z

Kornegay JN, Bogan DJ, Bogan JR, Dow JL, Wang J, Fan Z, Liu N, Warsing LC, Grange RW, Ahn M, Balog-Alvarez CJ, Cotton SW, Willis MS, Brinkmeyer-Langford C, Zhu H, Palandra J, Morris CA, Styner MA, Wagner KR (2016) Dystrophin-deficient dogs with reduced myostatin have unequal muscle growth and greater joint contractures. Skelet Muscle 6:14. https://doi.org/10.1186/s13395-016-0085-7

Krueger DD, Bear MF (2011) Toward fulfilling the promise of molecular medicine in fragile X syndrome. Annu Rev Med 62:411–429. https://doi.org/10.1146/annurev-med-061109-134644

La Croix NC (2005) Ocular manifestations of systemic disease in cats. Clin Tech Small Anim Pract 20(2):121–128. https://doi.org/10.1053/j.ctsap.2004.12.017

Lancaster MA, Knoblich JA (2014) Organogenesis in a dish: modeling development and disease using organoid technologies. Science 345(6194):1247125. https://doi.org/10.1126/science.1247125

Landis SC, Amara SG, Asadullah K, Austin CP, Blumenstein R, Bradley EW, Crystal RG, Darnell RB, Ferrante RJ, Fillit H, Finkelstein R, Fisher M, Gendelman HE, Golub RM, Goudreau JL, Gross RA, Gubitz AK, Hesterlee SE, Howells DW, Huguenard J, Kelner K, Koroshetz W, Krainc D, Lazic SE, Levine MS, Macleod MR, McCall JM, Moxley RT, Narasimhan K, Noble LJ, Perrin S, Porter JD, Steward O, Unger E, Utz U, Silberberg SD (2012) A call for transparent reporting to optimize the predictive value of preclinical research. Nature 490(7419):187–191. https://doi.org/10.1038/nature11556

Lau CH, Tin C, Suh Y (2020) CRISPR-based strategies for targeted transgene knock-in and gene correction. Fac Rev 9:20. https://doi.org/10.12703/r/9-20

Lee G, Papapetrou EP, Kim H, Chambers SM, Tomishima MJ, Fasano CA, Ganat YM, Menon J, Shimizu F, Viale A, Tabar V, Sadelain M, Studer L (2009) Modelling pathogenesis and treatment of familial dysautonomia using patient-specific iPSCs. Nature 461(7262):402–406. https://doi.org/10.1038/nature08320

Lee HJ, Kang TW, Kim BR (2018) Effects of diaphragm and deep abdominal muscle exercise on walking and balance ability in patients with hemiplegia due to stroke. J Exerc Rehabil 14(4):648–653. https://doi.org/10.12965/jer.1836252.126

Lengauer T, Pfeifer N, Kaiser R (2014) Personalized HIV therapy to control drug resistance. Drug Discov Today Technol 11:57–64. https://doi.org/10.1016/j.ddtec.2014.02.004

Leong XF, Ng CY, Jaarin K (2015) Animal models in cardiovascular research: hypertension and atherosclerosis. Biomed Res Int 2015:528757. https://doi.org/10.1155/2015/528757

Lieschke GJ, Currie PD (2007) Animal models of human disease: Zebrafish swim into view. Nat Rev Genet 8(5):353–367. https://doi.org/10.1038/nrg2091

Lissner LJ, Wartchow KM, Toniazzo AP, Gonçalves CA, Rodrigues L (2021) Object recognition and Morris water maze to detect cognitive impairment from mild hippocampal damage in rats:

a reflection based on the literature and experience. Pharmacol Biochem Behav 210:173273. https://doi.org/10.1016/j.pbb.2021.173273

Liu WH, Völse K, Senft D, Jeremias I (2021) A reporter system for enriching CRISPR/Cas9 knockout cells in technically challenging settings like patient models. Sci Rep 11:12649. https://doi.org/10.1038/s41598-021-91760-9

Liu-Chittenden Y, Huang B, Shim JS, Chen Q, Lee SJ, Anders RA, Liu JO, Pan D (2012) Genetic and pharmacological disruption of the TEAD-YAP complex suppresses the oncogenic activity of YAP. Genes Dev 26(12):1300–1305. https://doi.org/10.1101/gad.192856.112

Lobe CG, Nagy A (1998) Conditional genome alteration in mice. BioEssays 20(3):200–208. https://doi.org/10.1002/(SICI)1521-1878(199803)20:3<200::AID-BIES3>3.0.CO;2-V

Lochmüller H, Badowska DM, Thompson R, Knoers NV, Aartsma-Rus A, Gut I, Wood L, Harmuth T, Durudas A, Graessner H, Schaefer F, Riess O, RD-Connect Consortium, NeurOmics consortium, & EURenOmics Consortium (2018) RD-Connect, NeurOmics and EURenOmics: Collaborative European initiative for rare diseases. Eur J Hum Genet 26(6):778–785. https://doi.org/10.1038/s41431-018-0115-5

Long C, Li H, Tiburcy M, Rodriguez-Caycedo C, Kyrychenko V, Zhou H, Zhang Y, Min YL, Shelton JM, Mammen PPA, Liaw NY, Zimmermann WH, Bassel-Duby R, Schneider JW, Olson EN (2018) Correction of diverse muscular dystrophy mutations in human engineered heart muscle. Sci Adv 4(1):eaap9004. https://doi.org/10.1126/sciadv.aap9004

Loonstra A, Vooijs M, Beverloo HB, Allak BA, van Drunen E, Kanaar R, Berns A, Jonkers J (2001) Growth inhibition and DNA damage induced by Cre recombinase in mammalian cells. Proc Natl Acad Sci U S A 98(16):9209–9214. https://doi.org/10.1073/pnas.161269798

Lu B, Vogel H (2009) Drosophila models of neurodegenerative diseases. Annu Rev Pathol 4:315–342. https://doi.org/10.1602/neurorx.2.3.438

Lusic H, Grinstaff MW (2013) X-ray-computed tomography contrast agents. Chem Rev 113(3):1641–1666. https://doi.org/10.1021/cr200358s

Madisen L, Zwingman TA, Sunkin SM, Oh SW, Zariwala HA, Gu H, Ng LL, Palmiter RD, Hawrylycz MJ, Jones AR, Lein ES, Zeng H (2010) A robust and high-throughput Cre reporting and characterization system for the whole mouse brain. Nat Neurosci 13(1):133–140. https://doi.org/10.1038/nn.2467

Maguire AM, Simonelli F, Pierce EA, Pugh EN Jr, Mingozzi F, Bennicelli J, Banfi S, Marshall KA, Testa F, Surace EM, Rossi S, Lyubarsky A, Arruda VR, Konkle B, Stone E, Sun J, Jacobs J, Dell'Osso L, Hertle R, Ma JX, Redmond TM, Zhu X, Hauck B, Zelenaia O, Shindler KS, Maguire MG, Wright JF, Volpe NJ, McDonnell JW, Auricchio A, High KA, Bennett J (2008) Safety and efficacy of gene transfer for Leber's congenital amaurosis. N Engl J Med 358(21):2240–2248. https://doi.org/10.1056/NEJMoa0802315

Malakar Y, Lacey J, Twine NA, McCrea R, Bauer DC (2023) Balancing the safeguarding of privacy and data sharing: perceptions of genomic professionals on patient genomic data ownership in Australia. Eur J Hum Genet 28:1–7. https://doi.org/10.1038/s41431-022-01273-w

Mali P, Yang L, Esvelt KM, Aach J, Guell M, DiCarlo JE, Norville JE, Church GM (2013) RNA-guided human genome engineering via Cas9. Science (New York, NY) 339(6121):823–826. https://doi.org/10.1126/science.1232033

Malovini A, Illario M, Iaccarino G, Villa F, Ferrario A, Roncarati R, Anselmi CV, Novelli V, Cipolletta E, Leggiero E, Orro A, Rusciano MR, Milanesi L, Maione AS, Condorelli G, Bellazzi R, Puca AA (2011) Association study on long-living individuals from Southern Italy identifies rs10491334 in the CAMKIV gene that regulates survival proteins. Rejuvenation Res 14(3):283–291. https://doi.org/10.1089/rej.2010.1114

Mangiarini L, Sathasivam K, Seller M, Cozens B, Harper A, Hetherington C, Lawton M, Trottier Y, Lehrach H, Davies SW, Bates GP (1996) Exon 1 of the HD gene with an expanded CAG repeat is sufficient to cause a progressive neurological phenotype in transgenic mice. Cell 87(3):493–506. https://doi.org/10.1016/S0092-8674(00)81369-0

Mansour SL, Thomas KR, Capecchi MR (1988) Disruption of the proto-oncogene int-2 in mouse embryo-derived stem cells: a general strategy for targeting mutations to non-selectable genes. Nature 336(6197):348–352. https://doi.org/10.1038/336348a0

Marí-Beffa M, Mesa-Román AB, Duran I (2021) Zebrafish models for human skeletal disorders. Front Genet 12:675331. https://doi.org/10.3389/fgene.2021.675331
Massoud TF, Gambhir SS (2003) Molecular imaging in living subjects: seeing fundamental biological processes in a new light. Genes Dev 17(5):545–580. https://doi.org/10.1101/gad.1047403
McCarty CA, Chisholm RL, Chute CG, Kullo IJ, Jarvik GP, Larson EB, Li R, Masys DR, Ritchie MD, Roden DM, Struewing JP, Wolf WA, eMERGE Team (2011) The eMERGE network: a consortium of biorepositories linked to electronic medical records data for conducting genomic studies. BMC Med Genet 4(1):13. https://doi.org/10.1186/1755-8794-4-13
McCauley MD, Wehrens XH (2009) Animal models of arrhythmogenic cardiomyopathy. Dis Model Mech 2(11–12):563–570. https://doi.org/10.1242/dmm.002840
McKinney AJ, Holmen SL (2011) Animal models of melanoma: a somatic cell gene delivery mouse model allows rapid evaluation of genes implicated in human melanoma. Chin J Cancer 30(3):153–162. https://doi.org/10.5732/cjc.011.10007
McLeod C, Hobson-West P (2015) Opening up animal research and science-society relations? A thematic analysis of transparency discourses in the United Kingdom. Public Underst Sci 25(1):791–806. https://doi.org/10.1177/0963662515586320
Mei S, Zhang K (2021) A machine learning framework for predicting drug–drug interactions. Sci Rep 11:17619. https://doi.org/10.1038/s41598-021-97193-8
Mekada K, Abe K, Murakami A, Nakamura S, Nakata H, Moriwaki K, Obata Y, Yoshiki A (2009) Genetic differences among C57BL/6 substrains. Exp Anim 58(2):141–149. https://doi.org/10.1538/expanim.58.141
Mellman I, Coukos G, Dranoff G (2011) Cancer immunotherapy comes of age. Nature 480(7378):480–489. https://doi.org/10.1038/nature10673
Mendell JR, Al-Zaidy S, Shell R, Arnold WD, Rodino-Klapac LR, Prior TW, Lowes L, Alfano L, Berry K, Church K, Kissel JT, Nagendran S, L'Italien J, Sproule DM, Wells C, Cardenas JA, Heitzer MD, Kaspar A, Corcoran S, Braun L, Likhite S, Carlos Miranda C, Meyer K, Foust KD, Burghes AHM, Kaspar BK (2017) Single-dose gene-replacement therapy for spinal muscular atrophy. N Engl J Med 377(18):1713–1722. https://doi.org/10.1056/NEJMoa1706198
Mertens J, Paquola ACM, Ku M, Hatch E, Böhnke L, Ladjevardi S, McGrath S, Campbell B, Lee H, Herdy JR, Gonçalves JT, Toda T, Kim Y, Winkler J, Yao J, Hetzer MW, Gage FH (2015) Directly reprogrammed human neurons retain aging-associated transcriptomic signatures and reveal age-related nucleocytoplasmic defects. Cell Stem Cell 17(6):705–718. https://doi.org/10.1016/j.stem.2015.09.001
Meurs KM, Fox PR, Norgard M, Spier AW, Lamb A, Koplitz SL, Baumwart RD (2007a) A prospective genetic evaluation of familial dilated cardiomyopathy in the Doberman pinscher. J Vet Intern Med 21(5):1016–1020. https://doi.org/10.1892/0891-6640(2007)21[1016:apgeof]2.0.co;2
Meurs KM, Norgard MM, Ederer MM, Hendrix KP, Kittleson MD (2007b) A substitution mutation in the myosin binding protein C gene in ragdoll hypertrophic cardiomyopathy. Genomics 90(2):261–264. https://doi.org/10.1016/j.ygeno.2007.04.007
Miller MA, Zachary JF (2017) Mechanisms and morphology of cellular injury, adaptation, and death. Pathol Basis Vet Dis 2–43:e19. https://doi.org/10.1016/B978-0-323-35775-3.00001-1
Miller C, Abdo Z, Ericsson A, Elder J, VandeWoude S (2018) Applications of the FIV model to study HIV pathogenesis. Viruses 10(4):206. https://doi.org/10.3390/v10040206
Mistry PK, Cappellini MD, Lukina E, Ozsan H, Mach Pascual S, Rosenbaum H, Helena Solano M, Spigelman Z, Villarrubia J, Watman NP, Massenkeil G (2011) A reappraisal of Gaucher disease-diagnosis and disease management algorithms. Am J Hematol 86(1):110–115. https://doi.org/10.1002/ajh.21888
Mojic M, Takeda K, Hayakawa Y (2017) The dark side of IFN-γ: its role in promoting cancer immunoevasion. Int J Mol Sci 19(1):89. https://doi.org/10.3390/ijms19010089
Monani UR (2005) Spinal muscular atrophy: a deficiency in a ubiquitous protein; a motor neuron-specific disease. Neuron 48(6):885–896. https://doi.org/10.1016/j.neuron.2005.12.001
Monani UR, Sendtner M, Coovert DD, Parsons DW, Andreassi C, Le TT, Jablonka S, Schrank B, Rossoll W, Prior TW, Morris GE, Burghes AH (2000) The human centromeric survival

motor neuron gene (SMN2) rescues embryonic lethality in Smn(−/−) mice and results in a mouse with spinal muscular atrophy. Hum Mol Genet 9(3):333–339. https://doi.org/10.1093/hmg/9.3.333

Morrice JR, Gregory-Evans CY, Shaw CA (2018) Animal models of amyotrophic lateral sclerosis: a comparison of model validity. Neural Regen Res 13(12):2050–2054. https://doi.org/10.4103/1673-5374.241445

Moy SS, Nadler JJ, Perez A, Barbaro RP, Johns JM, Magnuson TR, Piven J, Crawley JN (2004) Sociability and preference for social novelty in five inbred strains: an approach to assess autistic-like behavior in mice. Genes Brain Behav 3(5):287–302. https://doi.org/10.1111/j.1601-1848.2004.00076.x

Mukherjee P, Roy S, Ghosh D, Nandi SK (2022) Role of animal models in biomedical research: a review. Lab Anim Res 38(1):18

Nachury MV, Loktev AV, Zhang Q, Westlake CJ, Peränen J, Merdes A, Slusarski DC, Scheller RH, Bazan JF, Sheffield VC, Jackson PK (2007) A core complex of BBS proteins cooperates with the GTPase Rab8 to promote ciliary membrane biogenesis. Cell 129(6):1201–1213. https://doi.org/10.1016/j.cell.2007.03.053

Nahrendorf M, Swirski FK (2016) Abandoning M1/M2 for a network model of macrophage function. Circ Res 119(3):414–417. https://doi.org/10.1161/CIRCRESAHA.116.309194

Naldini L, Blömer U, Gallay P, Ory D, Mulligan R, Gage FH, Verma IM, Trono D (1996) In vivo gene delivery and stable transduction of nondividing cells by a lentiviral vector. Science (New York, NY) 272(5259):263–267. https://doi.org/10.1126/science.272.5259.263

Narfström K (1999) Hereditary and congenital ocular disease in the cat. J Feline Med Surg 1(3):135–141. https://doi.org/10.1016/S1098-612X(99)90202-4

Nestler EJ, Hyman SE (2010) Animal models of neuropsychiatric disorders. Nat Neurosci 13(10):1161–1169. https://doi.org/10.1038/nn.2647

Neureither F, Ziegler K, Pitzer C, Frings S, Möhrlen F (2017) Impaired motor coordination and learning in mice lacking anoctamin 2 calcium-gated chloride channels. Cerebellum (London, England) 16(5–6):929–937. https://doi.org/10.1007/s12311-017-0867-4

Newhouse-Oisten MK, Peck KM, Conway AA, Frieder JE (2017) Ethical considerations for interdisciplinary collaboration with prescribing professionals. Behav Anal Pract 10(2):145–153. https://doi.org/10.1007/s40617-017-0184-x

Ney DM, Gleason ST, van Calcar SC, MacLeod EL, Nelson KL, Etzel MR, Rice GM, Wolff JA (2009) Nutritional management of PKU with glycomacropeptide from cheese whey. J Inherit Metab Dis 32(1):32–39. https://doi.org/10.1007/s10545-008-0952-4

Nickel M, Schulz A (2022) Natural history studies in NCL and their expanding role in drug development: experiences from CLN2 disease and relevance for clinical trials. Front Neurol 13. https://doi.org/10.3389/fneur.2022.785841

NIH (National Institutes of Health) (2021) Rare diseases and related terms. https://rarediseases.info.nih.gov/diseases/pages/31/faqs-about-rare-diseases

NIH (National Institutes of Health) (2023) New data management and data sharing policy. https://sharing.nih.gov/

Nosek BA, Alter G, Banks GC, Borsboom D, Bowman SD, Breckler SJ, Buck S, Chambers CD, Chin G, Christensen G, Contestabile M, Dafoe A, Eich E, Freese J, Glennerster R, Goroff D, Green DP, Hesse B, Humphreys M, Ishiyama J, Karlan D, Kraut A, Lupia A, Mabry P, Madon TA, Malhotra N, Mayo-Wilson E, McNutt M, Miguel E, Levy Paluck E, Simonsohn U, Soderberg C, Spellman BA, Turitto J, VandenBos G, Vazire S, Wagenmakers EJ, Wilson R, Yarkoni T (2015) SCIENTIFIC STANDARDS. Promoting an open research culture. Science (New York, NY) 348(6242):1422–1425. https://doi.org/10.1126/science.aab2374

NRC (National Research Council) (2011) Guide for the care and use of laboratory animals, 8th edn. National Academies Press. https://doi.org/10.17226/12910

O'Connor JP, Rose CJ, Waterton JC, Carano RA, Parker GJ, Jackson A (2015) Imaging intratumor heterogeneity: role in therapy response, resistance, and clinical outcome. Clin Cancer Res 21(2):249–257. https://doi.org/10.1158/1078-0432.CCR-14-0990

Ohmann C, Banzi R, Canham S, Battaglia S, Matei M, Ariyo C, Becnel L, Bierer B, Bowers S, Clivio L, Dias M, Druml C, Faure H, Fenner M, Galvez J, Ghersi D, Gluud C, Groves T, Houston P, Demotes-Mainard J (2017) Sharing and reuse of individual participant data from clinical trials: principles and recommendations. BMJ Open 7(12):e018647. https://doi.org/10.1136/bmjopen-2017-018647

Olivier M, Asmis R, Hawkins GA, Howard TD, Cox LA (2019) The need for multi-omics biomarker signatures in precision medicine. Int J Mol Sci 20(19):4781. https://doi.org/10.3390/ijms20194781

Ormandy EH, Schuppli CA (2014) Public attitudes toward animal research: a review. Animals 4(3):391–408. https://doi.org/10.3390/ani4030391

Orso G, Martinuzzi A, Rossetto MG, Sartori E, Feany M, Daga A (2005) Disease-related phenotypes in a Drosophila model of hereditary spastic paraplegia are ameliorated by treatment with vinblastine. J Clin Investig 115(11):3026–3034. https://doi.org/10.1172/JCI24694

Palmer EE, Howell K, Scheffer IE (2021) Natural history studies and clinical trial readiness for genetic developmental and epileptic encephalopathies. Neurotherapeutics 18(3):1432–1444. https://doi.org/10.1007/s13311-021-01133-3

Palmiter RD, Brinster RL, Hammer RE, Trumbauer ME, Rosenfeld MG, Birnberg NC, Evans RM (1982) Dramatic growth of mice that develop from eggs microinjected with metallothionein-growth hormone fusion genes. Nature 300(5893):611–615. https://doi.org/10.1038/300611a0

Pao W, Chmielecki J (2010) Rational, biologically based treatment of EGFR-mutant non-small-cell lung cancer. Nat Rev Cancer 10(11):760–774. https://doi.org/10.1038/nrc2947

Park IH, Lerou PH, Zhao R, Huo H, Daley GQ (2008) Generation of human-induced pluripotent stem cells. Nat Protoc 3(7):1180–1186. https://doi.org/10.1038/nprot.2008.92

Patel K, Tarkin J, Serruys PW, Tenekecioglu E, Foin N, Zhang YJ, Crake T, Moon J, Mathur A, Bourantas CV (2017) Invasive or non-invasive imaging for detecting high-risk coronary lesions? Expert Rev Cardiovasc Ther 15(3):165–179. https://doi.org/10.1080/14779072.2017.1297231

Paul A, Anand R, Karmakar SP, Rawat S, Bairagi N, Chatterjee S (2021) Exploring gene knockout strategies to identify potential drug targets using genome-scale metabolic models. Sci Rep 11:213. https://doi.org/10.1038/s41598-020-80561-1

Piedrahita JA, Williams JK (2017) Animal models in tissue engineering. Part I. Tissue Eng Part C Methods 23(11):641–642. https://doi.org/10.1089/ten.TEC.2017.0402

Pinnapureddy AR, Stayner C, McEwan J, Baddeley O, Forman J, Eccles MR (2015) Large animal models of rare genetic disorders: sheep as phenotypically relevant models of human genetic disease. Orphanet J Rare Dis 10:107. https://doi.org/10.1186/s13023-015-0327-5

Pittet MJ, Weissleder R (2011) Intravital imaging. Cell 147(5):983–991. https://doi.org/10.1016/j.cell.2011.11.004

Pittet MJ, Grimm J, Berger CR, Tamura T, Wojtkiewicz G, Nahrendorf M, Romero P, Swirski FK, Weissleder R (2007) In vivo imaging of T cell delivery to tumors after adoptive transfer therapy. Proc Natl Acad Sci U S A 104(30):12457–12461. https://doi.org/10.1073/pnas.0704460104

Piwowar HA, Vision TJ (2013) Data reuse and the open data citation advantage. PeerJ 1:e175. https://doi.org/10.7717/peerj.175

Platt RJ, Chen S, Zhou Y, Yim MJ, Swiech L, Kempton HR, Dahlman JE, Parnas O, Eisenhaure TM, Jovanovic M, Graham DB, Jhunjhunwala S, Heidenreich M, Xavier RJ, Langer R, Anderson DG, Hacohen N, Regev A, Feng G, Sharp PA, Zhang F (2014) CRISPR-Cas9 knockin mice for genome editing and cancer modelling. Cell 159(2):440–455. https://doi.org/10.1016/j.cell.2014.09.014

Proschan MA (2009) Sample size re-estimation in clinical trials. Biometrical J 51(2):348–357. https://doi.org/10.1002/bimj.200800266

Qiu S, Cai Y, Yao H, Lin C, Xie Y, Tang S, Zhang A (2023) Small molecule metabolites: discovery of biomarkers and therapeutic targets. Signal Transduct Target Ther 8:132. https://doi.org/10.1038/s41392-023-01399-3

Quadros RM, Miura H, Harms DW, Akatsuka H, Sato T, Aida T, Redder R, Richardson GP, Inagaki Y, Sakai D, Buckley SM, Seshacharyulu P, Batra SK, Behlke MA, Zeiner SA, Jacobi AM, Izu Y, Thoreson WB, Urness LD, Mansour SL, Ohtsuka M, Gurumurthy CB (2017) Easi-CRISPR:

a robust method for one-step generation of mice carrying conditional and insertion alleles using long ssDNA donors and CRISPR ribonucleoproteins. Genome Biol 18(1):92. https://doi.org/10.1186/s13059-017-1220-4

Raben N, Danon M, Gilbert AL, Dwivedi S, Collins B, Thurberg BL, Mattaliano RJ, Nagaraju K, Plotz PH (2003) Enzyme replacement therapy in the mouse model of Pompe disease. Mol Genet Metab 80(1–2):159–169. https://doi.org/10.1016/j.ymgme.2003.08.022

Raj A, van Oudenaarden A (2009) Nature, nurture, or chance: stochastic gene expression and its consequences. Cell 135(2):216–226. https://doi.org/10.1016/j.cell.2008.09.050

Ramazi S, Zahiri J (2021) Posttranslational modifications in proteins: resources, tools and prediction methods. Database (Oxford) 2021:baab012. https://doi.org/10.1093/database/baab012

Ramsey BW, Davies J, McElvaney NG, Tullis E, Bell SC, Dřevínek P, Griese M, McKone EF, Wainwright CE, Konstan MW, Moss R, Ratjen F, VX08-770-102 Study Group (2011) A CFTR potentiator in patients with cystic fibrosis and the G551D mutation. N Engl J Med 365(18):1663–1672. https://doi.org/10.1056/NEJMoa1105185

Rangarajan S, Walsh L, Lester W, Perry D, Madan B, Laffan M, Yu H, Vettermann C, Pierce GF, Wong WY, Pasi KJ (2017) AAV5-Factor VIII gene transfer in severe hemophilia A. N Engl J Med 377(26):2519–2530. https://doi.org/10.1056/NEJMoa1708483

Rashid ST, Corbineau S, Hannan N, Marciniak SJ, Miranda E, Alexander G, Huang-Doran I, Griffin J, Ahrlund-Richter L, Skepper J, Semple R, Weber A, Lomas DA, Vallier L (2010) Modeling inherited metabolic disorders of the liver using human induced pluripotent stem cells. J Clin Invest 120(9):3127–3136. https://doi.org/10.1002/hep.24143

Rath A, Olry A, Dhombres F, Brandt MM, Urbero B, Ayme S (2012) Representation of rare diseases in health information systems: the Orphanet approach to serve a wide range of end users. Hum Mutat 33(5):803–808. https://doi.org/10.1002/humu.22078

Ratjen F, Döring G (2003) Cystic fibrosis. Lancet 361(9358):681–689. https://doi.org/10.1016/S0140-6736(03)12567-6

Reeve BB, Hays RD, Bjorner JB, Cook KF, Crane PK, Teresi JA, Thissen D, Revicki DA, Weiss DJ, Hambleton RK, Liu H, Gershon R, Reise SP, Lai JS, Cella D, PROMIS Cooperative Group (2007) Psychometric evaluation and calibration of health-related quality of life item banks: plans for the patient-reported outcomes measurement information system (PROMIS). Med Care 45(5 Suppl. 1):S22–S31. https://doi.org/10.1097/01.mlr.0000250483.85507.04

Ribitsch I, Baptista PM, Lange-Consiglio A, Melotti L, Patruno M, Jenner F, Schnabl-Feichter E, Dutton LC, Connolly DJ, van Steenbeek FG, Dudhia J, Penning LC (2020) Large animal models in regenerative medicine and tissue engineering: to do or not to do. Front Bioeng Biotechnol 8:2020. https://doi.org/10.3389/fbioe.2020.00972

Robinson V, Jennings M, Working Group (2004) Refinement and reduction in the production of genetically modified mice: sixth report of the BVAAWF/FRAME/RSPCA/UFAW Joint Working Group on Refinement. Altern Lab Anim 32(Suppl. 1A):373–375. https://doi.org/10.1177/026119290403201s61

Roche DG, Kruuk LE, Lanfear R, Binning SA (2015) Public data archiving in ecology and evolution: how well are we doing? PLoS Biol 13(11):e1002295. https://doi.org/10.1371/journal.pbio.1002295

Rogers CS, Stoltz DA, Meyerholz DK, Ostedgaard LS, Rokhlina T, Taft PJ, Rogan MP, Pezzulo AA, Karp PH, Itani OA, Kabel AC, Wohlford-Lenane CL, Davis GJ, Hanfland RA, Smith TL, Samuel M, Wax D, Murphy CN, Rieke A, Whitworth K, Uc A, Starner TD, Brogden KA, Shilyansky J, McCray PB Jr, Zabner J, Prather RS, Welsh MJ (2008) Disruption of the CFTR gene produces a model of cystic fibrosis in newborn pigs. Science 321(5897):1837–1841. https://doi.org/10.1126/science.1163600

Rongvaux A, Takizawa H, Strowig T, Willinger T, Eynon EE, Flavell RA, Manz MG (2013) Human hemato-lymphoid system mice: current use and future potential for medicine. Annu Rev Immunol 31:635–674. https://doi.org/10.1146/annurev-immunol-032712-095921

Rosenthal N, Brown S (2007) The mouse ascending: perspectives for human-disease models. Nat Cell Biol 9(9):993–999. https://doi.org/10.1038/ncb437

Ross LF, Saal HM, David KL, Anderson RR, American Academy of Pediatrics, & American College of Medical Genetics and Genomics (2013) Technical report: ethical and policy issues in genetic testing and screening of children. Genet Med 15(3):234–245. https://doi.org/10.1038/gim.2012.176

Roth-Cline M, Nelson R (2014) The ethical principle of scientific necessity in pediatric research. Am J Bioethics 14:14–15. https://doi.org/10.1080/15265161.2014.964874

Roth-Cline M, Nelson R (2015) Ethical considerations in conducting pediatric and neonatal research in clinical pharmacology. Curr Pharm Des 21(39):5619–5635. https://doi.org/10.2174/1381612821666150901105146

Rubin BP, Antonescu CR, Scott-Browne JP, Comstock ML, Gu Y, Tanas MR, Ware CB, Woodell J (2005) A knock-in mouse model of gastrointestinal stromal tumor harboring kit K641E. Cancer Res 65(15):6631–6639. https://doi.org/10.1158/0008-5472.CAN-05-0891

Rubinstein WS, Maglott DR, Lee JM, Kattman BL, Malheiro AJ, Ovetsky M, Hem V, Gorelenkov V, Song G, Wallin C, Husain N, Chitipiralla S, Katz KS, Hoffman D, Jang W, Johnson M, Karmanov F, Ukrainchik A, Denisenko M, Fomous C, Hudson K, Ostell JM (2013) The NIH genetic testing registry: a new, centralized database of genetic tests to enable access to comprehensive information and improve transparency. Nucleic Acids Res 41(Database issue):D925–D935. https://doi.org/10.1093/nar/gks1173

Russell WMS, Burch RL (eds) (1959) The principles of humane experimental technique. Universities Federation for Animal Welfare, UFAW, Wheathampstead, UK. Methuen & Co. Limited, London. pp. 1–238. doi: https://doi.org/10.5694/j.1326-5377.1960.tb73127.x

Saam JR, Gordon JI (1999) Inducible gene knockouts in the small intestinal and colonic epithelium. J Biol Chem 274(52):38071–38082. https://doi.org/10.1074/jbc.274.53.38071

Sackett DL, Rosenberg WM, Gray JA, Haynes RB, Richardson WS (1996) Evidence based medicine: what it is and what it isn't. Br Med J (Clin Res Ed) 312(7023):71–72. https://doi.org/10.1136/bmj.312.7023.71

Sakai C, Ijaz S, Hoffman EJ (2018) Zebrafish models of neurodevelopmental disorders: past, present, and future. Front Mol Neurosci 11:294. https://doi.org/10.3389/fnmol.2018.00294

Salih S, Elliyanti A, Alkatheeri A, AlYafei F, Almarri B, Khan H (2023) The role of molecular imaging in personalized medicine. J Personal Med 13(2):369. https://doi.org/10.3390/jpm13020369

Sams-Dodd F (2005) Target-based drug discovery: is something wrong? Drug Discov Today 10(2):139–147. https://doi.org/10.1016/S1359-6446(04)03316-1

Sánchez-Danés A, Richaud-Patin Y, Carballo-Carbajal I, Jiménez-Delgado S, Caig C, Mora S, Di Guglielmo C, Ezquerra M, Patel B, Giralt A, Canals JM, Memo M, Alberch J, López-Barneo J, Vila M, Cuervo AM, Tolosa E, Consiglio A, Raya A (2012) Disease-specific phenotypes in dopamine neurons from human iPS-based models of genetic and sporadic Parkinson's disease. EMBO Mol Med 4(5):380–395. https://doi.org/10.1002/emmm.201200215

Sanjak J, Zhu Q, Math EA (2023) Clustering rare diseases within an ontology-enriched knowledge graph. bioRxiv: Bioinformatics. https://doi.org/10.1101/2023.02.15.528673.2023.02.15.528673

Sauer B (1998) Inducible gene targeting in mice using the Cre/lox system. Methods 14(4):381–392. https://doi.org/10.1006/meth.1998.0593

Scannell JW, Blanckley A, Boldon H, Warrington B (2012) Diagnosing the decline in pharmaceutical R&D efficiency. Nat Rev Drug Discov 11(3):191–200. https://doi.org/10.1038/nrd3681

Scarfe L, Taylor A, Sharkey J, Harwood R, Barrow M, Comenge J, Beeken L, Astley C, Santeramo I, Hutchinson C, Ressel L, Smythe J, Austin E, Levy R, Rosseinsky MJ, Adams DJ, Poptani H, Park BK, Murray P, Wilm B (2018) Non-invasive imaging reveals conditions that impact distribution and persistence of cells after in vivo administration. Stem Cell Res Ther 9(1):332. https://doi.org/10.1186/s13287-018-1076-x

Schee Genannt Halfmann S, Mählmann L, Leyens L, Reumann M, Brand A (2017) Personalized medicine: what's in it for rare diseases? Adv Exp Med Biol 1031:387–404. https://doi.org/10.1007/978-3-319-67144-4_22

Schipper H, Clinch J, McMurray A, Levitt M (1984) Measuring the quality of life of cancer patients: the functional living index-cancer: development and validation. J Clin Oncol 2(5):472–483. https://doi.org/10.1200/JCO.1984.2.5.472

Schnorrer F, Schönbauer C, Langer C, Dietzl G, Novatchkova M, Schernhuber K, Fellner M, Azaryan A, Radolf M, Stark A, Keleman K, Dickson BJ (2010) Systematic genetic analysis of muscle morphogenesis and function in Drosophila. Nature 464(7286):287–291. https://doi.org/10.1038/nature08799

Schriml LM, Arze C, Nadendla S, Chang YW, Mazaitis M, Felix V, Feng G, Kibbe WA (2012) Disease ontology: a backbone for disease semantic integration. Nucleic Acids Res 40(Database issue):D940–D946. https://doi.org/10.1093/nar/gkr972

Schuck PF, Malgarin F, Cararo JH, Cardoso F, Streck EL, Ferreira GC (2015) Phenylketonuria pathophysiology: on the role of metabolic alterations. Aging Dis 6(5):390–399. https://doi.org/10.14336/AD.2015.0827

Sehnert AJ, Huq A, Weinstein BM, Walker C, Fishman M, Stainier DY (2002) Cardiac troponin T is essential in sarcomere assembly and cardiac contractility. Nat Genet 31(1):106–110. https://doi.org/10.1038/ng875

Seok J, Warren HS, Cuenca AG, Mindrinos MN, Baker HV, Xu W, Richards DR, McDonald-Smith GP, Gao H, Hennessy L, Finnerty CC, López CM, Honari S, Moore EE, Minei JP, Cuschieri J, Bankey PE, Johnson JL, Sperry J, Nathens AB, Billiar TR, West MA, Jeschke MG, Klein MB, Gamelli RL, Gibran NS, Brownstein BH, Miller-Graziano C, Calvano SE, Mason PH, Cobb JP, Rahme LG, Lowry SF, Maier RV, Moldawer LL, Herndon DN, Davis RW, Xiao W, Tompkins RG (2013) Genomic responses in mouse models poorly mimic human inflammatory diseases. Proc Natl Acad Sci U S A 110(9):3507–3512. https://doi.org/10.1073/pnas.1222878110

Servadio M, Vanderschuren LJ, Trezza V (2015) Modeling autism-relevant behavioral phenotypes in rats and mice: do 'autistic' rodents exist? Behav Pharmacol 26(6):522–540. https://doi.org/10.1097/FBP.0000000000000163

Shah N, Coathup V, Teare H, Forgie I, Giordano GN, Hansen TH, Groeneveld L, Hudson M, Pearson E, Ruetten H, Kaye J (2019) Motivations for data sharing-views of research participants from four European countries: a DIRECT study. Eur J Hum Genet 27(5):721–729. https://doi.org/10.1038/s41431-019-0344-2

Shalem O, Sanjana NE, Hartenian E, Shi X, Scott DA, Mikkelson T, Heckl D, Ebert BL, Root DE, Doench JG, Zhang F (2014) Genome-scale CRISPR-Cas9 knockout screening in human cells. Science (New York, NY) 343(6166):84–87. https://doi.org/10.1126/science.1247005

Shedlovsky A, McDonald JD, Symula D, Dove WF (1993) Mouse models of human phenylketonuria. Genetics 134(4):1205–1210. https://doi.org/10.1093/genetics/134.4.1205

Sherman RE, Anderson SA, Dal Pan GJ, Gray GW, Gross T, Hunter NL, LaVange L, Marinac-Dabic D, Marks PW, Robb MA, Shuren J, Temple R, Woodcock J, Yue LQ, Califf RM (2016) Real-world evidence—what is it and what can it tell us? N Engl J Med 375(23):2293–2297. https://doi.org/10.1056/NEJMsb1609216

Silverman JL, Turner SM, Barkan CL, Tolu SS, Saxena R, Hung AY, Sheng M, Crawley JN (2011) Sociability and motor functions in Shank1 mutant mice. Brain Res 1380:120–137. https://doi.org/10.1016/j.brainres.2010.09.026

Singh M, Ferrara N (2012) Modeling and predicting clinical efficacy for drugs targeting the tumor milieu. Nat Biotechnol 30(7):648–657. https://doi.org/10.1038/nbt.2286

Singh VK, Seed TM (2021) How necessary are animal models for modern drug discovery? Expert Opin Drug Discovery 16(12):1391–1397. https://doi.org/10.1080/17460441.2021.1972255

Smith J, Cheater F, Bekker H (2015) Parents' experiences of living with a child with a long-term condition: a rapid structured review of the literature. Health Expect 18(4):452–474. https://doi.org/10.1111/hex.12040

Snouwaert JN, Brigman KK, Latour AM, Malouf NN, Boucher RC, Smithies O, Koller BH (1992) An animal model for cystic fibrosis made by gene targeting. Science 257(5073):1083–1088. https://doi.org/10.1126/science.257.5073.1083

Soldner F, Hockemeyer D, Beard C, Gao Q, Bell GW, Cook EG, Hargus G, Blak A, Cooper O, Mitalipova M, Isacson O, Jaenisch R (2009) Parkinson's disease patient-derived induced pluripotent stem cells free of viral reprogramming factors. Cell 136(5):964–977. https://doi.org/10.1016/j.cell.2009.02.013

Spindler SR, Hartenstein V (2010) The Drosophila neural lineages: a model system to study brain development and circuitry. Dev Genes Evol 220(1–2):1–10. https://doi.org/10.1007/s00427-010-0323-7

Streck EL, Bussular FP, Wessler LB, Duarte MB, Rezende VL, Rodrigues MS, Torres CA, Lemos IS, Candiotto G, Gava FF, de Oliveira J, Valvassori SS (2021) Administration of branched-chain amino acids alters epigenetic regulatory enzymes in an animal model of Maple Syrup Urine Disease. Metab Brain Dis 36(2):247–254. https://doi.org/10.1007/s11011-020-00631-1

Swindle MM, Makin A, Herron AJ, Clubb FJ Jr, Frazier KS (2012) Swine as models in biomedical research and toxicology testing. Vet Pathol 49(2):344–356. https://doi.org/10.1177/0300985811402846

Szulc J, Wiznerowicz M, Sauvain MO, Trono D, Aebischer P (2006) A versatile tool for conditional gene expression and knockdown. Nat Methods 3(2):109–116. https://doi.org/10.1038/nmeth846

Takahashi K, Yamanaka S (2006) Induction of pluripotent stem cells from mouse embryonic and adult fibroblast cultures by defined factors. Cell 126(4):663–676. https://doi.org/10.1016/j.cell.2006.07.024

Tan J, Ong CK, Lim WK, Ng CC, Thike AA, Ng LM, Rajasegaran V, Myint SS, Nagarajan S, Thangaraju S, Dey S, Nasir ND, Wijaya GC, Lim JQ, Huang D, Li Z, Wong BH, Chan JY, McPherson JR, Cutcutache I, Poore G, Tay ST, Tan WJ, Putti TC, Ahmad BS, Iau P, Chan CW, Tang APH, Yong WS, Madhukumar P, Ho GH, Tan VKM, Wong CY, Hartman M, Ong KW, Tan BKT, Rozen SG, Tan P, Tan PH, Teh BT (2015) Genomic landscapes of breast fibroepithelial tumors. Nat Genet 47(11):1341–1345. https://doi.org/10.1038/leu.2016.13

Taruscio D, Arriola L, Baldi F, Barisic I, Bermejo-Sánchez E, Bianchi F, Calzolari E, Carbone P, Curran R, Garne E, Gatt M, Latos-Bieleńska A, Khoshnood B, Irgens L, Mantovani A, Martínez-Frías ML, Neville A, Rißmann A, Ruggeri S, Wellesley D, Dolk H (2014) European recommendations for primary prevention of congenital anomalies: a joined effort of EUROCAT and EUROPLAN projects to facilitate inclusion of this topic in the National rare diseases plan. Public Health Genom 17(2):115–123. https://doi.org/10.1159/000360602

Thierry G (2012) Neurophysiological examination methods: electrophysiology and neuroimaging. In: Cappa S et al (eds) Cognitive neurology: a clinical textbook. Oxford Academic, pp 19–38. https://doi.org/10.1093/acprof:oso/9780198569275.003.0003

Thistlethwaite J, Moran M, World Health Organization Study Group on Interprofessional Education and Collaborative Practice (2010) Learning outcomes for interprofessional education (IPE): literature review and synthesis. J Interprof Care 24(5):503–513. https://doi.org/10.3109/13561820.2010.483366

Thomas KR, Capecchi MR (1987) Site-directed mutagenesis by gene targeting in mouse embryo-derived stem cells. Cell 51(3):503–512. https://doi.org/10.1016/0092-8674(87)90646-5

Threadgill DW, Churchill GA (2012) Ten years of the collaborative cross. Genetics 190(2):291–294. https://doi.org/10.1534/g3.111.001891

Toutain PL, Ferran A, Bousquet-Mélou A (2010) Species differences in pharmacokinetics and pharmacodynamics. Handb Exp Pharmacol 199:19–48. https://doi.org/10.1007/978-3-642-10324-7_2

Troy CM, Jean YY (2015) Caspases: therapeutic targets in neurologic disease. Neurother J Am Soc Exp NeuroTher 12(1):42–48. https://doi.org/10.1007/s13311-014-0307-9

Tsien JZ, Chen DF, Gerber D, Tom C, Mercer EH, Anderson DJ, Mayford M, Kandel ER, Tonegawa S (1996) Subregion- and cell type-restricted gene knockout in mouse brain. Cell 87(7):1317–1326. https://doi.org/10.1016/s0092-8674(00)81826-7

Turner RT, Maran A, Lotinun S, Hefferan T, Evans GL, Zhang M, Sibonga JD (2001) Animal models for osteoporosis. Rev Endocr Metab Disord 2(1):117. https://doi.org/10.1023/a:1010067326811

Umscheid CA, Margolis DJ, Grossman CE (2011) Key concepts of clinical trials: a narrative review. Postgrad Med 123(5):194–204. https://doi.org/10.3810/pgm.2011.09.2475

van der Worp HB, Howells DW, Sena ES, Porritt MJ, Rewell S, O'Collins V, Macleod MR (2010) Can animal models of disease reliably inform human studies? PLoS Med 7(3):e1000245. https://doi.org/10.1371/journal.pmed.1000245

Van Rompay K (2017) Tackling HIV and AIDS: contributions by non-human primate models. Lab Anim 46:259–270. https://doi.org/10.1038/laban.1279

Varga OE, Hansen AK, Sandøe P, Olsson IA (2010) Validating animal models for preclinical research: a scientific and ethical discussion. Altern Lab Anim 38(3):245–248. https://doi.org/10.1177/026119291003800309

Vashi N, Justice MJ (2019) Treating Rett syndrome: from mouse models to human therapies. Mamm Genome 30(5–6):90–110. https://doi.org/10.1007/s00335-019-09793-5

Veit G, Velkov T, Xu H, Vadeboncoeur N, Bilodeau L, Matouk E, Lukacs GL (2021) A precision medicine approach to optimize modulator therapy for rare CFTR folding mutants. J Pers Med 11(7):643. https://doi.org/10.3390/jpm11070643

Vidal-Puig A, Jimenez-Liñan M, Lowell BB, Hamann A, Hu E, Spiegelman B, Flier JS, Moller DE (1996) Regulation of PPAR gamma gene expression by nutrition and obesity in rodents. J Clin Invest 97(11):2553–2561. https://doi.org/10.1172/JCI118703

Villa-Bellosta R (2019) ATP-based therapy prevents vascular calcification and extends longevity in a mouse model of Hutchinson-Gilford progeria syndrome. Proc Natl Acad Sci U S A 116(47):23698–23704. https://doi.org/10.1073/pnas.1910972116

Vissers LE, de Ligt J, Gilissen C, Janssen I, Steehouwer M, de Vries P, van Lier B, Arts P, Wieskamp N, del Rosario M, van Bon BW, Hoischen A, de Vries BB, Brunner HG, Veltman JA (2010) A de novo paradigm for mental retardation. Nat Genet 42(12):1109–1112. https://doi.org/10.1038/ng.712

Vollmer TL, Sorensen PS, Selmaj K, Zipp F, Havrdova E, Cohen JA, Sasson N, Gilgun-Sherki Y, Arnold DL, On behalf of the BRAVO Study Group (2014) A randomized placebo-controlled phase III trial of oral laquinimod for multiple sclerosis. J Neurol 261(4):773–783. https://doi.org/10.1007/s00415-014-7264-4

Vujovic S, Henderson S, Presneau N, Odell E, Jacques TS, Tirabosco R, Boshoff C, Flanagan AM (2006) Brachyury, a crucial regulator of notochordal development, is a novel biomarker for chordomas. J Pathol 209(2):157–165. https://doi.org/10.1002/path.1969

Wang G, Xu Y, Wang Q, Chai Y, Sun X, Yang F, Zhang J, Wu M, Liao X, Yu X, Sheng X, Liu Z, Zhang J (2022) Rare and undiagnosed diseases: from disease-causing gene identification to mechanism elucidation. Fundamental Res 2(6):918–928. https://doi.org/10.1016/j.fmre.2022.09.002

Wangler MF, Yamamoto S, Bellen HJ (2015) Fruit flies in biomedical research. Genetics 199(3):639–653. https://doi.org/10.1534/genetics.114.171785

Weake V, Workman J (2010) Inducible gene expression: diverse regulatory mechanisms. Nat Rev Genet 11:426–437. https://doi.org/10.1038/nrg2781

Weatherald J, Reis A, Sitbon O, Humbert M (2019) Pulmonary arterial hypertension registries: past, present and into the future. Eur Respir Rev 28(154):190128. https://doi.org/10.1183/16000617.0128-2019

Weissleder R (2001) A clearer vision for in vivo imaging. Nat Biotechnol 19(4):316–317. https://doi.org/10.1038/86684

Whitesides GM (2006) The origins and the future of microfluidics. Nature 442(7101):368–373. https://doi.org/10.1038/nature05058

Wong VW, Adams LA, de Lédinghen V, Wong GL, Sookoian S (2018) Noninvasive biomarkers in NAFLD and NASH—current progress and future promise. Nat Rev Gastroenterol Hepatol 15(8):461–478. https://doi.org/10.1038/s41575-018-0014-9

Wooden JI, Spinetta MJ, Nguyen T, O'Leary CI, Leasure JL (2021) A sensitive homecage-based novel object recognition task for rodents. Front Behav Neurosci 15:680042. https://doi.org/10.3389/fnbeh.2021.680042

Woolf SH (2008) The meaning of translational research and why it matters. JAMA 299(2):211–213. https://doi.org/10.1001/jama.2007.26

Wu SS, Li QC, Yin CQ, Xue W, Song CQ (2020) Advances in CRISPR/Cas-based gene therapy in human genetic diseases. Theranostics 10(10):4374–4382. https://doi.org/10.7150/thno.43360
Xu M, Motabar O, Ferrer M, Marugan JJ, Zheng W, Ottinger EA (2016) Disease models for the development of therapies for lysosomal storage diseases. Ann N Y Acad Sci 1371(1):15–29. https://doi.org/10.1111/nyas.13052
Yang Q, Vijayakumar A, Kahn BB (2018) Metabolites as regulators of insulin sensitivity and metabolism. Nat Rev Mol Cell Biol 19(10):654–672. https://doi.org/10.1038/s41580-018-0044-8
Yang KD, Belyaeva A, Venkatachalapathy S, Damodaran K, Katcoff A, Radhakrishnan A, Shivashankar GV, Uhler C (2021) Multi-domain translation between single-cell imaging and sequencing data using autoencoders. Nat Commun 12:31. https://doi.org/10.1038/s41467-020-20249-2
Yau TY, Molina O, Courey AJ (2020) SUMOylation in development and neurodegeneration. Development 147(6):dev175703. https://doi.org/10.1242/dev.175703
Yin H, Xue W, Chen S, Bogorad RL, Benedetti E, Grompe M, Koteliansky V, Sharp PA, Jacks T, Anderson DG (2014) Genome editing with Cas9 in adult mice corrects a disease mutation and phenotype. Nat Biotechnol 32(6):551–553. https://doi.org/10.1038/nbt.2884
Yitbarek D, Dagnaw GG (2022) Application of advanced imaging modalities in veterinary medicine: a review. Vet Med (Auckland, N.Z.) 13:117–130. https://doi.org/10.2147/vmrr.s367040
Yue R, Dutta A (2022) Computational systems biology in disease modelling and control, review and perspectives. NPJ Syst Biol Appl 8:37. https://doi.org/10.1038/s41540-022-00247-4
Zhang F, Cong L, Lodato S, Kosuri S, Church GM, Arlotta P (2011) Efficient construction of sequence-specific TAL effectors for modulating mammalian transcription. Nat Biotechnol 29(2):149–153. https://doi.org/10.1038/nbt.1775
Zhang D, Luo G, Ding X, Lu C (2012) Preclinical experimental models of drug metabolism and disposition in drug discovery and development. Acta Pharm Sin B 2:549–561. https://doi.org/10.1016/j.apsb.2012.10.004
Zhang L, Wan S, Jiang Y, Wang Y, Fu T, Liu Q, Cao Z, Qiu L, Tan W (2017) Molecular elucidation of disease biomarkers at the interface of chemistry and biology. J Am Chem Soc 139(7):2532–2540. https://doi.org/10.1021/jacs.6b10646
Zimmer AM, Pan YK, Chandrapalan T, Kwong RWM, Perry SF (2019) Loss-of-function approaches in comparative physiology: is there a future for knockdown experiments in the era of genome editing? J Exp Biol 222(Pt 7):jeb175737. https://doi.org/10.1242/jeb.175737
Zurlo J, Rudacille D, Goldberg AM (1996) The three Rs: the way forward. Environ Health Perspect 104(8):878–880. https://doi.org/10.1126/science.272.5267.1403

Research and Clinical Approaches to Undiagnosed Rare Genetic Disorders

5

Archana Rajavel and Raja Natesan Sella

Abstract

Rare genetic diseases are complex and heterogeneous. The diagnostic challenges of undiagnosed rare genetic disorders are due to their diversity and limited knowledge of their genetic origin. This chapter provides a comprehensive overview of the research and clinical strategies for identifying and treating undiagnosed rare diseases.

Keywords

Undiagnosed diseases · Diagnosis · Whole-exome sequencing

5.1 Introduction

Undiagnosed rare diseases are a heterogeneous group of disorders that affect a small number of individuals. Characterizing and diagnosing these diseases are challenging as they have a wide range of clinical symptoms and are rare and heterogeneous. Even in developed countries, health systems neglect these diseases due to their incidence, but they should be considered major public health problems. Approximately 7000 rare diseases are reported worldwide (Haendel et al. 2020), and other undefined diseases come under rare undiagnosed disorders. Among 80% of diseases of a genetic cause, 60% have reduced life expectancy and affect 1 in 2000 people worldwide, but 6–8% of the general population is affected (Gagne et al. 2014). The status

A. Rajavel · R. N. Sella (✉)
Membrane Protein Interaction Laboratory, Department of Genetic Engineering, School of Bioengineering, SRM Institute of Science and Technology, Kattankulathur, Chengalpattu, Tamil Nadu, India
e-mail: rajan3@srmist.edu.in

M. Umair et al. (eds.), *Rare Genetic Disorders*,
https://doi.org/10.1007/978-981-99-9323-9_5

of undiagnosed rare diseases (URDs) in India is a growing concern, with estimates suggesting that there are approximately 70 million people in India living with rare diseases.

Most rare and undiagnosed disease-affected patients are children with symptoms, and physicians help them manage the severity of symptoms using existing treatment. Approximately 30% of people die before age 5, and less than 3% receive USFDA-approved therapies (Ferreira 2019). Currently, available genetic tests and methods cannot diagnose the genetic disease, and these types of diseases are named undiagnosed rare genetic disorders. URD patients and their families spend an average of 5–8 years of a diagnostic odyssey to identify the genetic condition and disease management strategies, design individualized treatment options, and prevent the condition early in the family (Global Commission to End the Diagnostic Odyssey for Children with a Rare Disease 2018; Black et al. 2015). The misdiagnosis of the genetic condition puts an economic and emotional burden on the family. In most cases, the siblings of patients with hereditary disorders or undiagnosed diseases are often subjected to social discrimination due to fear of spreading contagious illnesses. Diagnosis, inheritance, hazards, and chances for social inclusion must be determined for rare disease patients. Most genetically-related rare diseases cause dysfunctions in multi-organ systems and cost higher healthcare resources. Patients with URDs face significant challenges, including inadequate diagnostic tools, limited treatment options, and healthcare services. In the early 1960s, diagnosing RDs was done by biochemical parameters and cytogenetic methods. Emerging technologies for diagnosing and treating rare diseases, including advancements in high-throughput sequencing technologies, have provided an effective tool to diagnose rare diseases. Whole-genome sequencing, whole-exome sequencing, and RNA sequencing have enabled the identification of causative mutations for many rare disorders (de Ligt et al. 2012). Due to limited resources, the pharma industry shows little interest in manufacturing and marketing drugs for orphan diseases. Researchers and scientists may have limited knowledge of these diseases, limiting effective treatment development. This chapter discusses the challenges associated with undiagnosed rare diseases and emerging solutions to address these challenges.

5.2 Ethical Considerations in Undiagnosed Rare Diseases

Undiagnosed rare diseases present unique ethical considerations for patients, their families, and healthcare professionals. Obtaining informed consent is essential for any medical treatment. Still, in the case of rare diseases, patients may need help understanding the implications of participating in clinical trials or genetic testing. It is important to ensure that patients have access to accurate and understandable information about their condition, the potential risks and benefits of testing or treatment, and implications of any results (Coors et al. 2017). Due to the rarity of these conditions, the privacy of patients and their families must be carefully protected to prevent stigmatization and discrimination. Patients must be assured that their genetic information will not be used against them or shared without consent. Patients with rare diseases often face significant challenges in accessing healthcare, including difficulty obtaining a diagnosis, lack of effective treatments, and high care costs. It is important to ensure that these patients receive the same level of care and attention as patients with more common conditions and that resources are allocated fairly (Barrera and Galindo 2010). Healthcare professionals must balance the potential benefits of testing and treatment against the risks and potential harm to the patient. It is important to consider the individual needs and circumstances of each patient, as well as the broader implications of testing and treatment for the patient and their family. Patients with URDs may have unique wishes or beliefs about their treatment and care, which healthcare professionals should respect. It is important to involve patients and their families in decision-making and to support them in making choices that align with their values and goals.

5.3 Challenges Faced by Undiagnosed Rare Disease Patients

5.3.1 Lack of Healthcare Access

In developing countries, accessing healthcare is one of the major challenges in identifying undiagnosed rare disease populations earlier. Even though patients with rare diseases live in countries with state-of-the-art facilities, they still struggle for basic healthcare. These struggles come from the root of rarity, the heterogeneity in understanding the disease, the lack of awareness and knowledge of the clinicians on rare diseases, and the difficulty in identifying them earlier and providing proper care. Some diagnostic tests for rare diseases may be expensive, making it difficult for patients to access them. One of the major challenges in diagnosing rare diseases in India is the need for more access to genetic testing and expertise (Bogart et al. 2022). Only a few specialized centers in India offer genetic testing and counseling, and these services are often expensive and not covered by insurance. It is difficult for patients to receive an accurate diagnosis and access appropriate treatment. Currently, genomic technologies are used as diagnostic tools for identifying rare diseases, but only a few people among the global population have access to the new technologies.

5.3.2 Difficulty in Diagnostic Result Interpretation

Rare disease patients have difficulty interpreting the test reports data, which delays the definitive diagnosis. The genetic variations of unknown significance (VuS) lead to an unknown effect on the patient's health. The findings explain that partial clinical phenotype and novel disease gene connections are all determinant factors. The VuS is a major diagnostic problem for RGDs, causing unclear outcomes (Lopes et al. 2018; Marwaha et al. 2022). Genetic testing, like next-generation sequencing, can detect atypical manifestations of recognized RGDs, with 5–7% of cases being secondary causes of complicated clinical presentations. Disease-gene connections are constantly evolving, with over 300 discoveries each year. Clinical laboratories will identify compelling variations in genes of unknown significance and identify other families with variants in the same GUS and overlapping symptoms. The existing diagnostic paradigm cannot identify novel disease-gene connections despite the potential to discover 6000–13,000 Mendelian circumstances.

5.3.3 Current Diagnostic Paradigm

Patients suspected of RGD often acquire a diagnosis during the first few tests. Still, some may need to be diagnosed due to technological limitations, inadequate understanding of the coding genome, and pathological genetic mechanisms. Clinicians should understand that even after doing the most accurate genetic test for their patient, they cannot completely rule out a certain RGD (Hartley et al. 2020). Microarrays, NGS, and bioinformatic workflows that align genomic data to a reference sequence have technical restrictions. Studying the diagnostic testing strategy for a patient with an undetected RGD is important. Targeted genetic testing has been performed on millions of people with suspected RGD, but access to data is difficult due to security and privacy concerns (Marwaha et al. 2022). To fully utilize the potential of collective data, strategies must be created to de-silo existing and future data. Current NGS and informatics analyses cannot detect genetic variation linked to non-Mendelian inheritance, such as tissue-specific somatic mosaicism, epigenetic changes, oligogenic inheritance, or gene-environment interactions.

5.3.4 Challenges in Rare Disease Research

Researchers who work on rare diseases have more career opportunities to explore and face more challenges. Funding for rare disease research is often limited, as the resources available to support research are typically much lower than those available for common diseases. There may be a lack of appropriate diagnostic tests or biomarkers for many rare diseases, making it difficult to diagnose these diseases. The limited rare disease patient population makes it challenging to design specialized studies and conduct large-scale clinical trials. The underlying pathophysiology mechanism is poorly understood, making it challenging to develop effective treatments (Griggs et al. 2009). It is difficult to access data on rare disease patients due to privacy concerns. The limited number of patients who participated in a clinical study may lead to a lack of knowledge in understanding the diseases and developing treatment strategies.

Adequate funding and proper training for expertise from the government and funding agencies help to address the above challenges in rare diseases. Advances in multi-omics approaches and artificial intelligence are also helping to improve diagnosis rates for rare undiagnosed disorders, but further research is needed to fully realize the potential of these technologies (Stoller 2018). Encourage biopharmaceutical companies to develop orphan drug products for an underserved population.

5.4 Clinical Approaches to Undiagnosed Rare Diseases

A thorough patient history and physical examination can help identify unusual or characteristic features of a rare disease. Genetic testing can help identify mutations or other genetic variants associated with a rare disease. Whole-genome or whole-exome sequencing, gene panel testing, and other techniques have been used to diagnose URDs. A multidisciplinary team, including physicians, genetic counselors, and other healthcare professionals, can work together to identify and treat rare diseases. This team works together, coordinating care, sharing data and resources, and developing personalized treatment plans, as shown in Fig. 5.1. Clinical trials can be used to test new therapies such as gene therapies, precision medicine, and other targeted approaches for rare diseases. Patient advocacy groups can provide resources, support, and advocacy for patients and their families affected by rare diseases by providing access to clinical trials, financial assistance, and social support.

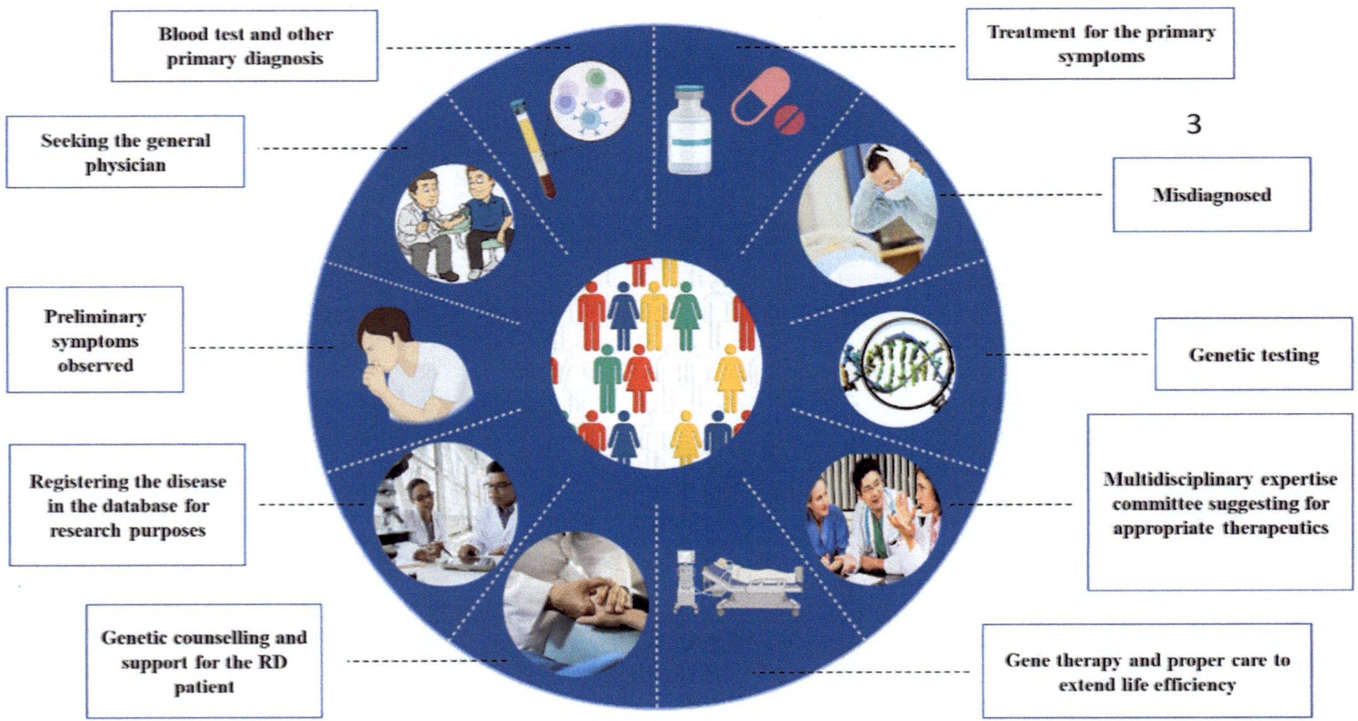

Fig. 5.1 Illustrative representation of rare disease patient journey from the diagnosis to management of the disease

5.5 Diagnosis of Undiagnosed Rare Diseases

The first-line tests for patients with congenital anomalies (CAs), developmental delay (DD), and intellectual disability (ID) are karyotyping, fluorescence in situ hybridization (FISH), and chromosomal microarray analysis (CMA) (Marwaha). If the patient is suspected of having a chromosomal disorder, go with chromosomal microarray tests (e.g., fragile × chromosome); otherwise, specific diagnostic tests need to be done. Still, chromosome analysis plays a role in diagnosing URDs and investigates balanced chromosomal rearrangements, mosaic abnormalities, cryptic deletions, and large copy number variants.

Sequencing is the laboratory test that helps to determine the order of the bases in the genome and protein-coding region of an organism. Now, the American College of Medical Genetics and Genomics (ACMG) recommends genome sequencing (GS) and exome sequencing (ES) as the second-tier tests for the identification of structural variants, intronic variants, tandem repeats, and disease-causing pathogenic variants (Gonzaga-Jauregui et al. 2012). ES can analyze protein-coding, untranslated, and intron-exon boundaries at a low cost. However, they are limited by low complexity, specificity of the capturing probe, and nonuniform coverage. ES reveals only partial information about the pathogenic variant and phenotype of the patient (Retterer et al. 2016). GS-mediated diagnosis identified large and small CNVs, balanced chromosomal rearrangements, and low-coverage regions.

5.6 Advances in Research on Undiagnosed Rare Diseases

There is a lot of exciting research on undiagnosed rare diseases, and researchers are progressing toward improving diagnosis and treatment for these patients. Undiagnosed rare diseases are an area of active research, and various ongoing efforts exist to address this problem.

5.6.1 Multi-Omics Approach

Genomics and advances in gene technologies have revolutionized our ability to diagnose rare diseases. Researchers use whole-genome, whole-exome, and targeted gene sequencing to identify disease-causing mutations in previously undiagnosed patients. Whole-exome sequencing (WES) and whole-genome sequencing (WGS) are techniques that can identify rare genetic variants, including single nucleotide variants (SNVs) and small insertions or deletions (indels), that may be responsible for the disease (Wenger et al. 2019). By comparing the patient's DNA sequence to reference genomes, researchers can identify rare or absent variants in the general population and prioritize these variants for further study. Scientists can use techniques such as CRISPR-Cas9 gene editing to introduce specific genetic variants into cell lines or animal models and study the resulting phenotype (Lek et al. 2020). These modern molecular biology tools help elucidate the disease's underlying mechanisms and identify potential therapeutic targets. Genomics helps to identify genetic variants, study their functional impact, and identify potential biomarkers.

Transcriptomics involves analyzing an individual's entire set of expressed genes (transcriptome), identifying differentially expressed genes, and alternative splicing (Kremer et al. 2017). Transcriptomics identifies potential biomarkers or therapeutic targets by comparing the transcriptomes of patients with undiagnosed rare diseases to those of healthy individuals or individuals with known genetic diseases. Researchers can identify differentially expressed genes that may be associated with the disease. The modifications of DNA and associated proteins can alter gene expression without changing the underlying DNA sequence. Epigenetic modifications include DNA methylation, histone modifications, and noncoding RNA molecules. DNA methylation is a common epigenetic modification that involves adding a methyl group to the DNA molecule, often leading to the repression of gene expression (Sadikovic et al. 2020) and identifying epigenetic modifications associated with specific genetic variants in URD patient samples that carry the functionally impact variants and identifying potential therapeutic targets (Aref-Eshghi et al. 2019). Epigenomics studies also study the effects of environmental exposures on gene expression and contribute to disease development.

Proteomics studies proteins, including their structure, function, and interactions. Proteomics can play an important role in studying undiagnosed rare genetic diseases, as it can identify differences in protein expression or activity associated with the disease. It can study undiagnosed rare genetic diseases by analyzing protein expression patterns in patient samples. This can be done using mass spectrometry-based techniques, such as liquid chromatography-tandem mass spectrometry (LC-MS/MS), which can identify and quantify proteins in complex mixtures. By comparing protein expression patterns in patient samples to those in healthy controls or other disease states, researchers can identify potential biomarkers of the disease, which may aid in diagnosis or treatment (Grabowski et al. 2019). Techniques such as co-immunoprecipitation or yeast two-hybrid assays allow researchers to identify and study protein-protein interactions. It can be used to study posttranslational modifications (PTMs) of proteins (Ebert et al. 2022), which can significantly impact protein function. Phosphorylation is a common posttranslational modification that regulates protein activity and signaling pathways. By studying PTMs in patient samples, researchers can identify differences that may be associated with the disease and identify potential therapeutic targets.

Metabolomics is an emerging technology increasingly used in rare disease research by identifying metabolic abnormalities that may be associated with the disease. Metabolomics studies small molecules called metabolites, which are the end products of cellular metabolism (Tarailo-Graovac et al. 2016). The analysis of metabolic profiles in patient samples can be done using techniques such as mass spectrometry or nuclear magnetic resonance spectroscopy. By comparing metabolic profiles in patient samples to those in healthy controls or other disease states, researchers can identify potential biomarkers of the disease, which may aid in diagnosis or treatment (Abela et al. 2016). By identifying metabolic pathways in patient samples, researchers can gain insights into the underlying mechanisms of the disease and identify potential therapeutic targets.

These technologies are used to identify protein biomarkers and metabolic pathways associated with undiagnosed rare diseases. Examples of successful applications of proteomics and metabolomics in undiagnosed rare diseases include the identification of biomarkers for disorders such as Niemann-Pick disease (Percival et al. 2020), Fabry disease (Ducatez et al. 2021), and Gaucher disease (Menkovic et al. 2020).

5.6.2 Role of Artificial Intelligence (AI) in URD Research

AI tools such as machine learning and natural language processing are being used to mine electronic health records and identify patients with similar symptoms and clinical features, for example, the symptoms checker website (Piñol et al. 2017), as shown in Table 5.1. These approaches can help identify new disease-causing genes and accelerate diagnosis by suggesting likely candidate genes for further investigation. AI algorithms can be used to analyze genomic and clinical data to identify disease-causing genes and variants. It will also predict the likelihood of a patient having a rare disease based on their symptoms and medical history. AI algorithms also identify potential drug targets and predict the effectiveness of new therapies like URSAHD (unveiling RNA sample annotation for human diseases) (Lee et al. 2019). AI analyzes large-scale genomic and clinical datasets to identify patient subgroups that may benefit from a particular treatment. AI-based genetic devices, including Xrare and radio phenomics, help to analyze patient data (Li et al. 2019), such as medical images and clinical measurements, to monitor disease progression and response to treatment. Using the data, clinicians could adjust treatment plans

Table 5.1 Artificial intelligence tool for rare disease identification

AI tool	Description	Reference
SISH	Self-supervised pathology image search tool for identifying rare diseases	Chen et al. (2022)
Deep Variant	Deep neural network tool which identifies the SNP and small indel variants from NGS data	Poplin et al. (2018)
DeepGestalt	An open-source deep learning tool which identifies rare genetic syndromes	Gurovich et al. (2018)
HPO2Vec	HPO embeddings with heterogeneous knowledge resources for human phenotype ontology	Shen et al. (2019)
DIDA	A deep learning tool for predicting the pathogenicity of the genes associated with rare genetic disorders	Gazzo et al. (2016)
DREAM-RD	Artificial intelligence and machine learning algorithm-based drug repurposing tools for rare diseases which identifies drug candidates	Agastheeswaramoorthy and Sevilimedu (2020)

and improve patient outcomes. AI can match patients with rare diseases to clinical trials (Blumenrath et al. 2020) and research studies based on their genetic and clinical characteristics and help accelerate the development of new therapies and treatments for rare diseases. AI-powered telemedicine platforms can connect patients with rare diseases to specialists and resources from anywhere globally. AI can revolutionize diagnosis (Akobeng 2007), treatment, and management of rare genetic disorders by leveraging large-scale data analysis and predictive modeling. While there are still challenges to overcome, such as data privacy and standardization, AI is promising to improve outcomes for patients with rare diseases (Hasani et al. 2022).

5.6.3 Understanding Disease Mechanisms in Model Organisms

Models for rare genetic diseases are important tools for understanding the underlying biology of these conditions and developing new therapies and treatments. Studying model organisms is the gold standard for understanding the pathogenic genes and loci of the URDs. Animal models, such as mice and zebrafish (Phillips and Westerfield 2014), are commonly used in rare disease research to study the underlying biology of the disease and test potential therapies. Animal models can be genetically engineered to mimic human disease mutations or bred to naturally develop similar symptoms to those seen in humans with the disease. Cellular models (Rath and Felbor 2021), such as induced pluripotent stem cells (iPSCs) and organoids, are used to study the cellular mechanisms of rare diseases. iPSCs are generated by reprogramming adult cells to become stem cells, which can then be differentiated into various cell types for study (Sabitha et al. 2021; Anderson and Francis 2018). Organoids are 3D structures that mimic the architecture and function of human organs, allowing researchers to study disease mechanisms in a more realistic setting (Heydari et al. 2021). In silico models, such as computer simulations and artificial intelligence, are used to analyze large amounts of genomic and clinical data to identify disease-causing genes, pathways, and potential drug targets (Frederiksen et al. 2022). These models can also be used to predict the effectiveness of potential therapies and identify patient subgroups that may benefit from a particular treatment. Patient-derived models, such as patient-derived xenografts (PDXs) and patient-derived cell lines, are generated from patient samples and can be used to study the genetic and molecular characteristics of the disease. By combining multiple models and approaches, researchers can gain a more comprehensive understanding of the disease and identify new avenues for treatment and intervention. Sometimes, these models fail to resemble the diseases of humans (Bhimani et al. 2020). It is a priority to construct viable models to address the clinical issues in rare diseases, including high population prevalence, low prevalence in single diseases, scarcity of clinical samples, and traditional screening.

5.7 Rare Disease Registries

A patient registry is a database of information about patients with a specific disease or condition, which can be used to track the natural history of the disease, identify potential treatments, and facilitate research collaborations. Patient registries can be particularly valuable for undiagnosed rare diseases because they allow researchers to collect and analyze data on a larger number of patients than might be available at a single institution (Jansen-Van Der Weide et al. 2018). Registries can also help identify patterns and commonalities among patients, aiding in the diagnosis and treatment of the disease. Several patient registries specifically focus on undiagnosed rare diseases, including the Undiagnosed Diseases Network (UDN) in the United States (Brownstein et al. 2015) and the European Reference Network for Rare Endocrine Conditions (Endo-ERN) in Europe (Pereira and Hiort 2021). These registries typically involve a multidisciplinary team of experts who analyze patient data and identify potential genetic or other causes of the disease (Gliklich et al. 2014). These registries may be maintained by patient advocacy organizations, medical centers, or research groups and can be an important resource for patients and families seeking information about a particular disease.

5.8 Patient Match-Making Exchanger

Matchmaker Exchange is a collaborative platform that enables the sharing of rare disease patient data and facilitates the discovery of novel disease-gene associations (Philippakis et al. 2015). It is a federated network of databases that allows researchers and clinicians to search for patients with similar clinical and genetic characteristics across different institutions and countries. Matchmaker Exchange operates on the principle of "Matchmaker Recruitment," where participating institutions upload de-identified patient data into their local databases (Dyke et al. 2017). When a researcher or clinician submits a query for a patient with specific clinical or genetic characteristics, the question is distributed across all the participating databases, including DECIPHER (Firth et al. 2009), GeneMatcher (Sobreira et al. 2015), and Phenomecentral (Buske et al. 2015). If a match is found, the researcher or clinician can contact the potential match's submitter to request additional information or collaborate on a research project. It has been instrumental in rare diseases where individual institutions may only have a small number of patients with a particular disease or genetic variant. By enabling data sharing across institutions, Matchmaker Exchange has facilitated the discovery of new disease-gene associations and improved the diagnosis and treatment of rare diseases.

5.9 Collaborative Networks

Collaborative networks such as the International Rare Diseases Research Consortium (IRDiRC), Global Rare Diseases Patient Registry Data Repository (GRDR), Undiagnosed Diseases Network (UDN), European Reference Networks (ERNs), and Rare Diseases Clinical Research Network (RDCRN) bring together researchers, clinicians, patient advocates, and industry partners to accelerate research into rare diseases. These networks aim to develop new treatments and improve diagnosis and care for patients with rare diseases (Radu et al. 2021). In India, there are several rare disease collaborative networks: the Indian Council of Medical Research (ICMR), the Organization for Rare Diseases India (ORDI) Alliance, the Indian Society for Clinical Research in Rare Diseases (ISCRD), the Center for Medical Genetics, Sir Ganga Ram Hospital, and Genomics for Understanding Rare Diseases: India Alliance Network (GUaRDIAN). This network was established to improve the diagnosis and management of rare genetic disorders in India (Julkowska et al. 2017). The network includes a consortium of medical institutions, academic centers, and patient advocacy groups.

Patient advocacy groups are playing an increasingly important role in rare disease research. These groups provide funding for research, facilitate patient engagement in research studies, and advocate for increased research funding and policy changes to improve the lives of patients with rare diseases (Dunkle et al. 2010). In India, there are few patient advocacy groups to support rare disease patients, including ORDI, ISCRD, Lysosomal Storage Disorders Support Society (LSDSS), Muscular Dystrophy Foundation India (MDFI), and Hemophilia Federation India (HFI). These organizations provide various services, including awareness, multidisciplinary care, advocacy, support, and community-building events for rare disease patients and their families (Rajasimha et al. 2014).

5.10 Research Gaps in URDs

Despite significant progress in rare diseases, many research gaps must be addressed to improve diagnosis, treatment, and outcomes for patients with these conditions (Zhu et al. 2021). Many rare diseases are poorly understood, and there is limited knowledge of the underlying molecular mechanisms that cause these conditions. More research is needed to identify the genetic and environmental factors contributing to the development of rare diseases and to understand the biological pathways these conditions affect. There are currently no effective treatments for many rare diseases, and even when treatments are available, they may not be effective for all patients. More research is needed for new therapies for rare diseases, including gene, cell, and small-molecule drugs. Diagnosing rare diseases can be challenging, and there is often limited access to specialized diagnostic tools and expertise. More research is needed to develop new diagnostic tools and technologies, including genomics and proteomics, that can be used to identify rare diseases more quickly and accurately. Clinical trials for rare diseases are often small and underpowered, and there needs to be more data on the long-term safety and efficacy of treatments for these conditions. Patients with rare diseases face unique challenges, including difficulty obtaining a diagnosis, limited treatment options, and lack of access to specialized care. More research is needed to understand the patient's perspective on rare diseases, including their experiences with diagnosis, treatment, and maintenance, and to develop patient-centered approaches to rare disease research and care.

Addressing these research gaps will require increased funding and collaboration among researchers, clinicians, patient advocates, and policymakers. By working together to address these challenges, we can improve the lives of patients with rare diseases and make progress toward more effective diagnosis and treatment of these conditions.

5.11 Research Funds and Policies

Research funding and policies for rare diseases are essential to accelerate the development of new therapies, diagnostic tools, and support systems for patients with rare diseases (Khosla and Valdez 2018). The research funding and policies established for rare diseases are the Orphan Drug Act (ODA), passed by the U.S. Congress in 1983, providing financial incentives to pharmaceutical companies to develop rare diseases. The incentives include tax credits, grants, and exclusive marketing rights for 7 years. Rare Diseases Clinical Research Network (RDCRN), established by the National Institutes of Health (NIH) in 2003, is a network of research consortia that conduct clinical research on rare diseases. The RDCRN provides funding, support, and expertise to researchers studying rare diseases. The European Union established 24 ERNs in 2017, each focusing on a specific rare disease or a group of rare diseases. The ERNs are networks of healthcare providers and researchers across Europe collaborating on clinical research and treating rare diseases (Rodwell and Aymé 2015). National Organization for Rare Disorders (NORD) is a patient advocacy organization in the United States that provides funding, education, and advocacy for rare disease research. NORD provides grants for research on rare diseases and advocates for policies that support rare disease research. The National Center for Advancing Translational Sciences (NCATS), a part of the NIH, provides funding for research on rare diseases and supports the development of new diagnostic tools and treatments for these conditions. International Rare Diseases Research Consortium (IRDiRC) is a global consortium of research organizations, patient advocacy groups, and funding agencies that aims to accelerate research on rare diseases. The IRDiRC provides funding for rare disease research and establishes research priorities for the community (Lochmüller et al. 2017). These research funding and policy initiatives are critical for advancing the development of new therapies and diagnostic tools for rare diseases and improving support systems for patients with these conditions. By increasing investment in rare disease research, we can accelerate progress toward better diagnosis, treatment, and care for patients with rare diseases.

5.12 Management of URDS

Managing undiagnosed rare diseases is often complicated and challenging, as limited treatment options are often available. The different management strategies for these diseases include symptomatic treatment, supportive care, and experimental therapies (Choudhury and Chaube 2022). Rare diseases have genetic or inherited components, which may not be preventable in some cases, adding to the complexity. However, some measures can be taken to reduce the risk of developing certain rare diseases. Many rare diseases have symptoms that could be managed with medications or other therapies (Blaschke et al. 2021). Symptom management may also involve rehabilitation, physical therapy, and occupational therapy. Some rare diseases have no known cure, and patients may require palliative care to manage symptoms and improve their quality of life. Palliative care can include pain management, emotional and spiritual support, and end-of-life care (Adams et al. 2016). Genetic counseling can also help individuals and families understand their risk of developing a rare disease and guide family planning and reproductive options to reduce the risk of passing the inherited genetic mutations to the next generation (Miao et al. 2018). URDs may have lifestyle or environmental factors contributing to their development or exacerbating symptoms. Examples include avoiding exposure to certain toxins or chemicals, maintaining a healthy diet and exercise routine, and avoiding risky behaviors such as smoking or excessive alcohol consumption (Valdez et al. 2016). Certain rare diseases, such as meningitis and hepatitis B, could be prevented through vaccination. Regular health checkups and screenings may help detect rare diseases in their early stages, when treatment options may be more effective (Cox et al. 2022). Investing in research to better understand the underlying biology of rare diseases and develop new therapies can ultimately help prevent or mitigate the impact of rare diseases on individuals and society. It is important to note that prevention strategies may vary depending on the rare disease. Consultation with a healthcare provider is recommended for individualized guidance on reducing the risk of developing a rare disease (Chung et al. 2022).

5.13 Therapies for URDs

Therapy research in undiagnosed rare diseases is an active area of investigation, and there are ongoing efforts to develop new treatments for patients with undiagnosed rare diseases. Gene therapy is a promising approach for treating rare genetic disorders. Researchers are developing methods for delivering corrected or functional genes to patients with genetic mutations that cause rare diseases (Tambuyzer et al. 2020). Clinical gene therapy trials are underway for rare diseases, including spinal muscular atrophy, Duchenne muscular dystrophy, and Leber congenital amaurosis. Precision medicine involves tailoring treatments to individual patients based on their genetic and medical characteristics. Researchers use genomic and other diagnostic tools to identify patients with rare diseases who may benefit from targeted therapies or repurposed drugs. Repurposing existing drugs for new uses is a cost-effective approach to developing treatments for rare diseases (Roessler et al. 2021). Researchers are screening approved drug libraries to identify compounds that may effectively treat rare diseases (Bellomo et al. 2017). Stem cell therapy is another promising approach for treating rare diseases that affect specific cell types or tissues. Stem cells replace or repair damaged tissues in patients with rare diseases. Ongoing research on developing new drugs for rare diseases, traditional drug development approaches, and innovative strategies such as crowdsourcing drug development through patient advocacy groups are also gaining momentum.

Therapy research in URDs focuses on developing targeted and personalized treatments for individual patients, repurposing existing drugs, and developing new therapies. These efforts aim to improve the patient's quality of life and provide hope for those not diagnosed.

5.14 Conclusion and Future Direction

Undiagnosed rare genetic disorders significantly challenge patients, their families, and healthcare professionals. However, recent advances in genomics, bioinformatics, and other research areas have led to new approaches to diagnosis and treatment. Here are some of the conclusions and future directions for research and clinical practices for URDs:

1. Collaboration is the key: Due to the rarity and complexity of these conditions, collaboration among clinicians, researchers, and patients is essential. Sharing data and resources, developing standards for diagnosis and treatment, and establishing patient advocacy groups are the areas where there is a high scope for improvement.
2. Advances in genomics: Genomics has already significantly impacted the identification and diagnosis of rare genetic disorders and will likely continue to play a central role in future research and clinical approaches such as whole-genome and whole-exome sequencing, gene editing, and other techniques.
3. Integration of phenome and genome data: Integrating clinical data with genomic data can provide a complete picture of a patient's health and help identify new genetic variants and disease-causing mutations. New tools and technologies are to be developed for data analysis and interpretation.
4. Personalized medicine: Advances in genomics and other areas of research, such as the use of gene therapies, precision medicine, and different targeted approaches, are leading to the development of personalized treatment options for rare genetic disorders.
5. Ethical considerations: As research and clinical approaches to undiagnosed rare genetic disorders continue to evolve, it is important to consider the ethical implications, such as informed consent, privacy, equity, and access to care.

In conclusion, the field of rare genetic disorders is rapidly evolving, with new advances in genomics, bioinformatics, and personalized medicine providing new opportunities for diagnosis and treatment. Collaboration, clinical and genomic data integration, and ethical considerations will continue to be essential in advancing the field and improving patient and family outcomes.

References

Abela L, Simmons L, Steindl K, Schmitt B, Mastrangelo M, Joset P, Papuc M, Sticht H, Baumer A, Crowther LM, Mathis D, Rauch A, Plecko B (2016) N8-acetylspermidine as a potential plasma biomarker for Snyder-Robinson syndrome identified by clinical metabolomics. J Inherit Metab Dis 39(1):131–137. https://doi.org/10.1007/s10545-015-9876-y

Adams LS, Miller JL, Grady PA (2016) The spectrum of caregiving in palliative care for serious, advanced, rare diseases: key issues and research directions. J Palliat Med 19(7):698–705. https://doi.org/10.1089/jpm.2015.0464

Agastheeswaramoorthy K, Sevilimedu A (2020) Drug REpurposing using AI/ML tools—for Rare Diseases (DREAM-RD): a case study with Fragile X Syndrome (FXS). bioRxiv. https://doi.org/10.1101/2020.09.25.311142

Akobeng AK (2007) Understanding diagnostic tests 2: likelihood ratios, pre-and post-test probabilities and their use in clinical practice. Acta Paediatr Int J Paediatr 96(4):487–491. https://doi.org/10.1111/j.1651-2227.2006.00179.x

Anderson RH, Francis KR (2018) Modeling rare diseases with induced pluripotent stem cell technology. In: Molecular and cellular probes, vol 40. Academic Press, pp 52–59. https://doi.org/10.1016/j.mcp.2018.01.001

Aref-Eshghi E, Bend EG, Colaiacovo S, Caudle M, Chakrabarti R, Napier M, Brick L, Brady L, Carere DA, Levy MA, Kerkhof J, Stuart A, Saleh M, Beaudet AL, Li C, Kozenko M, Karp N, Prasad C, Siu VM et al (2019) Diagnostic utility of genome-wide DNA methylation testing in genetically unsolved individuals with suspected hereditary conditions. Am J Hum Genet 104(4):685–700. https://doi.org/10.1016/j.ajhg.2019.03.008

Barrera LA, Galindo GC (2010) Ethical aspects on rare diseases. Adv Exp Med Biol 686:493–511. https://doi.org/10.1007/978-90-481-9485-8_27

Bellomo F, Medina DL, De Leo E, Panarella A, Emma F (2017) High-content drug screening for rare diseases. J Inherit Metab Dis 40(4):601–607. https://doi.org/10.1007/s10545-017-0055-1

Bhimani J, Ball K, Stebbing J (2020) Patient-derived xenograft models—the future of personalized cancer treatment. Br J Cancer 122(5):601–602. https://doi.org/10.1038/s41416-019-0678-0

Black N, Martineau F, Manacorda T (2015) Diagnostic odyssey for rare diseases: exploration of potential indicators. PIRU Publication. www.piru.ac.uk

Blaschke CRK, Hartig JP, Grimsley G, Liu L, Semmes OJ, Wu JD, Ippolito JE, Hughes-Halbert C, Nyalwidhe JO, Drake RR (2021) Direct N-glycosylation profiling of urine and prostatic fluid glycoproteins and extracellular vesicles. Front Chem 9:734280. https://doi.org/10.3389/fchem.2021.734280

Blumenrath SH, Lee BY, Low L, Prithviraj R, Tagle D (2020) Tackling rare diseases: clinical trials on chips. Exp Biol Med 245(13):1155–1162. https://doi.org/10.1177/1535370220924743

Bogart K, Hemmesch A, Barnes E, Blissenbach T, Beisang A, Engel P, Tolar J, Schacker T, Schimmenti L, Brown N, Morrison K, Albright T, Klein M, Coleman J, Nelsen K, Blaylark R, LaFond K, Berkowitz S, Schultz KA et al (2022) Healthcare access, satisfaction, and health-related quality of life among children and adults with rare diseases. Orphanet J Rare Dis 17(1):196. https://doi.org/10.1186/s13023-022-02343-4

Brownstein CA, Holm IA, Ramoni R, Goldstein DB, Bacino C, Lee B, Eng C, Veeraghavan N, Bernick D, Brownstein C, Holm I, Kohane I, McCray A, Splinter K, Shashi V, Goldstein D, Beggs A, Loscalzo J, MacRae C et al (2015) Data sharing in the undiagnosed diseases network. HumMutat 36(10):985–988. https://doi.org/10.1002/humu.22840

Buske OJ, Schiettecatte F, Hutton B, Dumitriu S, Misyura A, Huang L, Hartley T, Girdea M, Sobreira N, Mungall C, Brudno M (2015) The Matchmaker Exchange API: automating patient matching through the exchange of structured phenotypic and genotypic profiles. Hum Mutat 36(10):922–927. https://doi.org/10.1002/humu.22850

Chen C, Lu MY, Williamson DFK, Chen TY, Schaumberg AJ, Mahmood F (2022) Fast and scalable search of whole-slide images via self-supervised deep learning. Nat Biomed Eng 6(12):1420–1434. https://doi.org/10.1038/s41551-022-00929-8

Choudhury MC, Chaube P (2022) Integrating rare disease management in public health programs in India: exploring the potential of National Health Mission. Orphanet J Rare Dis 17(1):43. https://doi.org/10.1186/s13023-022-02194-z

Chung CCY, Hong Kong Genome Project, Chu ATW, Chung BHY (2022) Rare disease emerging as a global public health priority. Front Public Health 10:1028545. https://doi.org/10.3389/fpubh.2022.1028545. PMID: 36339196; PMCID: PMC9632971

Coors M, Bauer L, Edwards K, Erickson K, Goldenberg A, Goodale J, Goodman K, Grady C, Mannino D, Wanner A, Wilson T, Yarborough M, Zirkle M (2017) Ethical issues related to clinical research and rare diseases. Transl Sci Rare Dis 2(3–4):175–194. https://doi.org/10.3233/trd-170013

Cox TM, Tylki-Szymańska A, Aymé S, Dooms M (2022) Editorial: prevention, diagnosis and treatment of rare disorders. Front Pharmacol 13:1026064. https://doi.org/10.3389/fphar.2022.1026064

de Ligt J, Willemsen MH, van Bon BWM, Kleefstra T, Yntema HG, Kroes T, Vulto-van Silfhout AT, Koolen DA, de Vries P, Gilissen C, del Rosario M, Hoischen A, Scheffer H, de Vries BBA, Brunner HG, Veltman JA, Vissers LELM (2012) Diagnostic exome sequencing in persons with severe intellectual disability. N Engl J Med 367(20):1921–1929. https://doi.org/10.1056/nejmoa1206524

Ducatez F, Mauhin W, Boullier A, Pilon C, Pereira T, Aubert R, Benveniste O, Marret S, Lidove O, Bekri S, Tebani A (2021) Parsing Fabry disease metabolic plasticity using metabolomics. J Pers Med 11(9):898. https://doi.org/10.3390/jpm11090898

Dunkle M, Pines W, Saltonstall PL (2010) Advocacy groups and their role in rare diseases research. Adv Exp Med Biol 686:515–525. https://doi.org/10.1007/978-90-481-9485-8_28

Dyke SOM, Knoppers BM, Hamosh A, Firth HV, Hurles M, Brudno M, Boycott KM, Philippakis AA, Rehm HL (2017) "Matching" consent to purpose: the example of the matchmaker exchange. Human Mutat 38(10):1281–1285. https://doi.org/10.1002/humu.23278

Ebert T, Tran N, Schurgers L, Stenvinkel P, Shiels PG (2022) Ageing—oxidative stress, PTMs and disease. Mol Asp Med 86:101099. https://doi.org/10.1016/j.mam.2022.101099

Ferreira CR (2019) The burden of rare diseases. Am J Med Genet A 179(6):885–892. https://doi.org/10.1002/ajmg.a.61124

Firth HV, Richards SM, Bevan AP, Clayton S, Corpas M, Rajan D, Van Vooren S, Moreau Y, Pettett RM, Carter NP (2009) DECIPHER: database of chromosomal imbalance and phenotype in humans using ensembl resources. Am J Hum Genet 84(4):524–533. https://doi.org/10.1016/j.ajhg.2009.03.010

Frederiksen SD, Avramović V, Maroilley T, Lehman A, Arbour L, Tarailo-Graovac M (2022) Rare disorders have many faces: in silico characterization of rare disorder spectrum. Orphanet J Rare Dis 17(1):1–8. https://doi.org/10.1186/s13023-022-02217-9

Gagne JJ, Thompson L, O'Keefe K, Kesselheim AS (2014) Innovative research methods for studying treatments for rare diseases: methodological review. BMJ 349:g6802. https://doi.org/10.1136/bmj.g6802

Gazzo AM, Daneels D, Cilia E, Bonduelle M, Abramowicz M, Van Dooren S, Smits G, Lenaerts T (2016) DIDA: a curated and annotated digenic diseases database. Nucleic Acids Res 44(D1):D900–D907. https://doi.org/10.1093/nar/gkv1068

Gliklich R, Dreyer N, Leavy M (eds) (2014) Registries for evaluating patient outcomes: a user's guide. Third edition. Two volumes. (Prepared by the Outcome DEcIDE Center [Outcome

Sciences, Inc., a Quintiles company] under Contract No. 290 2005 00351 TO7.) AHRQ Publication No. 13(14)-EHC111. Agency for Healthcare Research and Quality, Rockville, MD. April 2014. http://www.effectivehealthcare.ahrq.gov/registries-guide-3.cfm

Global Commission to End the Diagnostic Odyssey for Children with a Rare Disease (2018) Ending the diagnostic Odyssey global commission year one report for children with a rare disease

Gonzaga-Jauregui C, Lupski JR, Gibbs RA (2012) Human genome sequencing in health and disease. Ann Rev Med 63:35–61. https://doi.org/10.1146/annurev-med-051010-162644

Grabowski P, Hesse S, Hollizeck S, Rohlfs M, Behrends U, Sherkat R, Tamary H, Ünal E, Somech R, Patiroğlu T, Canzar S, Der Werff V, Ten Bosch J, Klein C, Rappsilber J (2019) Proteome analysis of human neutrophil granulocytes from patients with monogenic disease using data-independent acquisition. Mol Cell Proteomics 18(4):760–772. https://doi.org/10.1074/mcp.RA118.001141

Griggs RC, Batshaw M, Dunkle M, Gopal-Srivastava R, Kaye E, Krischer J, Nguyen T, Paulus K, Merkel PA (2009) Clinical research for rare disease: opportunities, challenges, and solutions. Mol Genet Metab 96(1):20–26. https://doi.org/10.1016/j.ymgme.2008.10.003

Gurovich Y, Hanani Y, Bar O, Fleischer N, Gelbman D, Basel-Salmon L, Krawitz P, Kamphausen SB, Zenker M, Bird LM, Gripp KW (2018) DeepGestalt—identifying rare genetic syndromes using deep learning. http://arxiv.org/abs/1801.07637

Haendel M, Vasilevsky N, Unni D, Bologa C, Harris N, Rehm H, Hamosh A, Baynam G, Groza T, McMurry J, Dawkins H, Rath A, Thaxon C, Bocci G, Joachimiak MP, Köhler S, Robinson PN, Mungall C, Oprea TI (2020) How many rare diseases are there? Nat Rev Drug Discov 19(2):77–78. https://doi.org/10.1038/d41573-019-00180-y

Hartley T, Lemire G, Kernohan KD, Howley HE, Adams DR, Boycott KM (2020) New diagnostic approaches for undiagnosed rare genetic diseases. Ann Rev Genomics Hum Genet 21:351–372. https://doi.org/10.1146/annurev-genom-083118

Hasani N, Farhadi F, Morris MA, Nikpanah M, Rhamim A, Xu Y, Pariser A, Collins MT, Summers RM, Jones E, Siegel E, Saboury B (2022) Artificial intelligence in medical imaging and its impact on the rare disease community: threats, challenges and opportunities. PET Clin 17(1):13–29. https://doi.org/10.1016/j.cpet.2021.09.009

Heydari Z, Moeinvaziri F, Agarwal T, Pooyan P, Shpichka A, Maiti TK, Timashev P, Baharvand H, Vosough M (2021) Organoids: a novel modality in disease modeling. In Bio-Design Manuf 4(4):689–716. https://doi.org/10.1007/s42242-021-00150-7

Jansen-Van Der Weide MC, Gaasterland CMW, Roes KCB, Pontes C, Vives R, Sancho A, Nikolakopoulos S, Vermeulen E, Van Der Lee JH (2018) Rare disease registries: potential applications towards impact on development of new drug treatments. Orphanet J Rare Dis 13(1):154. https://doi.org/10.1186/s13023-018-0836-0

Julkowska D, Austin CP, Cutillo CM, Gancberg D, Hager C, Halftermeyer J, Jonker AH, Lau LPL, Norstedt I, Rath A, Schuster R, Simelyte E, Van Weely S (2017) The importance of international collaboration for rare diseases research: a European perspective. Gene Ther 24(9):562–571. https://doi.org/10.1038/gt.2017.29

Khosla N, Valdez R (2018) A compilation of national plans, policies and government actions for rare diseases in 23 countries. Intractable Rare Dis Res 7(4):213–222. https://doi.org/10.5582/irdr.2018.01085. International Advancement Center for Medicine and Health Research

Kremer LS, Bader DM, Mertes C, Kopajtich R, Pichler G, Iuso A, Haack TB, Graf E, Schwarzmayr T, Terrile C, Koňaříkova E, Repp B, Kastenmüller G, Adamski J, Lichtner P, Leonhardt C, Funalot B, Donati A, Tiranti V et al (2017) Genetic diagnosis of Mendelian disorders via RNA sequencing. Nat Commun 8:1582. https://doi.org/10.1038/ncomms15824

Lee Y-S, Krishnan A, Oughtred R, Rust J, Chang CS, Ryu J, Kristensen VN, Dolinski K, Theesfeld CL, Troyanskaya OG (2019) A computational framework for genome-wide characterization of the human disease landscape. Cell Syst 8(2):152–162. https://doi.org/10.1016/j.cels.2018.12.010

Lek A, Zhang Y, Woodman KG, Huang S, DeSimone AM, Cohen J, Ho V, Conner J, Mead L, Kodani A, Pakula A, Sanjana N, King OD, Jones PL, Wagner KR, Lek M, Kunkel LM (2020) Applying genome-wide CRISPR-Cas9 screens for therapeutic discovery in facioscapulo-

humeral muscular dystrophy. Sci Transl Med 12(536):eaay0271. https://doi.org/10.1126/scitranslmed.aay0271
Li Q, Zhao K, Bustamante CD, Ma X, Wong WH (2019) Xrare: a machine learning method jointly modeling phenotypes and genetic evidence for rare disease diagnosis. Genet Med 21(9):2126–2134. https://doi.org/10.1038/s41436
Lochmüller H, TorrentFarnell J, Le Cam Y, Jonker AH, Lau LP, Baynam G, Kaufmann P, Dawkins HJ, Lasko P, Austin CP, Boycott KM (2017) The International Rare Diseases Research Consortium: policies and guidelines to maximize impact. Eur J Hum Genet 25(12):1293–1302. https://doi.org/10.1038/s41431-017-0008-z
Lopes MT, Koch VH, Sarrubbi-Junior V, Gallo PR, Carneiro-Sampaio M (2018) Difficulties in the diagnosis and treatment of rare diseases according to the perceptions of patients, relatives and health care professionals. Clinics (Sao Paulo) 73:e68. https://doi.org/10.6061/clinics/2018/e68
Marwaha S, Knowles JW, Ashley EA (2022) A guide for the diagnosis of rare and undiagnosed disease: beyond the exome. In: Genome Med 14(1):1–22. Ltd, BioMed Central. https://doi.org/10.1186/s13073-022-01026-w
Menkovic I, Boutin M, Alayoubi A, Mercier FE, Rivard GÉ, Auray-Blais C (2020) Identification of a reliable biomarker profile for the diagnosis of Gaucher disease type 1 patients using a mass spectrometry-based metabolomic approach. Int J Mol Sci 21(21):1–21. https://doi.org/10.3390/ijms21217869
Miao H, Zhou J, Yang Q, Liang F, Wang D, Ma N, Gao B, Du J, Lin G, Wang K, Zhang Q (2018) Long-read sequencing identified a causal structural variant in an exome-negative case and enabled preimplantation genetic diagnosis. Hereditas 155:32. https://doi.org/10.1186/s41065-018-0069-1
Percival BC, Gibson M, Wilson PB, Platt FM, Grootveld M (2020) Metabolomic studies of lipid storage disorders, with special reference to Niemann-Pick type C disease: a critical review with future perspectives. In: Int J Mol Sci 21(7):2533. AG, MDPI. https://doi.org/10.3390/ijms21072533
Pereira AM, Hiort O (2021) Introduction to Endo-ERN–scope and mission. Endocrine 71(3):537–538. https://doi.org/10.1007/s12020-020-02602-z
Philippakis AA, Azzariti DR, Beltran S, Brookes AJ, Brownstein CA, Brudno M, Brunner HG, Buske OJ, Carey K, Doll C, Dumitriu S, Dyke SOM, den Dunnen JT, Firth HV, Gibbs RA, Girdea M, Gonzalez M, Haendel MA, Hamosh A et al (2015) The matchmaker exchange: a platform for rare disease gene discovery. Hum Mutat 36(10):915–921. https://doi.org/10.1002/humu.22858
Phillips JB, Westerfield M (2014) Zebrafish models in translational research: tipping the scales toward advancements in human health. DMM Dis Models Mech 7(7):739–743. https://doi.org/10.1242/dmm.015545
Piñol M, Alves R, Teixidó I, Mateo J, Solsona F, Vilaprinyó E (2017) Rare disease discovery: an optimized disease ranking system. IEEE Trans Indust Inform 13(3):1184–1192. https://doi.org/10.1109/TII.2017.2686380
Poplin R, Chang PC, Alexander D, Schwartz S, Colthurst T, Ku A, Newburger D, Dijamco J, Nguyen N, Afshar PT, Gross SS, Dorfman L, McLean CY, Depristo MA (2018) A universal snp and small-indel variant caller using deep neural networks. Nat Biotechnol 36(10):983. https://doi.org/10.1038/nbt.4235
Radu R, Hernández-Ortega S, Borrega O, Palmeri A, Athanasiou D, Brooke N, Chapí I, Le Corvec A, Guglieri M, Perera-Lluna A, Garrido-Aguirre J, Ryll B, Escalera BN (2021) Global collaborative social network (Share4Rare) to promote citizen science in rare disease research: platform development study. JMIR Format Res 5(3):e22695. https://doi.org/10.2196/22695
Rajasimha HK, Shirol PB, Ramamoorthy P, Hegde M, Barde S, Chandru V, Ravinandan ME, Ramchandran R, Haldar K, Lin JC, Babar IA, Girisha KM, Srinivasan S, Navaneetham D, Battu R, Devarakonda R, Kini U, Vijayachandra K, Verma IC (2014) Organization for rare diseases India (ORDI)-addressing the challenges and opportunities for the Indian rare diseases' community. Genet Res 96:e009. https://doi.org/10.1017/S0016672314000111

Rath M, Felbor U (2021) Cellular models of genetic diseases. Med Genet 33(3):219–220. https://doi.org/10.1515/medgen-2021-2095

Retterer K, Juusola J, Cho MT, Vitazka P, Millan F, Gibellini F, Vertino-Bell A, Smaoui N, Neidich J, Monaghan KG, McKnight D, Bai R, Suchy S, Friedman B, Tahiliani J, Pineda-Alvarez D, Richard G, Brandt T, Haverfield E et al (2016) Clinical application of whole-exome sequencing across clinical indications. Genet Med 18(7):696–704. https://doi.org/10.1038/gim.2015.148

Rodwell C, Aymé S (2015) Rare disease policies to improve care for patients in Europe. Biochim Rev 121:201–219. https://doi.org/10.1016/j.neubiorev.2020.12.025

Roessler HI, Knoers NVAM, van Haelst MM, van Haaften G (2021) Drug repurposing for rare diseases. Trends Pharmacol Sci 42(4):255–267. https://doi.org/10.1016/j.tips.2021.01.003. Epub 2021 Feb 6. PMID: 33563480

Sabitha KR, Shetty AK, Upadhya D (2021) Patient-derived iPSC modeling of rare neurodevelopmental disorders: molecular pathophysiology and prospective therapies. Neurosci Biobehav Rev 121:201–219.https://doi.org/10.1016/j.neubiorev.2020.12.025. Epub 2020 Dec 25. PMID: 33370574; PMCID: PMC7962756

Sadikovic B, Levy MA, Aref-Eshghi E (2020) Functional annotation of genomic variation: DNA methylation episignatures in neurodevelopmental Mendelian disorders. Hum Mol Genet 29:R27–R32. https://doi.org/10.1093/hmg/ddaa144

Shen F, Peng S, Fan Y, Wen A, Liu S, Wang Y, Wang L, Liu H (2019) HPO2Vec+: leveraging heterogeneous knowledge resources to enrich node embeddings for the human phenotype ontology. J Biomed Inform 96:103246. https://doi.org/10.1016/j.jbi.2019.103246

Sobreira N, Schiettecatte F, Valle D, Hamosh A (2015) GeneMatcher: a matching tool for connecting investigators with an interest in the same gene. Hum Mutat 36(10):928–930. https://doi.org/10.1002/humu.22844

Stoller JK (2018) The challenge of rare diseases. Chest 153(6):1309–1314. https://doi.org/10.1016/j.chest.2017.12.018

Tambuyzer E, Vandendriessche B, Austin CP, Brooks PJ, Larsson K, Miller Needleman KI, Valentine J, Davies K, Groft SC, Preti R, Oprea TI, Prunotto M (2020) Therapies for rare diseases: therapeutic modalities, progress and challenges ahead. Nat Rev Drug Discov 19(2):93–111. https://doi.org/10.1038/s41573-019-0049-9

Tarailo-Graovac M, Shyr C, Ross CJ, Horvath GA, Salvarinova R, Ye XC, Zhang L-H, Bhavsar AP, Lee JJY, Drögemöller BI, Abdelsayed M, Alfadhel M, Armstrong L, Baumgartner MR, Burda P, Connolly MB, Cameron J, Demos M, Dewan T et al (2016) Exome sequencing and the management of neurometabolic disorders. N Engl J Med 374(23):2246–2255. https://doi.org/10.1056/nejmoa1515792

Valdez R, Ouyang L, Bolen J (2016) Public health and rare diseases: oxymoron no more. Prev Chronic Dis 13(1):E05. https://doi.org/10.5888/pcd13.150491

Wenger AM, Peluso P, Rowell WJ, Chang PC, Hall RJ, Concepcion GT, Ebler J, Fungtammasan A, Kolesnikov A, Olson ND, Töpfer A, Alonge M, Mahmoud M, Qian Y, Chin CS, Phillippy AM, Schatz MC, Myers G, DePristo MA et al (2019) Accurate circular consensus long-read sequencing improves variant detection and assembly of a human genome. Nat Biotechnol 37(10):1155–1162. https://doi.org/10.1038/s41587-019-0217-9

Zhu Q, Nguyễn ĐT, Sheils T, Alyea G, Sid E, Xu Y, Dickens J, Mathé EA, Pariser A (2021) Scientific evidence based rare disease research discovery with research funding data in knowledge graph. Orphanet J Rare Dis 16(1):483. https://doi.org/10.1186/s13023-021-02120-9

Drug Discovery and Development for Rare Genetic Disorders

6

Mohamed Y. Zaky and Tasneem Abaza

Abstract

A rare disease is any condition that has an extremely low prevalence on an individual basis, and most rare diseases are genetic disorders. Owing to the limited market size, expensive demand, and perhaps low financial return, research and development of rare disorder therapies have only recently increased internationally, in several domains including small-molecule pharmaceuticals and biologics. There is a considerable gap between fundamental research and patient unfulfilled demands for rare disease treatment development due to the complicated etiology and varying symptoms. Due to the regular developments in the methodology and techniques of drug discovery research, the development of drugs for the treatment of rare disorders can now be accelerated. Disorder organizations and research institutes across the world are working to better study rare diseases. The recent drug development methodologies for orphan diseases are discussed. Genome and pharmacogenetics studies provide interpretation for the disease etiology and treatment. Regarding the road map of small-molecule drug development, building up a compound library and bioassay development, virtual and experimental screening, hit confirmation and lead generation, preclinical and clinical studies, and drug registration are explained, and drug pharmacokinetics, formulation, and repositioning are illustrated. Biologics are considered, including protein replacement, oligonucleotide, antibody, and cell therapy.

M. Y. Zaky (✉)
Molecular Physiology Division, Department of Zoology, Faculty of Science, Beni-Suef University, Beni-Suef, Egypt
e-mail: mohamedzaki448@science.bsu.edu.eg

T. Abaza
Biotechnology and Biomolecular Chemistry Program, Faculty of Science, Cairo University, Giza, Egypt

M. Umair et al. (eds.), *Rare Genetic Disorders*,
https://doi.org/10.1007/978-981-99-9323-9_6

Keywords
Rare disorder · Drug development · Small molecule · Biologics

6.1 Introduction

Rare disorders are permanent, severe, and gradually disabling; those diseases have the potential to reduce life expectancy and threaten patients' life. A disease is classified as "rare" in the United States if it affects fewer than 200,000 persons and in the European Union if its frequency is less than 5 per 10,000. Moreover, in certain geographies, some diseases are considered to be "rare" but not in others. For instance, Tay–Sachs disorder, a rare genetic condition that can be inherited from one parent to the offspring and results from the absence of an enzyme that contributes to the breakdown of the fatty gangliosides, is rare in the community at large but among Ashkenazi Jews, the frequency of carriers is 4/100 (Haendel et al. 2020). Although rare diseases have a slight influence on public health, a huge number of people are suffering around the world as more than 200 million individuals are estimated to be affected by them globally. According to the US National Institutes of Health (NIH), 25–30 million Americans are reportedly affected, and about 7000 rare disorders are documented.

Orphanet, the portal for rare diseases and orphan drugs, has data on 6172 distinct rare disorders, of which 71.9% are genetic (Nguengang Wakap et al. 2020). Most rare genetic diseases are driven by a genetic mutation, which alters the intact function of a single gene. Rare genetic diseases (RGDs) are also called Mendelian or monogenic diseases. Even though RGDs are progressive, early diagnosis can reduce or retard some of their long-term effects, for example, by newborn screening in which each baby is examined for a series of health disorders at 1 or 2 days of age. Then the diagnosed cases can be controlled by standard and/or targeted therapies in an optimal way. Patients with rare diseases sometimes go through an "odyssey" before receiving the proper diagnosis. Several years may pass before a disease is diagnosed, during which time it may develop. The identification of some rare genetic disorders is difficult because of the great disease variety, overlapping or heterogeneous phenotypes, an enormous volume of additional data, or complex nomenclature, i.e., eponyms or archaic names such as Bourneville-Pringle syndrome (TSC) for tuberous sclerosis complex. Even for trained dermatologists, detecting genetic skin conditions remains difficult. However, a conclusive molecular diagnosis eliminates the requirement for additional diagnostic procedures, makes it easier for patients to reach the right medical resources, offers accurate recurrence-risk information, lowers the risk of prognosis, encourages impacted families to make reproductive decisions, and provides the patient's family with psychosocial advantages. Molecular mechanisms linking genetic abnormalities to their disease phenotypes are elusive. The currently available and constantly evolving sophistication of genome-wide expression analysis methods, like genealogical proteomics, comparing the proteomes of cells from individuals with rare diseases to those from their congenitally normal relatives to understand the mechanisms behind these disorders, and interactome proteomics, interacting proteins in stable complexes in a biological system, is mostly achieved by affinity-based techniques (Zlatic et al. 2018; Gokhale et al. 2012a, b, 2019; Comstra et al. 2017), which assists in conducting an inclusive investigation for these disorders to understand deeply how abnormal proteins contribute to the pathogenesis of rare diseases.

Improved mechanistic insights into the genes and enzymes linked to rare genetic diseases offer a special potential for the development of orphan drugs, drugs used to treat, prevent, or diagnose rare conditions. Some eminent examples are novel anticancer therapeutic opportunities that have been created by improvements in the detection of synthetic lethal connections between rare disease genes and oncogenes and tumor suppressor genes and by the fast progression of small-molecule inhibitors against enzymes that contribute to DNA deterioration response and repair (Bhattacharjee and Nandi 2018).

6.2 Etiology

Studying the etiology of rare genetic diseases is fundamental for the development of unique and effective prognostic, diagnostic, and therapeutic techniques. Rare genetic diseases are correlated with mutations in specific genes; this means that eventually, genes alter because of abnormalities in DNA; hence, the possibility of having a genetic disorder increases. Optimally, the mutation that leads to a disorder is called a pathogenic variant, and a number of them can be transmitted from parents to the next generations demonstrating the reason why some rare diseases run in families. For a deep understanding of how a mutation in one gene led to an RGD, it is guileful to remember that genes' DNA contains instructions for manufacturing proteins responsible for cell complex interactions to maintain vital functions and support their health. Vascular Ehlers-Danlos syndrome is a prime example of RGD; it is caused by a mutation in the COL3A1 gene leading to a defect in collagen III. This disorder is remarked by fine, friable, and translucent skin and laceration of internal organs like the intestine and arteries, resulting in reduced lifespan (Royce and Steinmann 2003). Cystic fibrosis (CF) is another illustrative example; this heterogeneous resistive genetic condition develops due to a mutation in the cystic fibrosis transmembrane conductance regulator (CFTR) gene. The CFTR protein functions as a channel in healthy cells, enabling the release of chloride and other ions. However, this protein is damaged in CF patients, preventing the cells from releasing chloride. As a result, the cells' salt balance is off, and thick, sticky mucus is produced. The gastrointestinal tract and the respiratory system are mainly damaged, but later various organs are as well, posing a life-threatening hazard (Knowles and Durie 2002; Rafeeq and Murad 2017). It is crucial to remember that genetics is only one aspect of the problem. Diet and several environmental elements can combine with genetic factors to develop or enhance disease (Graham Jr. and Shaw 2005; Kim and Leventhal 2015).

6.3 Challenges in Rare Disease Treatment

Despite the severity of most rare genetic disorders, there are few efficient treatments available since the drug industry cannot afford the costs of research and development and selling such medications (Boycott et al. 2013). On the other hand, rare illnesses frequently have insufficient knowledge of the natural history of the diseases, which is essential for planning and powering research. This is typically due to the rareness of those disorders, and in many situations, there may only be a few international experts who have provided care for several patients (Kempf et al. 2018).

The complexity of some rare genetic illnesses makes it more difficult to create efficient therapies. For instance, a scaly skin condition known as congenital ichthyosis has more than 30 recognized variants with overlapping clinical characteristics linked to several gene abnormalities (Dunoyer 2011). Over 200 missense mutations in the NPC1 gene cause Niemann-Pick disease type C (NPC), and they all share a common clinical phenotype (Runz et al. 2008). Even if the individuals are diagnosed with the same condition, their varied variants will unavoidably result in varying degrees of effectiveness from the same therapy. So, this inspires scientists to go forward with multi-omics analysis and precision medicine to get accurate diagnoses and treatments.

Moreover, clinical trial results with insufficient patient enrollment yield no statistically significant results. Clinical study models require regular in-person visits at study locations and need to be reevaluated since they might be resource-intensive for patients. In addition, there may only be a few treatment centers for rare diseases in the nation or perhaps the entire globe, making research participation unreasonably burdensome for the participants (Kempf et al. 2018). This issue has been inventively solved in several studies by validating outcome measures for at-home measurements and employing internet-based data collection while utilizing centralized expert evaluators to reduce evaluation variability. For example, to examine the pathological conditions of juvenile Batten disease, the University of Rochester Batten Center (URBC) performs both remote examinations (Kempf et al. 2018; Cialone et al. 2011).

6.4 Treatment Options for Genetic Disorders

Improved mechanistic understanding of the defects linked to rare disorders offers a special chance for the discovery of therapies. The treatment of rare genetic disorders includes small-molecule drugs and biologics such as protein replacement therapy, oligonucleotide therapy, antibody therapy, and cell therapy. Small molecules are generally the most commonly used pharmacological framework for diseases, and they remain appealing as medications due to their numerous routes of administration, regulated dosage, stability, scalability, and overall low cost. Despite long-held worries that the frequency at which small-molecule therapies enter the market is decreasing, modern testing methods and developments in structural biology, computerized screening, and synthetic chemistry facilitate finding and designing unique biologically active compounds (Scannell et al. 2012). Since presently licensed medications target less than 700 of an estimated 3000 disease-associated proteins encoded in the human genome, there is also an enormous opportunity to enhance the understanding of the understudied genes as therapeutic targets (Santos et al. 2017; Rodgers et al. 2018). The process of small-molecule drug discovery often contains a screening step to test the potential of those molecules to fit with the previously determined protein objective. Moreover, further investigations of linked molecular pathways can detect a proper targetable protein for small-molecule drugs other than the product of defective DNA if it is not druggable (Schreiber 2000).

6.5 Identification of Target Molecules

The target identification step determines a gene, protein, receptor, or enzyme (therapeutic agent), which performs a crucial function in the disorder. The perfect targets have a range of characteristics; the pivotal one is to be druggable, which means that the suspected drug molecule can access and bind to that target, and thus a biological response is quantified both in vitro and in vivo. Following that, researchers report the therapeutic properties of the target.

Conventional genetic analysis of human subgroups with rare genetic conditions was effective in identifying molecular targets in specific cases. For instance, Bartter's syndromes are rare genetic conditions that cause defects in kidney functions and are associated with metabolic alkalosis, hypokalemia, hypotension, and other symptoms. Those syndromes are caused due to genetic variants in four genes: the class 1 SLC12A1 gene, the class 2 KCNJ1 gene, the class 3 C1CNKB gene, and the class 4 BSND gene (Gamba 2005).

With the availability of biological databases and the development of molecular sciences, the druggable genetic targets in the human genome are identified successfully; thus, molecular target-based drug development has emerged as the preferred method (Eder et al. 2014). Entire genome or exome sequencing provides important prospects for determining the specific source of rare disorders. Direct biochemical approaches, genetic interactions, or computational inference can all be used to identify targets. Furthermore, in several situations, more than one approach could be done to adequately describe on-target and off-target impacts and explain pathways of small-molecule activity (Schenone et al. 2013). Biochemical approaches include identifying the possible target molecules or the small molecules using a suitable label, incubating the two entities, and detecting binding directly, generally after a wash phase (Burdine and Kodadek 2004). Genetic manipulation could also be employed to find proteins of interest by altering suspected targets in cells based on the concept of genetic modifiers (enhancers or suppressors), hence affecting small-molecule sensitivity (Zheng et al. 2004). Target assumptions can be developed using computational inference, which evaluates small-molecule effects to those of known reference compounds or genetic alterations (Weinstein et al. 1997; Hughes et al. 2000; Young et al. 2008). Once a disorder target has been identified, it must be thoroughly validated, which means emphasizing that this target is the best for the corresponding disorder, through a range of methods from applying in vitro techniques to the use of whole-animal models. Although each strategy is acceptable in its own terms, a multi-validation strategy considerably increases confidence in the observed conclusion.

6.6 Bioassay Development

Assay development is a critical stepin of the drug discovery and development pipeline. After the disorder target has been identified and validated, a bioassay should be carried out to measure the impact of drug candidates on different scopes, biochemically, molecularly, and cellularly, and to evaluate their therapeutic activities. In vitro tests have become common techniques for screening bioactive candidates with the introduction of molecular biology protocols, recombinant proteins, and modified cell lines that express a specific protein or a reporter system. Two prime approaches are used in the step of compound screening; the first is based on culturing stable mammalian cell lines that are characterized by overexpressing the preidentified target, and the second relies on the purification of overexpressed recombinant protein, which is used later in a biochemical assay. Comparatively, nowadays, more studies are directed towards the utilization of cell-based methods (Dunne et al. 2009). Cell-based protocols have been used to study target categories such as ion channels, membrane receptors, and nuclear receptors, and they typically produce a functional report as a result of drug candidate biological action (Michelini et al. 2010). For instance, screening assays at G-protein-coupled receptors (GPCRs), like GTPγS binding assays, have been designed to assess the binding affinity of a radiolabeled or fluorescent-labeled ligand to the receptor for evaluating the change in guanine nucleotide at the G-protein level (DeLapp et al. 2012). Moreover, biochemical assays which have been applied to enzyme and receptor targets frequently quantify the compound's affinity for the target protein. Both methods of testing have been effectively employed to determine hit and candidate compounds. Furthermore, the comparative advantages of biochemical and cell-based tests have been widely argued and addressed previously (Moore and Rees 2001). An animate example is the ELISA assay, which is used to detect protein-protein interactions (PPIs) that are essential in cell signal transductions; they play a significant role in both normal and pathological cell functioning (Arkin et al. 2012).

Determination of the used assay is based on a range of factors such as the biological nature of the drug target protein, the available facilities in the laboratory, the expertise of the scientists in that laboratory, whether the target molecule is an inhibitor or activator, and whether it is large-scale or small-scale screening. Moreover, the chosen protocol should follow some considerations like the assay pharmacological relevance: research should be conducted using known ligands with activity at the target under investigation, if available, to evaluate if the test pharmacology is indicative of disorder status and to demonstrate that the test is able to detect drug candidates with both the necessary strength and action mechanisms (Hughes et al. 2011); assay reproducibility: it is a prerequisite in a compound screening setting that the assay is repeatable across assay plates, screen days, and, in a research project that may continue for many years, over the whole drug development project (Hughes et al. 2011); assay expense: whether the screening process is conducted on the small scale with a small number of compounds or on an industrial large scale, used reagents and volumes are chosen to keep assay expenses to a minimum; assay sensitivity to chemicals: organic solvents such as dimethyl sulfoxide (DMSO) or ethanol are commonly used to dissolve chemicals; therefore, assays must be designed to be insensitive to the amounts of solvents utilized in the experiment (Hughes et al. 2011); and assay quality: this is commonly assessed by the screening window coefficient known as Z' factor, which is a statistical factor that refers to the assay robustness and estimated from the sample means and sample standard deviations (Z'' = 1-(3*SD(total signal) + 3*SD(basal signal))/(Total signal—Basal signal)); the Z factor has a value between 0 and 1 (Inglese et al. 2007; Zhang et al. 1999); an assay with a Z factor more than 0.4 is regarded as sufficiently powerful for drug screening, while many research teams prefer to work with assays with a Z factor larger than 0.6 (Hughes et al. 2011); furthermore, pharmacological controls within each experiment also serve to verify assay quality, which means that several precautions should be considered; the reference compounds' pharmacology is within set limitations, and conducting simple protocols with fewer steps, stable reagents, few washing steps or transferring from plate to plate, and optimal use for instruments in each step (Hughes et al. 2011).

6.7 Compound Library

A typical step in the discovery and development of drug candidates is the construction of a compound library; it is a pool of approximately 10^6–10^7 lead candidates, in a physical form, collected to be tested against the disease target molecule in vitro, and then their biological activity is evaluated. A variety of chemical libraries in massive pharmaceutical corporations were created from internally synthesized compounds; therefore, those collections are relatively limited. Recently, the volume and variety of available compound libraries in the industry have increased. Initial screening of 500,000–3,000,000 candidates for lead identification for one therapeutic target has become common practice in both the pharmaceutical industry and academic research. After the fluctuation of combinatorial chemistry (Kodadek 2011), researchers in medicinal chemistry have been progressively directed toward creating virtual candidate libraries, which means saving the chemical structures of the lead candidates on the computers instead of storing many reagents in laboratories. Before the synthesis and experimental testing, virtual screening may be done with such virtual compound libraries to reduce the number of compounds that are considered prospective leads for less time, effort, and money consumption (van Hilten et al. 2019). There are static libraries containing all the distinct virtual biomolecules that can be within a given group of boundaries. The chemical universe database GDB-17 is now the most powerful example, with 116 billion virtual compounds containing up to 17 atoms of C, N, O, S, and halogens within the basic chemical principles (Reymond 2015; Ruddigkeit et al. 2012; van Hilten et al. 2019). On the other hand, dynamic virtual libraries provide more effective virtual screening as full enumeration is not necessary. Frequently, ensuring synthetic possibility is the focus; there are two methods to do this: first, by utilizing the building blocks that are produced when existing molecules are retro-synthesized using powerful chemical reactions, and second, through the simulation of well-known chemistry employed in combinatorial libraries, which is frequently followed in pharmaceutical firms (van Hilten et al. 2019).

6.8 Virtual Screening

Virtual screening (VS) has developed as a useful computational tool in drug development for screening huge libraries of small molecules for novel hits with desirable features, which can subsequently be verified experimentally. Virtual screening's objective, like that of other computational techniques, is not to replace in vitro or in vivo experiments, but to quicken the investigation process, minimize the number of compounds to be examined practically, and justify their selection. Furthermore, in order to reduce consumed time, budget, effort, and reagents, virtual screening has become particularly desirable in pharmaceutical industries and academic institutions. There are several approaches for virtual screening; quantitative structure-activity relationship (QSAR) analysis, a ligand-based drug design tool created by Hansch and Fujita (1964), is the most potent one because of its high and rapid throughput and excellent hit rate (Neves et al. 2018). QSAR is an effective approach used to create mathematical models that use regression and classification methods, respectively, to find a statistically significant correlation between the molecular structure and a constant (pIC50, pEC50, Ki, etc.) or categorical (active, inactive, toxic, nontoxic, etc.) biological/toxicological property (Cherkasov et al. 2014; Neves et al. 2018). The first stage in developing a QSAR model is to obtain pertinent chemogenomic data from databases and the literature, and then computing the chemical descriptors on several levels of molecular structure representation. The moment that it is developed and validated, QSAR models are utilized to portend the biological characteristics of unique compounds. Although experimental testing of computational hits is not a requirement of the QSAR approach, it is extremely desirable and ought to be undertaken as the last validation of established systems (Neves et al. 2018).

6.9 Experimental Screening

High-throughput screening (HTS) is a quintessential stage in the process of drug discovery and development in which biochemical or cellular events are evaluated against drug candidates hundreds of thousands of times in a reproducible and quick manner. Automated high-throughput screening systems typically include incubators for controlling temperature, humidity, and gas flow rate; liquid handlers for dispensing cells and chemicals; sensitive plate readers for monitoring test outcomes; a mechanical arm system; and a software application to connect all these elements. Robotic screens were first created and employed in pharmaceutical businesses but have lately been adopted by academic researchers at universities and research organizations. A throughput of 500,000 to 1 million wells per day may be obtained with the automated screening system. HTS aims to quickly find active chemicals that influence a certain target, pathway, or biochemical or cellular event. The results of an HTS experiment serve as the foundation for pharmaceutical research and development, resulting in lead compounds with adequate physicochemical characteristics for therapeutic significance (Avery et al. 2010). HTS experiments are often done in 96-, 384-, or 1536-well microtiter plates, whereas classical HTS analyzes each drug in a compound library at a given concentration, usually 10 μM. On the other hand, a quantitative high-throughput screening (qHTS) platform is used to examine chemicals at different concentrations and the concentration-response curves for each screened drug are created immediately. qHTS has recently gained prominence in toxicology since it defines the biological effects of compounds more thoroughly and reduces the rates of misleading results (Attene-Ramos et al. 2014). The screening results are entered into a database and examined with informatics tools. The elementary screening hits are usually picked based on some parameters; for instance, for a single concentration screening, the inhibition should be greater than 50%, or for qHTS, the inhibitory concentration of 50% response (IC50) is less than 5 μM and the efficacy is greater than 70% (Inglese et al. 2006; Sun et al. 2017). The primary hits include both real positives and false positives, such as autofluorescent and other nonspecific chemicals. In the hit confirmation step, these false-positive and nonspecific chemicals must be identified and deleted.

6.10 Hit Confirmation

Hits can come from a variety of routes, but a real candidate of the predicted hit compound should show bioactivity in a primary biological experiment. Nevertheless, it is critical to demonstrate that the detected activity is a result of the anticipated machine and is not due to artifacts. Both secondary and tertiary tests are performed to verify the selectivity and activity of the hit compounds. Frequent or promiscuous hits are subsequently excluded from additional consideration. Employing a counter-screening assay, where hits are assessed for their activity against a different member of the target family under the same experimental conditions, is one method of identifying false positives, and if the hit exhibits similar activity, it is most certainly a false positive. Moreover, small-molecule precipitate and aggregation formation might result in false-positive results, and a counter screen helps to remove them from further processing (Keserű and Makara 2006). In order to get rid of various misleading signals, it has been suggested that color quenching corrections in scintillation proximity assays (SPAs) (Park et al. 1999), particle count measurements for varying sample concentrations, and two-dimensional fluorescence intensity distribution analysis for fluorescence interference be performed (Heyse et al. 2005).

Identification of substances that might induce activity based on harmful pathways must also be taken into account because they could result in toxicity. Therefore, toxic chemicals are often removed using a cytotoxicity test that uses the same concentrations and incubation period as the primary screening. Throughout the hit confirmation stage, medicinal chemists use biological data to order and classify hits into groups, gaining preliminary insight into structure-activity relationships (SARs) among group members. Furthermore, they also scrutinize the feasibility of chemical synthesis. Despite all this investigation, only a few lead compounds have been identified and given priority.

6.11 Lead Generation and Optimization

Lead generation, also known as the hit-to-lead stage, includes optimizing selected hits from a broad series to develop leads. For lead generation, three to five chemical groups are usually selected, and analogous chemicals are assessed to produce a quantitative SAR toward target selectivity, activity, physicochemical features, pharmacokinetics, ADME (absorption, distribution, metabolism, and excretion) aspects, and toxicity potential. During this step, following combinatorial chemistry, which is described as the systematic and repetitive covalent connection of a set of different building blocks of various structures to one another to yield a large array of diverse molecular entities (Pandeya 2005), the synthesis of leads is performed where small molecules or peptides are both generated efficiently in vast quantities; in some cases, up to a million compounds simultaneously, and sometimes up to a million chemicals, can be processed at once. Lead identification flow involves several steps. In the early stages of lead generation, it is wise to evaluate a compound's potential for toxicity first; this is achieved using various in vitro tests based on human cell lines; cytotoxicity experiments to examine how chemicals affect cell viability (Riss et al. 2011); hepatoxicity test by employing many platforms, including cell lines of hepatic cells and suspensions of isolated liver cells, liver slices, or subcellular fractions (Gómez-Lechón et al. 2010); human ether-a-go-go-related gene (hERG) inhibition experiment to determine the potential for QT interval prolonging of the drugs under study utilizing hERG-overexpressing cell lines (Pollard et al. 2008); and in vitro micronucleus assay to evaluate the possibility of genotoxicities such as clastogenic activity, which means structural abnormality in chromosomes, and aneugenic activity, which means numerical chromosome abnormality (Kirsch-Volders et al. 2011). The next steps include the estimation of the specificity of compounds for the target by cell-free assay, assessment of the activity of compounds in the target's known animal orthologs since the substances' effectiveness must be tested in experimental animals in addition to cell-based or in vitro functional tests, and investigation of the physicochemical attributes of exemplary compounds from the collection to verify the compounds' drug-likeness for example; the route of drug administration that is preferred the most is the oral method; therefore, lead is more likely to have high membrane permeability and can be absorbed inside the body (Leeson 2012); solubility, in in vitro and in vivo levels, is required to be greater than

60 μg/mL in lead compounds (Kerns et al. 2008). Moreover, evaluation of ADME profiling includes assessing permeability by modeling intestinal absorption using, for instance, the colon cancer cell line (van Breemen and Li 2005), metabolic stability by human liver microsomes (Baranczewski et al. 2006), induction and inhibition of cytochrome P450 to determine if the lead can affect the metabolism of medications that are taken at the same time (Yan and Caldwell 2001), and plasma protein binding which affects drug distribution and pharmacological activity in general (Lambrinidis et al. 2015).

Lead optimization aims to provide preclinical development candidates by modifying the lead structure chemically to address its flaws. In general, the goal is to improve the physicochemical and ADME qualities and reduce the toxicity risks in order to find a drug that can be harmless and have good pharmacokinetics. It is crucial to show a clear relationship between a lead concentration in plasma and its pharmacodynamic effects; this information might then be used to foretell the dosage plan for the drug. Preclinical drug development is then initiated with the optimized lead molecule.

6.12 Preclinical Studies

Preclinical studies are typically carried out by physicians and scientists in different disciplines: metabolism experts, toxicologists, chemists, pharmacologists, process chemists, formulation, and regulatory specialists. In most cases, researchers work on a few optimized leads (7–10) before choosing a candidate for preclinical studies. Candidates for preclinical studies must typically fulfill some parameters: bioanalytically validated convenient pharmacokinetics, confirmed in vivo activity, sufficient safety margin and dose range, acceptable drug interaction record, and scale-up feasibility (Strovel et al. 2016). Once the candidate is selected, preclinical studies, which serve as a link between drug development and the start of clinical studies, are performed via diverse practices: active pharmaceutical ingredient (API) manufacture, dose design and formulation studies, analytical and bioanalytical techniques, pharmacokinetics and metabolism investigations, good manufacturing practice (GMP) manufacture, and safety pharmacology and toxicology studies (Steinmetz and Spack 2009). These activities guarantee the establishment of pharmacological activity in experimental animal models in order to define pharmacokinetics, ensure the lead compounds' therapeutic safety, and provide the foundation for an Investigational New Drug (IND) application to the US Food and Drug Administration (FDA) for approval to begin human clinical trials. Therefore, identifying and developing high-quality leads is crucial for enhancing drug development success rates in later phases of clinical trials.

6.13 Drug Pharmacokinetics

Pharmacokinetics (PK) means what the body does to the drug and describes the absorption, distribution, metabolism, and excretion (ADME) of leads into, through, and out of the body. A deep understanding of prospective drug pharmacokinetics is essential to deciding the dosage levels and frequency of administration for safety pharmacology and toxicology investigations, as well as to aid in the interpretation of toxicology study data for estimating the first-in-man dose (Singh 2018a). This is done through predictive ADME techniques that involve in silico models, physiochemical parameters, and in vitro studies of permeation and drug metabolism. Following an evaluation of these screening findings in conjunction with effectiveness results, drugs projected to have desirable PK characteristics are explored further utilizing in vivo animal models.

6.14 Drug Formulation

Drug formulation is the process of combining active pharmaceutical ingredients with other chemical constituents to make a drug product. The administration route should be clearly specified at this time. Drugs can be given enterally (oral, buccal, and rectal), parenterally by injectable methods (intravenous, intramuscular, and subcutaneous), topically, and by inhalation. Orally delivered drugs might be in the form of a tablet, capsule, suspension, or solution. When developing a drug product formulation, several aspects must be addressed; the formulation's components must be physically and chemically compatible with the active pharmaceutical ingredient, and capsule size, flavor, stability and best before date, and convenience of administration are all important considerations. Dissociation enhancers, which have been shown to increase the bioavailability of the active DS, are one type of component that formulas may include. In addition, in order to enhance swallowing, conceal an undesirable taste, safeguard chemicals during storage, enhance appearance, manage medication release over time, or direct it to certain areas of the gastrointestinal system, coatings may be applied to solid formulations, especially tablets (Steinmetz and Spack 2009).

6.15 Human Clinical Trials

The process of clinical study of a drug is focused on determining the indication, dosage range, and schedule at which the drug is safe and effective for defined use in a patient group; therefore, a clinical development strategy is designed to specify a set of clinical trials. A phase I trial is generally done on 10–100 healthy normal volunteers non-blindly to examine the safety and tolerability of a drug. The maximum tolerated dosage is defined, and pharmacokinetic and pharmacodynamic features are tracked. The phase II trial is the initial study that evaluates the drug's clinical efficacy; as a result, it is conducted on patients (50–500). This trial may have several goals, including setting the dosage schedule and examining the dose-response relationship, and as in phase I, the safety evaluation continues. Due to ineffectiveness or safety concerns, most clinical candidates get unsuccessful at this stage. Phase III studies, which typically involve a few hundred to a few thousand patients, validate the efficiency of the prospective drug in a broader population. The trial is carried out at many locations and contrasts the investigational drug with the most effective current therapy or the accepted standard of care for that particular condition. In order to identify less frequent adverse effects, safety is further evaluated in a wider pool. Phase III is typically randomized, controlled, double-blind, and considered the priciest and trickiest trial. If positive outcomes are attained, all available data is put into a document, and a New Drug Application is submitted for licensing the drug. Phase IV trials are basically to evaluate the drug in a real situation for extra follow-up investigations, which start after the drug is sold in order to find unusual or long-term negative impacts in a much broader population. This phase has a high influence including changing a drug's label, its warnings about interactions and contraindications, or even removing a product from the market (Singh 2018b).

6.16 Drug Registration and Pharmacovigilance

According to the U.S. Food and Drug Administration (FDA), if researchers have proof from preclinical and clinical studies that a drug is safe and effective for its particular purpose, the corporation can submit an application to commercialize the drug. The FDA review committee carefully goes over all of the drug's data that has been presented before deciding whether or not to approve it. A New Drug Application (NDA) tells the full story of a drug from preclinical to phase III clinical study results. Its purpose is to demonstrate drug safety and effectiveness for its intended use in the population studied. A determination on whether to approve the drug must be made by the review panel within 6–10 months in which every participant of the review board thoroughly examines the area of the application that relates to him or her. For instance, a pharmacologist evaluates the data from animal studies, while the medical officer and statistician review the clinical data. In addition, FDA investigators perform routine inspections at clinical trial sites. The organization searches for proof of data falsification, manipulation, or withholding. All evaluations and documents are compiled into an "action package" by the program manager, and it is used as the official record for FDA review; then the review committee offers a recommendation, and the final decision is made by the senior FDA official. It is required to produce and improve prescription information when the FDA decides that the drug has been demonstrated to be safe and effective for its intended purpose. This is known as "labeling"; the label precisely and objectively states the foundation for approval as well as the optimal way to utilize the drug. However, many times, unresolved concerns must be addressed before the drug may be licensed for commercialization. The FDA may ask the applicant to respond to inquiries based on existing data or to demand further research. At this stage, the applicant can decide whether to continue the work, and there are procedures for filing a formal appeal in the event that a developer disagrees with an FDA decision. Once the drug is approved and marketed, pharmacovigilance procedures are established. The World Health Organization (WHO) defines pharmacovigilance as the science and actions involved in the identification, evaluation, comprehension, and mitigation of side effects or any other drug-related issues (WHO 2002). Pharmacovigilance's overarching objective is to guarantee safe medication usage by reducing medication-related risks and optimizing benefits. Common adverse events will be covered by data collected during clinical trials; however, it is possible to miss unusual adverse occurrences. Therefore, it is crucial to keep an eye on safety during the post-approval time and during the whole lifetime of the drug to determine its true risk-benefit profile and take the appropriate precautions to reduce hazards (Jalali 2018).

6.17 Drug Repositioning

Drug repositioning involves finding and using already approved medications to treat new conditions. The drug repurposing target is to collect some approved pharmaceuticals to test for novel functions quickly. The method is similar to that of unique compound screening, with the exception that instead of a broad and diverse collection of compounds, approved medication libraries are used (Sun et al. 2016). Consequently, drug repositioning screening has become a successful alternative strategy for the quick discovery of novel drugs for rare disorders. Using FDA-approved drugs and clinical candidates for drug repositioning has significant advantages over the traditional new drug development approach. These approved drugs have been tested on people, and their toxicity and safety are usually well known (Huang et al. 2011).

6.18 Biologics

Biologics have increased in relevance as an effective therapy for a variety of disorders in recent years (Kinch 2015). They come in different forms: vaccines, blood, tissues, cells, nucleic acids, enzymes, peptides, and antibodies. The discovery and development of biologics differ from those of small-molecule drugs. Compared to small-molecule drugs, which are chemically manufactured and have defined structures, most biologics are complex products with no specified structures. Through modern biological technologies, they may be extracted from a variety of living origins, such as human tissues, animals, and microbes. Biologics have the potential to produce the most effective therapies for rare disorders that lack licensed medications. Some of these methods entail replacing, modifying, or introducing new versions of genes, proteins, or cells into the body of the patient.

For instance, protein defects may be the cause of a number of rare disorders; therefore, administrating the intact form of those proteins, extracted from animals or synthesized recombinantly, to patients is a method for their treatment, which is known as protein replacement therapy. Those recombinant proteins are produced in the human cell cultures to guarantee that posttranslational modifications occur appropriately. A prime example is the process of glycosylation that is required following protein translation of lysosomal enzyme produced for lysosomal storage disease patients; this process is necessary for the binding of the lysosomal enzyme to particular cell surface receptors, as the cells will not receive it if the binding is improper (Desnick and Schuchman 2012). There are other biological therapeutic approaches: oligonucleotide therapy which is designed to target disease-associated genes by interfering at the level of RNA using some techniques (Setten et al. 2019), cell therapy which involves transplanting cells produced from iPS cells like retinal cells for eye diseases and chimeric antigen receptor T cells for rare cancers, and antibody therapy, in which monoclonal antibodies control signaling pathways, enroll cells or proteins to target sites, deliver cytotoxins, or neutralize or modulate circulating factors (Tambuyzer et al. 2020).

6.19 Future Prospective

Clustered regularly interspaced short palindromic repeats associated with nuclease 9 (CRISPR-Cas9) systems are convenient genomic and epigenomic editing methods. Nowadays, they can play a role in scientific research on genetic diseases as they have been utilized in several genomic screen investigations. CRISPR screenings allow for high-throughput analysis of gene roles in health and disease conditions. Unlike traditional RNAi screening, CRISPR screening has fewer off-target effects; can be applied in several forms like knockout, knockdown, and activation screens; and could target coding and noncoding sequences of the genome. Soon, this powerful screen technology will have the potential to revolutionize functional genomic investigations and genetic studies (Xue et al. 2016).

6.20 Conclusion

In conclusion, this chapter has manifested an explanation for the process of drug discovery and development with a focus on orphan diseases. Although research work on small-molecule drugs is currently the primary approach for developing medications for rare genetic diseases, biologics, including protein replacement, oligonucleotide, antibody, and cell therapy are anticipated to result in significant advances in treatment. The continuous advancement of researchers' understanding of the causes and mechanisms of rare diseases, in addition to rapid progressions in advanced biological technology, will lead us to discovering required therapies.

References

Arkin MR, Glicksman MA, Fu H, Havel JJ, Du Y (2012) Inhibition of protein-protein interactions: non-cellular assay formats. In: Assay guidance manual. Eli Lilly & Company and the National Center for Advancing Translational Sciences, Bethesda, MD

Attene-Ramos MS, Austin CP, Xia M (2014) High throughput screening

Avery VM, Camp D, Carroll AR, Jenkins ID, Quinn RJ (2010) The identification of bioactive natural products by high throughput screening (HTS)

Baranczewski P, Stanczak A, Sundberg K, Svensson R, Wallin A, Jansson J, Garberg P, Postlind H (2006) Introduction to in vitro estimation of metabolic stability and drug interactions of new chemical entities in drug discovery and development. Pharmacol Rep 58(4):453

Bhattacharjee S, Nandi S (2018) Rare genetic diseases with defects in DNA repair: opportunities and challenges in orphan drug development for targeted cancer therapy. Cancers 10(9):298. https://doi.org/10.3390/cancers10090298

Boycott KM, Vanstone MR, Bulman DE, MacKenzie AE (2013) Rare-disease genetics in the era of next-generation sequencing: discovery to translation. Nat Rev Genet 14(10):681–691. https://doi.org/10.1038/nrg3555

Burdine L, Kodadek T (2004) Target identification in chemical genetics: the (often) missing link. Chem Biol 11(5):593–597

Cherkasov A, Muratov EN, Fourches D, Varnek A, Baskin II, Cronin M, Dearden J, Gramatica P, Martin YC, Todeschini R, Consonni V, Kuz'min VE, Cramer R, Benigni R, Yang C, Rathman J, Terfloth L, Gasteiger J, Richard A, Tropsha A (2014) QSAR modeling: where have you been? Where are you going to? J Med Chem 57(12):4977–5010. https://doi.org/10.1021/jm4004285

Cialone J, Augustine EF, Newhouse N, Vierhile A, Marshall FJ, Mink JW (2011) Quantitative telemedicine ratings in batten disease: implications for rare disease research. Neurology 77(20):1808–1811

Comstra HS, McArthy J, Rudin-Rush S, Hartwig C, Gokhale A, Zlatic SA, Blackburn JB, Werner E, Petris M, D'Souza P, Panuwet P, Barr DB, Lupashin V, Vrailas-Mortimer A, Faundez V (2017) The interactome of the copper transporter ATP7A belongs to a network of neurodevelopmental and neurodegeneration factors. eLife 6:e24722. https://doi.org/10.7554/eLife.24722

DeLapp NW, Gough WH, Kahl SD, Porter AC, Wiernicki TR (2012) GTPγS binding assays. Eli Lilly & Company and the National Center for Advancing Translational Sciences

Desnick RJ, Schuchman EH (2012) Enzyme replacement therapy for lysosomal diseases: lessons from 20 years of experience and remaining challenges. Annu Rev Genomics Hum Genet 13(1):307–335. https://doi.org/10.1146/annurev-genom-090711-163739

Dunne A, Jowett M, Rees S (2009) Use of primary human cells in high-throughput screens. Methods Mol Biol 565:239–257

Dunoyer M (2011) Accelerating access to treatments for rare diseases. Nat Rev Drug Discov 10(7):475–476. https://doi.org/10.1038/nrd3493

Eder J, Sedrani R, Wiesmann C (2014) The discovery of first-in-class drugs: origins and evolution. Nat Rev Drug Discov 13(8):577–587. https://doi.org/10.1038/nrd4336

Gamba G (2005) Molecular physiology and pathophysiology of electroneutral cation-chloride cotransporters. Physiol Rev 85(2):423–493

Gokhale A, Larimore J, Werner E, So L, Moreno-De-Luca A, Lese-Martin C, Lupashin VV, Smith Y, Faundez V (2012a) Quantitative proteomic and genetic analyses of the schizophrenia susceptibility factor dysbindin identify novel roles of the biogenesis of lysosome-related organelles complex 1. J Neurosci 32(11):3697–3711

Gokhale A, Perez-Cornejo P, Duran C, Hartzell HC, Faundez V (2012b) A comprehensive strategy to identify stoichiometric membrane protein interactomes. Cell Logist 2(4):189–196. https://doi.org/10.4161/cl.22717

Gokhale A, Hartwig C, Freeman AA, Bassell JL, Zlatic SA, Savas CS, Vadlamudi T, Abudulai F, Pham TT, Crocker A (2019) Systems analysis of the 22q11. 2 microdeletion syndrome converges on a mitochondrial interactome necessary for synapse function and behavior. J Neurosci 39(18):3561–3581

Gómez-Lechón MJ, Castell JV, Donato MT (2010) The use of hepatocytes to investigate drug toxicity. Methods Mol Biol 640:389–415

Graham JM Jr, Shaw GM (2005) Gene-environment interactions in rare diseases that include common birth defects. Birth Defects Res A Clin Mol Teratol 73(11):865–867. https://doi.org/10.1002/bdra.20193

Haendel M, Vasilevsky N, Unni D, Bologa C, Harris N, Rehm H, Hamosh A, Baynam G, Groza T, McMurry J, Dawkins H, Rath A, Thaxon C, Bocci G, Joachimiak MP, Köhler S, Robinson PN, Mungall C, Oprea TI (2020) How many rare diseases are there? Nat Rev Drug Discov 19(2):77–78. https://doi.org/10.1038/d41573-019-00180-y

Hansch C, Fujita T (1964) p-σ-π analysis. A method for the correlation of biological activity and chemical structure. J Am Chem Soc 86(8):1616–1626

Heyse S, Brodte A, Bruttger O, Duerr O, Freeman T, Jung T, Lindemann M, Ottl J, Rinn B (2005) Quantifying bioactivity on a large scale: quality assurance and analysis of multiparametric ultra-HTS data. J Assoc Lab Autom 10(4):207–212. https://doi.org/10.1016/j.jala.2005.05.003

Huang R, Southall N, Wang Y, Yasgar A, Shinn P, Jadhav A, Nguyen D-T, Austin CP (2011) The NCGC pharmaceutical collection: a comprehensive resource of clinically approved drugs enabling repurposing and chemical genomics. Sci Transl Med 3(80):80ps16. https://doi.org/10.1126/scitranslmed.3001862

Hughes TR, Marton MJ, Jones AR, Roberts CJ, Stoughton R, Armour CD, Bennett HA, Coffey E, Dai H, He YD, Kidd MJ, King AM, Meyer MR, Slade D, Lum PY, Stepaniants SB, Shoemaker DD, Gachotte D, Chakraburtty K, Simon J, Bard M, Friend SH (2000) Functional discovery via a compendium of expression profiles. Cell 102(1):109–126. https://doi.org/10.1016/S0092-8674(00)00015-5

Hughes J, Rees S, Kalindjian S, Philpott K (2011) Principles of early drug discovery. Br J Pharmacol 162(6):1239–1249. https://doi.org/10.1111/j.1476-5381.2010.01127.x

Inglese J, Auld DS, Jadhav A, Johnson RL, Simeonov A, Yasgar A, Zheng W, Austin CP (2006) Quantitative high-throughput screening: a titration-based approach that efficiently identifies biological activities in large chemical libraries. Proc Natl Acad Sci 103(31):11473–11478. https://doi.org/10.1073/pnas.0604348103

Inglese J, Johnson RL, Simeonov A, Xia M, Zheng W, Austin CP, Auld DS (2007) High-throughput screening assays for the identification of chemical probes. Nat Chem Biol 3(8):466–479. https://doi.org/10.1038/nchembio.2007.17

Jalali RK (2018) Chapter 26: Pharmacovigilance and drug safety. In: Vohora D, Singh G (eds) Pharmaceutical medicine and translational clinical research. Academic Press, Boston, pp 403–406

Kempf L, Goldsmith JC, Temple R (2018) Challenges of developing and conducting clinical trials in rare disorders. Am J Med Genet A 176(4):773–783. https://doi.org/10.1002/ajmg.a.38413

Kerns EH, Di L, Carter GT (2008) In vitro solubility assays in drug discovery. Curr Drug Metab 9(9):879–885. https://doi.org/10.2174/138920008786485100

Keserű GM, Makara GM (2006) Hit discovery and hit-to-lead approaches. Drug Discov Today 11(15):741–748. https://doi.org/10.1016/j.drudis.2006.06.016

Kim YS, Leventhal BL (2015) Genetic epidemiology and insights into interactive genetic and environmental effects in autism Spectrum disorders. Biol Psychiatry 77(1):66–74. https://doi.org/10.1016/j.biopsych.2014.11.001

Kinch MS (2015) An overview of FDA-approved biologics medicines. Drug Discov Today 20(4):393–398. https://doi.org/10.1016/j.drudis.2014.09.003

Kirsch-Volders M, Plas G, Elhajouji A, Lukamowicz M, Gonzalez L, Vande Loock K, Decordier I (2011) The in vitro MN assay in 2011: origin and fate, biological significance, protocols, high throughput methodologies and toxicological relevance. Arch Toxicol 85(8):873–899. https://doi.org/10.1007/s00204-011-0691-4

Knowles MR, Durie PR (2002) What is cystic fibrosis? https://doi.org/10.1056/NEJMe020070. https://www.nejm.org/doi/pdf/10.1056/nejme020070. Accessed 12 Mar 2023

Kodadek T (2011) The rise, fall and reinvention of combinatorial chemistry. Chem Commun 47(35):9757–9763. https://doi.org/10.1039/C1CC12102B

Lambrinidis G, Vallianatou T, Tsantili-Kakoulidou A (2015) In vitro, in silico and integrated strategies for the estimation of plasma protein binding. A review. Adv Drug Deliv Rev 86:27–45. https://doi.org/10.1016/j.addr.2015.03.011

Leeson P (2012) Chemical beauty contest. Nature 481(7382):455–456. https://doi.org/10.1038/481455a

Michelini E, Cevenini L, Mezzanotte L, Coppa A, Roda A (2010) Cell-based assays: fuelling drug discovery. Anal Bioanal Chem 398(1):227–238. https://doi.org/10.1007/s00216-010-3933-z

Moore K, Rees S (2001) Cell-based versus isolated target screening: how lucky do you feel? J Biomol Screen 6(2):69–74. https://doi.org/10.1177/108705710100600202

Neves BJ, Braga RC, Melo-Filho CC, Moreira-Filho JT, Muratov EN, Andrade CH (2018) QSAR-based virtual screening: advances and applications in drug discovery. Front Pharmacol 9

Nguengang Wakap S, Lambert DM, Olry A, Rodwell C, Gueydan C, Lanneau V, Murphy D, Le Cam Y, Rath A (2020) Estimating cumulative point prevalence of rare diseases: analysis of the Orphanet database. Eur J Hum Genet 28(2):165–173. https://doi.org/10.1038/s41431-019-0508-0

Pandeya SN (2005) Combinatorial chemistry: a novel method in drug discovery and its application. IJC-B Vol44B(02)

Park Y-W, Cummings RT, Wu L, Zheng S, Cameron PM, Woods A, Zaller DM, Marcy AI, Hermes JD (1999) Homogeneous proximity tyrosine kinase assays: scintillation proximity assay versus homogeneous time-resolved fluorescence. Anal Biochem 269(1):94–104. https://doi.org/10.1006/abio.1999.4029

Pollard CE, Valentin J-P, Hammond TG (2008) Strategies to reduce the risk of drug-induced QT interval prolongation: a pharmaceutical company perspective. Br J Pharmacol 154(7):1538–1543. https://doi.org/10.1038/bjp.2008.203

Rafeeq MM, Murad HAS (2017) Cystic fibrosis: current therapeutic targets and future approaches. J Transl Med 15(1):84. https://doi.org/10.1186/s12967-017-1193-9

Reymond J-L (2015) The chemical space project. Acc Chem Res 48(3):722–730. https://doi.org/10.1021/ar500432k

Riss TL, Moravec RA, Niles AL (2011) Cytotoxicity testing: measuring viable cells, dead cells, and detecting mechanism of cell death. In: Stoddart MJ (ed) Mammalian cell viability: methods and protocols. Humana Press, Totowa, NJ, pp 103–114

Rodgers G, Austin C, Anderson J, Pawlyk A, Colvis C, Margolis R, Baker J (2018) Glimmers in illuminating the druggable genome. Nat Rev Drug Discov 17(5):301–302. https://doi.org/10.1038/nrd.2017.252

Royce PM, Steinmann B (2003) Connective tissue and its heritable disorders: molecular, genetic, and medical aspects. Wiley

Ruddigkeit L, van Deursen R, Blum LC, Reymond J-L (2012) Enumeration of 166 billion organic small molecules in the chemical universe database GDB-17. J Chem Inf Model 52(11):2864–2875. https://doi.org/10.1021/ci300415d

Runz H, Dolle D, Schlitter AM, Zschocke J (2008) NPC-db, a Niemann-pick type C disease gene variation database. Hum Mutat 29(3):345–350. https://doi.org/10.1002/humu.20636

Santos R, Ursu O, Gaulton A, Bento AP, Donadi RS, Bologa CG, Karlsson A, Al-Lazikani B, Hersey A, Oprea TI, Overington JP (2017) A comprehensive map of molecular drug targets. Nat Rev Drug Discov 16(1):19–34. https://doi.org/10.1038/nrd.2016.230

Scannell JW, Blanckley A, Boldon H, Warrington B (2012) Diagnosing the decline in pharmaceutical R&D efficiency. Nat Rev Drug Discov 11(3):191–200. https://doi.org/10.1038/nrd3681

Schenone M, Dančík V, Wagner BK, Clemons PA (2013) Target identification and mechanism of action in chemical biology and drug discovery. Nat Chem Biol 9(4):232–240. https://doi.org/10.1038/nchembio.1199

Schreiber SL (2000) Target-oriented and diversity-oriented organic synthesis in drug discovery. Science 287(5460):1964–1969. https://doi.org/10.1126/science.287.5460.1964

Setten RL, Rossi JJ, Han S (2019) The current state and future directions of RNAi-based therapeutics. Nat Rev Drug Discov 18(6):421–446. https://doi.org/10.1038/s41573-019-0017-4

Singh G (2018a) Chapter 4: Preclinical drug development. In: Vohora D, Singh G (eds) Pharmaceutical medicine and translational clinical research. Academic Press, Boston, pp 47–63

Singh G (2018b) Chapter 5: Target product profile and clinical development plan. In: Vohora D, Singh G (eds) Pharmaceutical medicine and translational clinical research. Academic Press, Boston, pp 65–80

Steinmetz KL, Spack EG (2009) The basics of preclinical drug development for neurodegenerative disease indications. BMC Neurol 9(1):S2. https://doi.org/10.1186/1471-2377-9-S1-S2

Strovel J, Sittampalam S, Coussens NP, Hughes M, Inglese J, Kurtz A, Andalibi A, Patton L, Austin C, Baltezor M, Beckloff M, Weingarten M, Weir S (2016) Early drug discovery and development guidelines: for academic researchers, collaborators, and start-up companies. Eli Lilly & Company and the National Center for Advancing Translational Sciences

Sun W, Sanderson PE, Zheng W (2016) Drug combination therapy increases successful drug repositioning. Drug Discov Today 21(7):1189–1195. https://doi.org/10.1016/j.drudis.2016.05.015

Sun W, Zheng W, Simeonov A (2017) Drug discovery and development for rare genetic disorders. Am J Med Genet A 173(9):2307–2322. https://doi.org/10.1002/ajmg.a.38326

Tambuyzer E, Vandendriessche B, Austin CP, Brooks PJ, Larsson K, Miller Needleman KI, Valentine J, Davies K, Groft SC, Preti R, Oprea TI, Prunotto M (2020) Therapies for rare diseases: therapeutic modalities, progress and challenges ahead. Nat Rev Drug Discov 19(2):93–111. https://doi.org/10.1038/s41573-019-0049-9

van Breemen RB, Li Y (2005) Caco-2 cell permeability assays to measure drug absorption. Expert Opin Drug Metab Toxicol 1(2):175–185. https://doi.org/10.1517/17425255.1.2.175

van Hilten N, Chevillard F, Kolb P (2019) Virtual compound libraries in computer-assisted drug discovery. J Chem Inf Model 59(2):644–651. https://doi.org/10.1021/acs.jcim.8b00737

Weinstein JN, Myers TG, O'Connor PM, Friend SH, Fornace AJ, Kohn KW, Fojo T, Bates SE, Rubinstein LV, Anderson NL, Buolamwini JK, van Osdol WW, Monks AP, Scudiero DA, Sausville EA, Zaharevitz DW, Bunow B, Viswanadhan VN, Johnson GS, Wittes RE, Paull KD (1997) An information-intensive approach to the molecular pharmacology of cancer. Science 275(5298):343–349. https://doi.org/10.1126/science.275.5298.343

World Health Organization (2002) Safety of medicines: a guide to detecting and reporting adverse drug reactions: why health professionals need to take action (No. WHO/EDM/QSM/2002.2). World Health Organization

Xue H-Y, Ji L-J, Gao A-M, Liu P, He J-D, Lu X-J (2016) CRISPR-Cas9 for medical genetic screens: applications and future perspectives. J Med Genet 53(2):91–97. https://doi.org/10.1136/jmedgenet-2015-103409

Yan Z, Caldwell GW (2001) Metabolism profiling, and cytochrome P450 inhibition & induction in drug discovery. Curr Top Med Chem 1(5):403–425. https://doi.org/10.2174/1568026013395001

Young DW, Bender A, Hoyt J, McWhinnie E, Chirn G-W, Tao CY, Tallarico JA, Labow M, Jenkins JL, Mitchison TJ, Feng Y (2008) Integrating high-content screening and ligand-target prediction to identify mechanism of action. Nat Chem Biol 4(1):59–68. https://doi.org/10.1038/nchembio.2007.53

Zhang J-H, Chung TDY, Oldenburg KR (1999) A simple statistical parameter for use in evaluation and validation of high throughput screening assays. J Biomol Screen 4(2):67–73. https://doi.org/10.1177/108705719900400206

Zheng XS, Chan T-F, Zhou HH (2004) Genetic and genomic approaches to identify and study the targets of bioactive small molecules. Chem Biol 11(5):609–618. https://doi.org/10.1016/j.chembiol.2003.08.011

Zlatic SA, Vrailas-Mortimer A, Gokhale A, Carey LJ, Scott E, Burch R, McCall MM, Rudin-Rush S, Davis JB, Hartwig C, Werner E, Li L, Petris M, Faundez V (2018) Rare disease mechanisms identified by genealogical proteomics of copper homeostasis mutant pedigrees. Cell Syst 6(3):368–380.e6. https://doi.org/10.1016/j.cels.2018.01.008

Rare Genetic Disorders: Unraveling the Pathophysiology, Gene Mutations, and Therapeutic Advances in Fabry Disease and Marfan Syndrome

7

Goutam Biswas, Nithar Ranjan Madhu, Bhanumati Sarkar, Soumosish Paul, Hadi Erfani, and Qamre Alam

Abstract

Fabry disease and Marfan syndrome are two distinct rare genetic disorders, each with intricate pathophysiologies, characterized by multi-organ involvement and life-threatening complications. This comprehensive review summarizes current knowledge on the pathophysiological mechanisms, gene mutations, and therapeutic approaches for these conditions. Fabry disease stems from mutations in the GLA gene, leading to deficient α-galactosidase A enzyme activity and subsequent accumulation of globotriaosylceramide (Gb3) in various tissues. This lipid storage disorder manifests with neuropathic pain, skin lesions, and progressive kidney and heart dysfunction. Early diagnosis through enzyme assays or genetic testing is vital. Enzyme replacement therapy (ERT) offers an effective therapeutic strategy, alleviating symptoms and slowing disease progression. Supportive care and personalized interventions are integral components of Fabry disease management. Marfan syndrome is primarily associated with mutations in the

G. Biswas · N. R. Madhu (✉) · S. Paul
Department of Zoology, Acharya Prafulla Chandra College, New Barrackpore, West Bengal, India

B. Sarkar
Department of Botany, Acharya Prafulla Chandra College, New Barrackpore, West Bengal, India

H. Erfani
Department of Chemical Engineering, Central Tehran Branch, Islamic Azad University, Tehran, Iran

Q. Alam
Department of Molecular Genomics and Precision Medicine, ExpressMed Diagnostics and Research, Zinj, Kingdom of Bahrain

M. Umair et al. (eds.), *Rare Genetic Disorders*,
https://doi.org/10.1007/978-981-99-9323-9_7

FBN1 gene, encoding truncated or nonfunctional fibrillin-1. Altered fibrillin-1 structure disrupts connective tissue homeostasis, resulting in diverse clinical manifestations. Pathophysiology encompasses skeletal abnormalities, ocular lens dislocation, and cardiovascular complications, notably aortic root dilation. The abstracted mutation of FBN1 necessitates a multidisciplinary approach to management, including regular aortic monitoring, lifestyle modifications, and emerging medical therapies. Understanding the pathophysiology and genetic basis of these two rare genetic disorders is paramount for timely diagnosis and effective management. While therapeutic options are available, ongoing research continues to refine treatments and holds promise for improving the quality of life for affected individuals. This review underscores the importance of a comprehensive approach, genetic counseling, and multidisciplinary care to optimize outcomes in these complex genetic disorders.

Keywords

Therapeutics · Fabry · Marfan · ERT · FBN1 · α-Gal A

7.1 Introduction

Rare genetic disorders, often referred to as orphan diseases, stand as a distinct and captivating realm within the expansive landscape of medical conditions. These disorders, characterized by their low prevalence in the general population, present an intricate tapestry of scientific, clinical, and societal challenges. While individually uncommon, collectively, rare genetic disorders encompass many conditions, each with its own unique genetic underpinnings, clinical manifestations, and implications for affected individuals and their families (Schieppati et al. 2008). Understanding rare genetic disorders requires a multidisciplinary approach that combines genetics, molecular biology, clinical medicine, and supportive care. The advent of genomic medicine and advances in genetic sequencing technologies have facilitated the identification of causative genetic mutations and contributed to the growing knowledge base surrounding these conditions.

Within the intriguing landscape of rare diseases, two conditions, Fabry disease and Marfan syndrome, stand as compelling examples, each with its own distinct genetic origins, clinical profiles, and impact on individuals' lives. Despite their differences, both Fabry disease and Marfan syndrome share some commonalities in their diagnostic complexities, the need for early identification, and the multidisciplinary approaches required for their management. This review aims to explore these two distinct but equally captivating rare genetic disorders, delving into their genetic foundations, dissecting their clinical diversity, navigating diagnostic intricacies, evaluating therapeutic interventions, and highlighting recent advancements. By doing so, we can increase awareness and help improve patient care, advancing research and ultimately positively impacting the lives of those affected by these conditions.

7.2 Methods

We formulated a comprehensive search strategy for this review, utilizing various databases such as Google Scholar, PubMed, and Scopus. We also incorporated specific keywords "Fabry disease," "Marfan syndrome," and "therapeutics" to identify relevant literature. We established clear inclusion criteria in our selection process to determine which studies would be part of our review. These criteria considered factors like publication dates, study designs, and overall relevance to Fabry disease and Marfan syndrome. Having identified relevant literature, we proceeded to extract pertinent information from them. This encompassed data related to genetic mutations, clinical manifestations, diagnostic methodologies, treatment options, and recent research findings pertinent to both disorders. The selected papers were chosen to bolster the argument's credibility in this review.

7.3 Fabry Disease

Fabry disease is a genetic, lysosomal storage disorder that develops when the enzyme alpha-galactosidase A (EC 3.2.1.22) cannot effectively break down lipids into simple components. It is a rare genetic disorder caused by mutations in the GLA gene. This gene encodes the enzyme alpha-galactosidase A (α-Gal A). When functional α-Gal A is lacking due to these mutations, a fatty substance called globotriaosylceramide (Gb3 or GL-3) accumulates in various cells and tissues throughout the body, leading to the characteristic pathophysiology of Fabry disease (Wang et al. 2011). The Gb3 buildup has several significant consequences, and the patients suffer from vascular impairment, skin lesions, ocular opacities, liver and renal dysfunction, chronic discomfort, and cardiac abnormalities (Table 7.1). Fabry disease often involves a thorough clinical assessment of symptoms like skin changes (angiokeratomas), eye abnormalities, neurological symptoms, and family history, followed by laboratory tests for the enzyme (α-Gal A) activity assay and biomarkers like Gb3. Genetic testing confirms the diagnosis by identifying GLA gene mutations.

7.3.1 History and Occurrence

In 1898, two dermatologists, Johannes Fabry and William Anderson, identified and characterized "angiokeratoma corporis diffusum" (Anderson 1898; Fabry 1898). The disease was first diagnosed as a systemic vascular disorder and later as a dysfunction of glycolipid storage (Pompen et al. 1947). In 1963, it was revealed that the disease leads to the abnormally high content of the glycolipid ceramide trihexoside accumulation (known as globotriaosylceramide or Gb3) and galabiosylceramide in various cell types (Sweeley and Klionsky 1963). The error was caused by

Table 7.1 Key aspects of Fabry disease's pathophysiology highlight how the buildup of Gb3 due to α-Gal A deficiency affects various organs and systems in the body

Aspect of Fabry disease pathophysiology	Description
Underlying cause	Mutations in the GLA gene result in deficient alpha-galactosidase A (α-Gal A) enzyme
Accumulation of Gb3	Lack of functional α-Gal A leads to globotriaosylceramide (Gb3 or GL-3) buildup within cells, particularly in lysosomes
Endothelial dysfunction	Gb3 accumulation in endothelial cells causes dysfunction, contributing to vascular problems such as impaired blood flow and microvascular damage
Organ involvement	Gb3 deposits in organs and tissues, impacting the kidneys (progressive damage and failure), heart (hypertrophy, cardiac issues), nervous system (neuropathic pain, neurological symptoms), and skin (angiokeratomas)
Pain and neuropathy	Neuropathic pain is common due to Gb3 accumulation in nerve cells, damaging the peripheral nervous system
Cardiac complications	Gb3 buildup in cardiac tissue can lead to heart enlargement (hypertrophy), potentially causing heart failure and arrhythmias
Renal complications	Progressive Gb3 deposition in the kidneys results in kidney dysfunction, proteinuria, and eventual renal failure
Microvascular disease	Gb3 accumulation in small blood vessels contributes to microvascular disease, affecting various organs and tissues
Systemic inflammation	Gb3 buildup may trigger inflammation and immune responses, exacerbating tissue damage
Variable clinical presentation	The disease severity and presentation vary among individuals, even those with the same genetic mutation, ranging from mild to severe symptoms with variable onset

inadequate activity of the ceramide trihexosidase, which catalyzes the hydrolysis reaction of the terminal galactose unit from ceramide trihexoside or Gb3 (Brady et al. 1967). Data from linkage studies proved that the disease is associated with X-linked inheritance (Opitz et al. 1965). Fabry disease is caused by an absence of alpha-galactosidase A, which is a lysosomal enzyme (Bishop et al. 1988). This enzyme is called for by the GLA gene, which is located on region Xq22.1 of the X chromosome. The GLA gene comprises seven exons totaling approximately 12 Kb sequence that encodes a precursor protein of 50 kDa (429 amino acids). Proteolytic cleavage and processing of the precursor protein ultimately yielded a mature homodimeric glycoprotein of 46 kDa (398 amino acids) (Garman and Garboczi 2004). The most severe form of Fabry disease, known as the classic form, comes with a mutation that leads to a total loss of α-Gal A, whereas those who have missense mutations frequently have persistent enzymatic activity ranging from 2 to 25% (Desnick et al. 2001). According to studies, 1 individual in 20,000 females and

1 individual in 40,000 males suffer from Fabry disease. One documented report claimed that the occurrence of Fabry hemizygotes in the population is 1:117,000 (Meikle et al. 1999). A substantially high frequency of the condition was discovered while attempting to establish the incidence through newborn screening, such as in Italy, where 1 in 3100 had shown this disorder (Spada et al. 2006). The frequency of Fabry disease in the Netherlands was estimated to be 1:476,000 (Poorthuis et al. 1999). Research conducted in the UK revealed an occurrence of the incidence of 1 in 366,000. However, it was based on information gathered from individuals with low α-Gal A activity who were seen in UK clinics (MacDermot et al. 2001a, b). Initial symptoms often appear in childhood, usually in 4–8 years. In middle age, untreated individuals frequently acquire life-threatening problems. Untreated males have a life expectancy of around 20 years less than the average population, which is 50 years, with a sharp drop in survival after 35 years (MacDermot et al. 2001a, b). Renal failure and cardiovascular problems are the most common symptoms of Fabry disease. Several patient groups with cardiac, renal, or cerebrovascular problems without any known reasons have also been shown to suffer from Fabry disease. Its prevalence in patients receiving hemodialysis treatment with end-stage renal disorder ranges from 0.22 to 1.2% (Linthorst et al. 2003; Nakao et al. 2003). Vascular deposition typically follows enhanced endothelial proliferation, which can then result in vascular blockage, ischemia, and infarction. The vertebrobasilar arteries and the tiny cerebral vessels are most commonly affected. Plasminogen activator inhibitor (PAI) activities are often higher. At the same time, thrombomodulin levels become low in patients who suffered early stroke at younger ages, indicating that the Fabry disease condition supports a prothrombotic state (DeGraba et al. 2000).

7.3.2 Gene Mutations

Fabry disease is ranked as the second most common disease based on the spectrum of all lysosomal storage disorders. The precise incidence of this disorder is unclear, and the patients who are not diagnosed may cause the existing data to misrepresent the frequency of this disease. Because of the restricted availability of genetic testing, many Fabry disease patients receive incorrect diagnoses and have their symptoms misunderstood as being caused by other illnesses. Various mutations (nearly about 900) have been identified in the GAL gene. Various small- and large-scale deletions or insertions have been recognized, although point mutations, mainly missense or nonsense mutations, are more common (Schäfer et al. 2005; Baptista et al. 2010). Except for a small number of identified mutations at CpG dinucleotides, most mutations are unique in occurrence (Cooper and Youssoufian 1988).

Different strategies were employed to understand the mutation in the GAL A gene, such as SSCP analyses (Rodríguez-Marí et al. 2003), chemical cleavage of mismatches, and entire sequencing of the coding and flanking regions (Ashley et al. 2001). Among these, amplification by PCR for identification of gene rearrangements followed by DNA sequencing has proven to be the most reliable. It has emerged as the most accurate approach for detecting mutations in X-linked disorders (Eng et al. 1993). Another sensitive and reliable method of DHPLC screening proved to be fruitful in detecting gene mutations (Shabbeer et al. 2005). Heterozygous female individuals have random X chromosome inactivation, and it is difficult to forecast how their condition will develop. X chromosomal inactivation is greatly imbalanced and is found in 29% of female Fabry patients.

Classic forms of the illness are linked with mutations that result in the gene product's full function loss. In contrast, late variations and occasionally moderate phenotypes are associated with mutations that result in amino acid changes. The majority of attempts to link genetics and clinical presentation have failed. Even within one family, there is significant heterogeneity in the age of start, pace of development, and organ manifestations. Clinical presentation is also extremely diverse. Several genetic and environmental variables are expected to be significant for both men and women. Unfortunately, the exact variables that alter the course of the disease have not been thoroughly studied (Gibas et al. 2008).

7.3.3 Therapeutics

The current recommended treatment is continuous intravenous infusions every 2 weeks to restore the reduced enzyme. Enzyme replacement therapy (ERT) has exhibited good early outcomes in adult male patients, including reduced pain, better peripheral nerve conduction and sweat, decreased glycolipid accumulation in different organs and tissues, and decreased heart hypertrophy (Schiffmann et al. 2001; Eng et al. 2001). Enzyme replacement treatment entails the possibility of infusion-related responses including, rigors and chills, which are substantially more severe in males than in females. In some cases, infusion reactions lead to a life-threatening condition that is triggered by antidrug antibody (ADA) responses. Several studies have revealed that antidrug antibodies restrict the ERT enzymes by blocking the active site and neutralizing the effect of the medication (Lenders et al. 2016, 2018). In earlier therapeutics, intravenous administration of two enzymes, agalsidase-a (human fibroblast-produced enzyme) and agalsidase-b (developed in Chinese hamster ovary cells), at a biweekly dose of 0.2 mg/kg and 1 mg/kg, respectively, has demonstrated sufficient clearance of Gb-3 deposits from cells (Eng et al. 2001). In comparison to animal products, the use of plant-based biologics, such as pegunigalsidase alfa, a unique modified version of alpha-galactosidase A, produced in tobacco cells and chemically modified with polyethylene glycol (PEG), has been shown to lessen clearance and boost stability (Schiffmann et al. 2019). Moss-aGal is another variation of recombinant Gal A generated from plants. Moss-aGal was successfully absorbed by endothelial cells in vitro. In a phase I clinical investigation with seven females having GLA mutations given a single dosage of moss-aGal (0.2 mg/kg), the result revealed a short half-life of 14 min in plasma following a single infusion, and no major side effects were detected (Hennermann et al. 2019).

The disadvantage of ERT is that the enzyme must be administered intravenously, which might be difficult because it necessitates several cannulations. Moreover, biweekly intravenous delivery might be challenging, especially in younger patients. Hyperpyrexia, dyspnea, and rash are some of the common symptoms of infusion reactions that might happen. Also, ERT does not address some clinical issues, such as advancing cardiac fibrosis and white matter damage (Hanneman et al. 2018).

The only alternative therapy for Fabry disease presently licensed is chaperone therapy, which uses Migalastat. Migalastat or 1-deoxygalactonojirimycin is an iminosugar and potent inhibitor of α-Gal A. This consumable small molecule can only be administered to individuals with certain mutations and works to increase the enzyme activity of mutant α-Gal A enzyme (Hughes et al. 2017). Migalastat stimulates protein trafficking to the lysosome and enhances protein folding by attaching to the catalytic domain of damaged α-Gal A in the endoplasmic reticulum. Before chaperone therapy, determination tests for a patient's amenability, such as the GLP HEK assay, are required. These tests can determine whether or not the patient can undergo this type of medication (Benjamin et al. 2017).

Substance reduction therapy (SRT), another oral treatment for Fabry disease, is intended to reduce the production of metabolites that cannot be broken down because of the underlying enzyme deficiency. Several SRT compounds were created and evaluated for their efficiency, such as galactose derivative Lucerastat and ceramide-based Venglustat. Both of these compounds inhibit glucosylceramide synthase and help limit the buildup of Gb-3 by lowering the quantity of ceramide that is transformed into glycosphingolipid (Ashe et al. 2015; Guérard et al. 2018). Recent research revealed that circulation levels of globotriaosylceramide and other sphingolipids were decreased at dosages of 1000 mg twice a day (Welford et al. 2018).

Gene therapies hold out hope for a cure for various uncommon hereditary illnesses. The potential application of gene therapy in Fabry disease is currently attracting interest. Ex vivo and in vivo gene editing are both possible. Hematopoietic stem cells from the patient are taken with the ex vivo method. After receiving myeloablative therapy, these stem cells go through gene editing before being reinfused into the patient's body for engraftment. In the in vivo method, a gene editing vector is injected directly into the patient; thereafter, cells, such as liver cells, immediately go through gene editing in order to produce the protein that is lacking (Ruiz de Garibay et al. 2013). In recent clinical trials employing ex vivo technique, in which the patient's CD34+ hematopoietic stem cells were collected and transfected using recombinant lentiviruses (AVR-RD-01, Avrobio), the results showed higher levels of Gal A activity than normal CD34+ hematopoietic cells (Huang et al. 2017). A study with adeno-associated vector-mediated in vivo gene therapy was investigated in a mouse model deficient in Gal A enzymatic activity and having substantial Gb3/lyso-Gb3 accumulation in blood plasma and tissues. The experimental results demonstrated high levels of Gal A activity as hepatocytes produced therapeutic amounts of human α-Gal A. Also, there was a concurrently noticeable reduction in the built-up Gb3/lyso-Gb3 in critical tissues (Huston et al. 2019).

Another therapeutic strategy that is currently being investigated is mRNA therapy, which uses α-Gal mRNA to induce enzyme synthesis. The endogenous protein translation mechanism may be superior to ERT because it assures correct folding, glycosylation, and intracellular trafficking of α-Gal A. Hepatocytes, where the enzyme is made, released into the bloodstream, and absorbed by tissues, are the primary focus of mRNA treatment, which is encapsulated in lipid nanoparticles. In mice and nonhuman primates, constant enzyme synthesis following a single mRNA infusion led to a plasma half-life of 7.5 h for Gal A (DeRosa et al. 2019). Studies revealed that Gb3 and cholesterol homeostasis interact in several ways (Slotte 1999). Therefore, the depletion of intra-lysosomal storage decreases by an alternate method that targets cholesterol metabolism, which may be a useful strategy for modifying glycosphingolipid balance (Schueler et al. 2016).

In order to improve the health conditions of Fabry disease, several patients also require some other nonspecific treatment such as adjuvant medication, antiproteinuric agents, and antiplatelet drugs (Schuller et al. 2016; Tahir et al. 2007). The exact cure for Fabry disease has not been discovered, although significant efforts to enhance the patient's condition are ongoing. Treatment using chaperones may be appropriate for a limited fraction of patients with certain mutations. Integrating different techniques, like substrate reduction and ERT, may be helpful for others, but the high cost of each individual therapy remains a significant obstacle to this strategy. Early outcomes of mRNA-based therapeutics are encouraging, but it is challenging to conduct studies of long enough duration to draw reliable conclusions about their efficacy. Limited number of patients, heterogeneity and slow course of the illness, inadequate screening, fewer number of case reports, slow progression of the disease, and shorter life span of the patients are also challenges in conducting well-powered studies for treatment (Fig. 7.1).

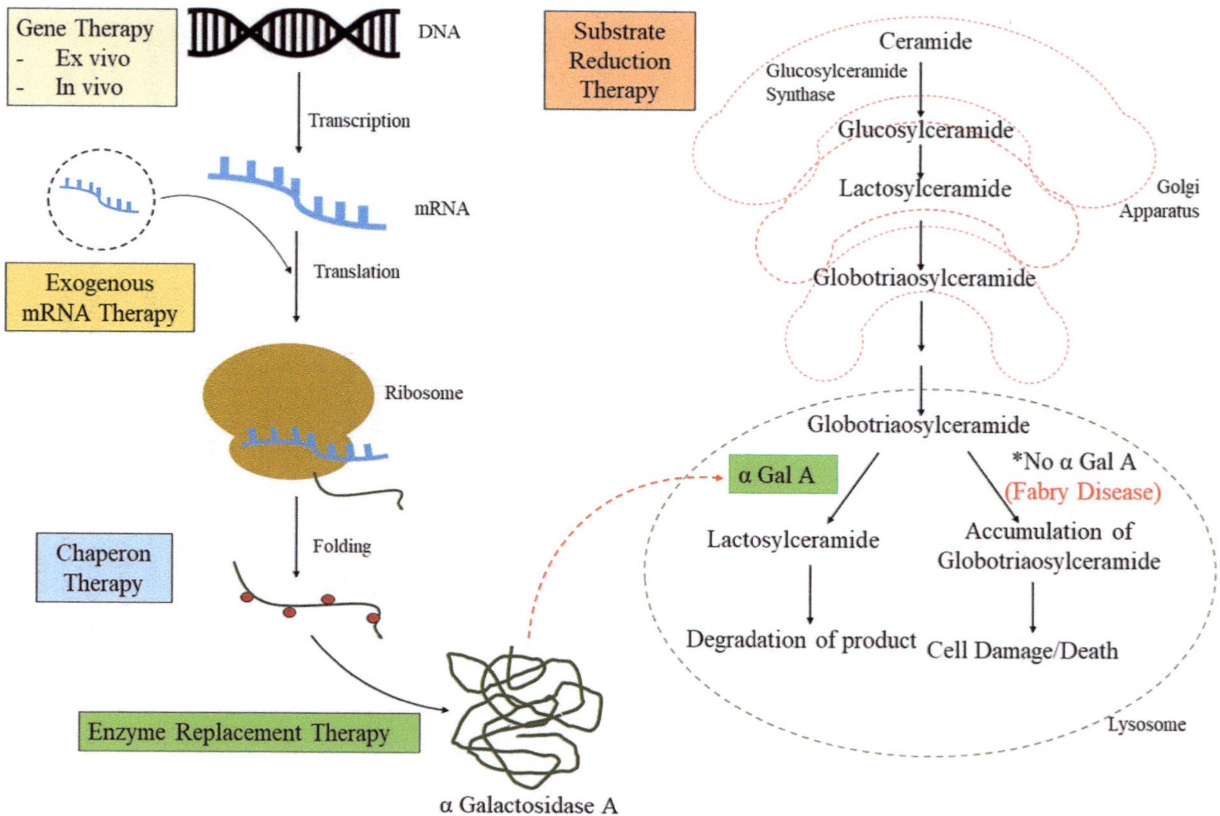

Fig. 7.1 Different therapeutic approaches in Fabry disease showing restoration of enzyme α-Gal A and the metabolic pathway (Adapted from Felis et al. 2020)

7.4 Marfan Syndrome

Marfan syndrome is an inherited autosomal dominant connective tissue disorder that affects the different systems in the body, including he neuromuscular, ocular, and circulatory systems. The disease is linked to the underlying mutation in the glycoprotein gene fibrillin-1 (FBN1) (Cañadas et al. 2010; Kumar and Agarwal 2014). Many symptoms of the typical Marfan syndrome appear around puberty or later, and serious consequences seldom ever appear before adulthood. However, prior to the development of open-heart surgery, Marfan patients typically passed away from acute aortic dissection or rupture, with an average life span of only 32 years. Today, treatment by specialized facilities has increased the life expectancy of Marfan patients by more than 60 years (Silverman et al. 1995). Understanding the intricate pathophysiology of Marfan syndrome is essential for diagnosis, management, and development of treatment strategies to mitigate its impact on affected individuals. A summary of the key aspects of Marfan syndrome's pathophysiology is given in Table 7.2.

Table 7.2 Key aspects of Marfan syndrome's pathophysiology show its effect on various body organs and systems

Aspect of Marfan syndrome pathophysiology	Description
Genetic basis	Marfan syndrome is primarily caused by mutations in the FBN1 gene, leading to abnormal fibrillin-1 protein production
Connective tissue abnormalities	Mutations in FBN1 result in the production of faulty fibrillin-1 protein, which compromises the integrity of connective tissues throughout the body
Cardiovascular involvement	Weakening of the aorta's connective tissue can lead to aortic root dilation, aortic aneurysms, and potentially life-threatening aortic dissections
Skeletal abnormalities	Altered connective tissue affects the skeleton, leading to characteristics like tall stature, long limbs, joint hypermobility, scoliosis, and chest deformities (pectus excavatum or carinatum)
Ocular features	Fibrillin-1 abnormalities can result in lens dislocation (ectopia lentis), myopia (nearsightedness), and an increased risk of retinal detachment
Pulmonary manifestations	Weakened lung tissue may cause spontaneous pneumothorax (collapsed lung) or other respiratory issues
Dural ectasia	Connective tissue weakness in the spine can lead to dural ectasia, a condition where the protective covering of the spinal cord expands abnormally
Variable clinical presentation	Symptoms and severity can vary widely among individuals with Marfan syndrome, depending on the specific mutations and their effects on connective tissues
Multisystem involvement	Marfan syndrome can affect multiple systems, including the cardiovascular, skeletal, ocular, pulmonary, and nervous systems, resulting in diverse clinical features
Aortic complications	Aortic dissection is a potentially life-threatening complication that demands vigilant monitoring and, in some cases, surgical intervention

Early diagnosis is crucial for properly managing and treating individuals with Marfan syndrome to prevent or mitigate potential complications, particularly those related to the cardiovascular system. These are the diagnostic criteria for Marfan syndrome as stated by the Ghent nosology. It defines diagnostic aspects into "major criteria," "minor criteria," "organ involvement," and manifestations that only exist in conjunction with other symptoms to form a "major" or "minor" criterion. While the Ghent nosology provides a structured framework for diagnosis, genetic testing is often an important component of the diagnostic process, especially for confirming the presence of a pathogenic FBN1 mutation (Dean 2007).

7.4.1 History and Occurrence

In 1896, French pediatrician Antoine Bernard-Jean Marfan first identified skeletal abnormalities in a five-and-a-half-year-old child and described the disease that ultimately went by the term Marfan syndrome. Forty years after publishing his first report, Marfan examined 150 instances that had been recorded with dolichostenomelia (the name he chose for abnormally long limbs). By then, it was evident that other tissues were also associated with the disease, as shown by congenital lens displacement (ectopia lentis) and mitral valve dysfunction. Marfan recognized that the inheritance pattern of this trait was in accordance with a Mendelian dominant characteristic because it impacted both sexes and several generations in families. In 1931, it was known that mesodermal tissues might have a developmental abnormality. The disease-causing defect on the aorta was characterized in 1943, while the severity of cardiovascular involvement was established in 1955 (Etter and Glover 1943; McKusick 1955). In 1972, it was summarized that aortic dilatation, myocardial infarction, and bacterial endocarditis all contributed to the patient's shorter life expectancy (Murdoch et al. 1972a, b). In 1991, the misssense mutation in the gene FBN1 on chromosome 15 (15q21.1) was discovered, which produces the glycoprotein fibrillin-1, an essential constituent of the extracellular microfibril (Dietz et al. 1991). In 1994, an issue of genetic heterogeneity was proposed. In FBN1-excluded individuals, a second locus for Marfan syndrome was mapped on chromosome 3p24.2-p25 through linkage analysis that produced Marfan syndrome type II (Collod et al. 1994). In 2004, it was revealed that this gene was linked with the production of transforming growth factor (TGF)-β receptor type II (Mizuguchi et al. 2004).

The overall prevalence of Marfan syndrome is frequently estimated as 1 per 10,000, and the incidence is considered to be 2–3 per 10,000 individuals, although these estimations depend on the identification of all afflicted and genetically susceptible individuals (Murdoch et al. 1972a, b; Judge and Dietz 2005). Obtaining numbers of its incidence and prevalence can occasionally be challenging because the clinical presentation of Marfan syndrome will not become obvious until later in life. The age range of 15–20 years had the highest age-specific frequency, suggesting that individuals with Marfan syndrome are likely to get a diagnosis within this age range due to more obvious clinical symptoms. However, in older age groups, the total frequency was 10.2 per 100,000 people, while the age group from 15 to 19 years had the greatest prevalence at 32.3 per 100,000 people (Chiu et al. 2014).

A study with 1013 patients bearing an FBN1 mutation revealed gender disparities in the onset of aortic illnesses in Marfan patients, in contrast to non-Marfan individuals. Men were more likely than women to develop acute aortic dissection type A (AADA), dilation of the ascending aorta, and other aortic events (Détaint et al. 2010). The Danish National MFS Registry supported this finding, showing that male patients had a considerably greater risk of aortic events at younger ages (Groth et al. 2017). Franken et al. (2016) found for the Dutch national registry CONCOR that gender was not related to the frequency of AADA or cardiovascular mortality but that males underwent aortic surgery more frequently than women out of 570 patients with Marfan syndrome (Franken et al. 2016).

7.4.2 Gene Mutation

In 1986, glycoprotein fibrillin-1 (350 kDa) was identified as one of the primary constituents of the extracellular microfibrils (Sakai et al. 1986). Fibrillin-1 is found in the matrix of both elastic and non-elastic tissues. This glycoprotein is encoded by the 230 kb long FBN1 gene, which has 65 exons (Corson et al. 1993). FBN1 mutations produce defective fibrillin-1 proteins that lead to poor connective tissue structure development. Among all the reported mutations in the FBN1 gene, missense mutations account for two-thirds and most frequently result in the substitution of cysteine amino acid. About 10% of all mutations documented are nonsense mutations, while 13% belong to small insertions, deletions, or duplication types. The occurrence of mutational errors during splicing is also documented for another 13% (Robinson et al. 2006).

Various mutations have been documented along the entire length of the FBN1 gene. Only 12% of all known FBN1 mutations are recurrent. Though not all FBN1 mutations have been reported or recorded, the real number of known mutations is probably significantly higher (Collod-Béroud et al. 2003). The exceptionally high penetration of FBN1 mutations is age-dependent, and there has never been a non-penetrance in hundreds of pedigrees. During zygote development, new or spontaneous mutations account for around 25% of instances (Tekin et al. 2007). However, de novo or spontaneous mutations, that occurred during gametogenesis or early embryogenesis, can also underlie the manifestation of Marfan syndrome. De novo mutations in the FBN1 gene occur stochastically and independently of parental genetic history, resulting in the condition's emergence in affected individuals without a familial predisposition (Fig. 7.2). Rare reports of gonadal mosaicism in people with Marfan syndrome exist. In unusual cases, the father's age was older when the child was conceived, which is consistent with the fact that sporadic Marfan is frequently caused by a novel mutation in a spermatogonium (Murdoch et al. 1972a, b).

The normal fibrillin-1 protein binds to another protein, TGF-β and serves as a connective protein that provides the structural foundation for tissues outside the cell. TGF-β has negative effects on vascular smooth muscle development and

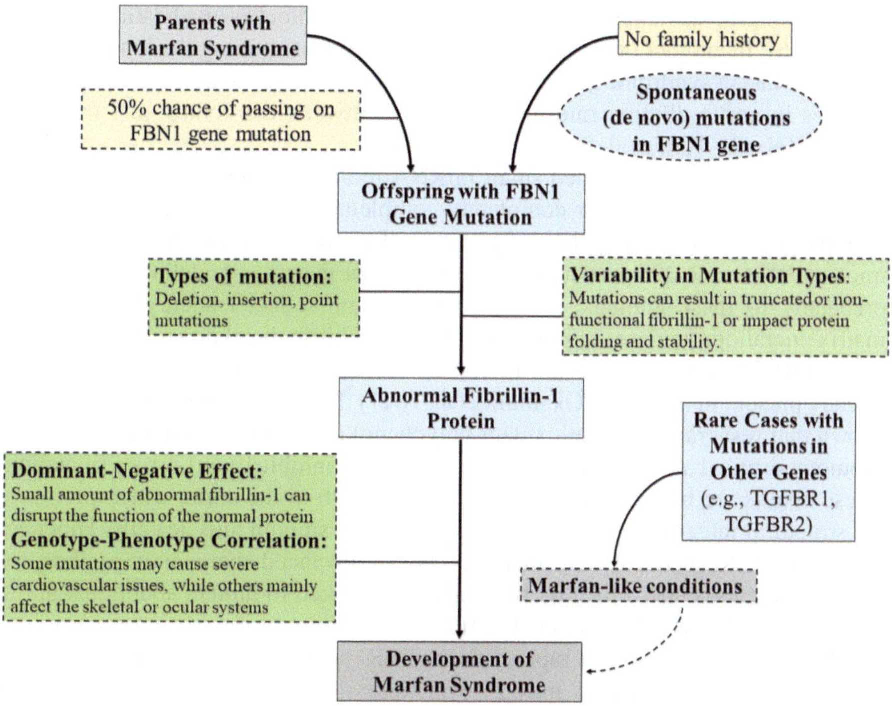

Fig. 7.2 Gene mutations related to Marfan syndrome

extracellular matrix stability. It is believed that mutant fibrillin causes elevated TGF-β in the lungs, heart valves, and aorta, which weakens the tissues and contributes to the symptoms of Marfan syndrome (Franken et al. 2015). However, other rare mutations and genetic factors can contribute to Marfan-like syndromes or related connective tissue disorders. Mutations in genes such as TGFBR1 and TGFBR2, which code for receptors in the TGF-β pathway, have been identified in individuals with conditions resembling Marfan syndrome. These mutations can disrupt the normal functioning of the TGF-β signaling pathway, leading to similar connective tissue abnormalities and clinical features seen in Marfan syndrome (Takeda et al. 2018).

7.4.3 Therapeutics

Beta-blockers, one of numerous blood pressure-lowering drugs, are considered the gold standard for treating congenital aneurysm development. TGF-β antagonists, such as neutralizing antibodies or angiotensin II receptor blockers (losartan, etc.), can stop increased TGF-β signaling. Losartan has shown exceptional promise in inhibiting aortic root development in a mouse model (Habashi et al. 2006). A study with randomized, single-blinded, controlled human studies does not seem to offer any meaningful advantages over conventional beta-blocker medication (Forteza et al. 2016). In another open-label research extension examining the long-term outcome of Marfan syndrome in patients randomly allocated to losartan or atenolol, no changes in aortic dilation rate were observed between the control and treatment groups (Teixido-Tura et al. 2018).

Endogenous single-stranded small miRNA molecules can suppress the expression of genes by partially or completely complementary binding to certain target regions of mRNA. During the early stages of aneurysm development in Marfan mice, miRNA miR-29b participates in extracellular matrix remodeling. When given prenatally to these animals, miR-29b suppression reduced aneurysms, lowered matrix metalloproteinase activity, and reduced elastin breakdown. On the other hand, miR-29b blockage did not inhibit the development of aneurysms that were already present in the aorta (Okamura et al. 2017). In a Marfan mouse model, it was shown that resveratrol, an antioxidant polyphenol derived from red wine known to promote sirtuin-1 activity and lengthen longevity, inhibited aortic root dilatation. These results are based on downregulating miR-29b in the aorta, which promotes the structural integrity of elastin and the survival of smooth muscle cells (Hibender et al. 2016). Resveratrol may therefore be an advanced therapeutic method for Marfan patients. Other miRNAs have also been linked to human abdominal aortic aneurysms and miR-29b (Iyer et al. 2017).

The mechanistic target of rapamycin (mTOR) signaling family, which has been discovered to be activated by miR-29 overexpression, is another newly discovered molecular target. mTOR is a highly conserved signaling mechanism that may control apoptosis, protein synthesis, and cell cycle in almost all mammalian cells (Laplante and Sabatini 2009; McCarthy 2013). Almost all mammalian cells can have their cell cycle, protein production, and apoptosis under control thanks to the highly conserved signaling mechanism known as mTOR. Rapamycin reduces aneurysm development in a rat calcium chloride-induced thoracic aortic aneurysm model through mTOR-mediated downregulation of various proinflammatory mediators (Cao et al. 2017).

The NOX family of NADPH oxidases and oxidative stress have recently been considered in Marfan syndrome research. Reactive oxygen species (ROS) are potentially responsible for the etiology of aortic aneurysms (McCormick et al. 2007). The high levels of ROS seen in smooth muscle cells and macrophages were shown to be brought on by enhanced TGF-β signaling caused by NADPH oxidase-4 activation, which contributes to the aortopathy (Sturrock et al. 2006; Manea et al. 2008). Studies involving NOX4-deficient Marfan mouse model and cultured vascular smooth muscle cells from affecting Marfan patients showed that NOX4 has an effect on the development of aortic dilation in Marfan and also on the organization and structure of the aortic tunica media (Jiménez-Altayó et al. 2018).

Due to two significant constraints, there have only been a small number of studies on the applicability of gene therapy for aneurysm stabilization. First, an appropriate vector technology is needed for effective and long-lasting gene transfer into the vascular wall during gene therapy. Secondly, unlike genetic defects in the gene encoding, fibrillin has a dominant-negative effect on the vessel wall, disrupting the formation of fibrillin multimers. Therefore, reexpressing fibrillin alone may not be enough to have a therapeutic impact (Byers 2004). Adenoviral vectors have been used a lot to deliver therapeutic genes into the vasculature. Still, the main issue is the intense inflammatory response caused by the vector within the target tissue. Even the use of second-generation viral constructs fails to increase the vascular transduction efficiency of adenoviral vectors. Furthermore, it was established that one of the primary reasons for temporary gene overexpression in the vascular system after adenovirus injection was tissue immunological response (Flugelman et al. 1992; Parker et al. 2013; Seppelt et al. 2016).

A problem in gene therapy of Marfan syndrome is that mutant fibrillin-1 has a dominant-negative impact on the vessel wall, as was previously addressed. Therefore, it is expected that lowering mutant fibrillin expression levels will improve the architecture of elastin. In fact, it was demonstrated that an RNA-based approach based on overexpressing antisense hammerhead ribozymes that target mutant fibrillin-1 efficiently inhibits protein deposition. However, only in vitro studies were used to evaluate this strategy's effectiveness (Kilpatrick and Phylactou 1998). Further research into this strategy in vivo is necessary to see whether it affects clinical outcomes.

Next-generation RNA sequencing (RNA-Seq), which reveals the presence and amount of RNA in a biological sample at a specific time, is an important step in discovering novel therapeutic targets. This enables the discovery of new targets that may affect MFS therapy. RNA-Seq enables a more precise gene and transcript expression assessment compared to microarray technology. RNA-Seq analysis of the aortic tissues of mgR/mgR mice that were 9 weeks old and compared to wild-type mice revealed unique aorta-specific pathways implicated in the pathophysiology of the disease (Bhushan et al. 2019).

Although the pathophysiology of Marfan syndrome is not entirely understood, it is thought that fibrillin-1 gene mutations have a dominating detrimental impact via overactive TGF-β signaling pathways. While there is no cure for Marfan syndrome, various treatments and therapies are available to manage its symptoms and improve the quality of life for individuals with the condition. The treatment and management of Marfan syndrome typically involve a multidisciplinary approach, with input from various medical specialists (Fig. 7.3). Early diagnosis and appropriate management can help prevent or minimize the complications associated with Marfan syndrome and improve the overall quality of life for affected individuals. The most potentially fatal sign of Marfan syndrome is cardiovascular abnormalities, particularly aortic root dilatation and mitral valve prolapse since these individuals are susceptible to abrupt aortic dissection. Prophylactic surgery of the aortic root in suffering patients continues to be the only treatment strategy that, although not curing the underlying illness, results in a meaningful extension of life expectancy in Marfan syndrome. The effectiveness of beta-blockers, the gold standard for decreasing the expansion

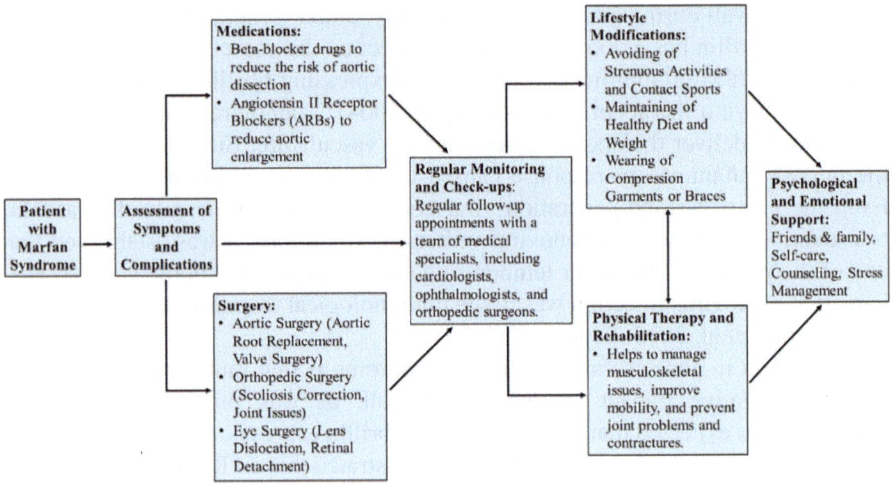

Fig. 7.3 A comprehensive scenario for Marfan syndrome treatment and management

of the aorta, and angiotensin type II receptor blockers like losartan on clinical outcomes has not been obvious in clinical research. In the future, there may be potential methods for increasing the effectiveness of vascular gene transfer and identifying new therapeutic targets to treat at least the vascular illness of suffering individuals.

7.5 Conclusion

Despite their genetic disparities, both Fabry disease and Marfan syndrome share common diagnostic challenges, necessitating early identification to mitigate disease progression. Furthermore, the management of these conditions is multifaceted. Fabry disease demands enzyme replacement therapy (ERT) and emerging gene therapy approaches to alleviate symptoms and slow disease progression. In contrast, Marfan syndrome necessitates meticulous cardiovascular surveillance, lifestyle modifications, and, in some cases, surgical interventions to prevent life-threatening complications. Additionally, there is a need for more personalized therapies and improved management strategies to address the complexities of Fabry disease and Marfan syndrome effectively. Recent advancements in genetics and medical research have broadened our understanding of these rare diseases, offering hope for improved diagnostic accuracy and innovative therapies. Ongoing research can continue to advance our understanding of these disorders and improve treatment options, offering hope for better outcomes and quality of life for individuals affected by Marfan syndrome and Fabry disease. Through the collective efforts of the medical community, researchers, advocacy groups, and affected individuals and families, the landscape of rare genetic disorders can be transformed from one of obscurity to one of understanding, support, and hope.

References

Anderson WA (1898) A case of angiokeratoma. Br J Dermatol 18:113–117

Ashe KM, Budman E, Bangari DS, Siegel CS, Nietupski JB, Wang B et al (2015) Efficacy of enzyme and substrate reduction therapy with a novel antagonist of glucosylceramide synthase for Fabry disease. Mol Med 21:389–399. https://doi.org/10.2119/molmed.2015.00088

Ashley GA, Shabbeer J, Yasuda M, Eng CM, Desnick RJ (2001) Fabry disease: twenty novel α-galactosidase a mutations causing the classical phenotype. J Hum Genet 46:192–196. https://doi.org/10.1007/s100380170088

Baptista MV, Ferreira S, Pinho-e-Melo T, Carvalho M, Cruz VT, Carmona C et al (2010) Mutations of the GLA gene in young patients with stroke: the PORTYSTROKE study—screening genetic conditions in Portuguese young stroke patients. Stroke 41(3):431–436. https://doi.org/10.1161/STROKEAHA.109.570499

Benjamin ER, Della Valle MC, Wu X, Katz E, Pruthi F, Bond S et al (2017) The validation of pharmacogenetics for the identification of Fabry patients to be treated with migalastat. Genet Med 19(4):430–438. https://doi.org/10.1038/gim.2016.122

Bhushan R, Altinbas L, Jäger M, Zaradzki M, Lehmann D, Timmermann B et al (2019) An integrative systems approach identifies novel candidates in Marfan syndrome-related pathophysiology. J Cell Mol Med 23(4):2526–2535. https://doi.org/10.1111/jcmm.14137

Bishop DF, Kornreich R, Desnick RJ (1988) Structural organization of the human alpha-galactosidase A gene: further evidence for the absence of a 3′ untranslated region. Proc Natl Acad Sci U S A 85(11):3903–3907. https://doi.org/10.1073/pnas.85.11.3903

Brady RO, Gal AE, Bradley RM, Martensson E, Warshaw AL, Laster L (1967) Enzymatic defect in Fabry's disease: ceramidetrihexosidase deficiency. N Engl J Med 276(21):1163–1167. https://doi.org/10.1056/NEJM196705252762101

Byers PH (2004) Determination of the molecular basis of Marfan syndrome: a growth industry. J Clin Invest 114(2):161–163. https://doi.org/10.1172/JCI22399

Cañadas V, Vilacosta I, Bruna I, Fuster V (2010) Marfan syndrome. Part 1: pathophysiology and diagnosis. Nat Rev Cardiol 7(5):256–265. https://doi.org/10.1038/nrcardio.2010.30

Cao J, Wu Q, Geng L, Chen X, Shen W, Wu F, Chen Y (2017) Rapamycin inhibits CaCl2-induced thoracic aortic aneurysm formation in rats through mTOR-mediated suppression of proinflammatory mediators. Mol Med Rep 16(2):1911–1919. https://doi.org/10.3892/mmr.2017.6844

Chiu HH, Wu MH, Chen HC, Kao FY, Huang SK (2014) Epidemiological profile of Marfan syndrome in a general population: a national database study. Mayo Clin Proc 89(1):34–42. https://doi.org/10.1016/j.mayocp.2013.08.022

Collod G, Babron MC, Jondeau G, Coulon M, Weissenbach J, Dubourg O et al (1994) A second locus for Marfan syndrome maps to chromosome 3p24. 2–p25. Nat Genet 8(3):264–268. https://doi.org/10.1038/ng1194-264

Collod-Béroud G, Le Bourdelles S, Ades L, Ala-Kokko L, Booms P, Boxer M et al (2003) Update of the UMD-FBN1 mutation database and creation of an FBN1 polymorphism database. Hum Mutat 22(3):199–208. https://doi.org/10.1002/humu.10249

Cooper DN, Youssoufian H (1988) The CpG dinucleotide and human genetic disease. Hum Genet 78:151–155. https://doi.org/10.1007/BF00278187

Corson GM, Chalberg SC, Dietz HC, Charbonneau NL, Sakai LY (1993) Fibrillin binds calcium and is coded by cDNAs that reveal a multidomain structure and alternatively spliced exons at the 5′ end. Genomics 17(2):476–484. https://doi.org/10.1006/geno.1993.1350

Dean J (2007) Marfan syndrome: clinical diagnosis and management. Eur J Hum Genet 15(7):724–733. https://doi.org/10.1038/sj.ejhg.5201851

DeGraba T, Azhar S, Dignat-George F, Brown E, Boutière B, Altarescu G, McCarron, Schiffmann R (2000) Profile of endothelial and leukocyte activation in Fabry patients. Ann Neurol 47(2):229–233. https://doi.org/10.1002/1531-8249(200002)47:2%3C229::AID-ANA13%3E3.0.CO;2-T

DeRosa F, Smith L, Shen Y, Huang Y, Pan J, Xie H et al (2019) Improved efficacy in a Fabry disease model using a systemic mRNA liver depot system as compared to enzyme replacement therapy. Mol Ther 27(4):878–889. https://doi.org/10.1016/j.ymthe.2019.03.001

Desnick RJ, Ioannou YA, Eng CM (2001) α-galactosidase A deficiency: Fabry disease. In: Scriver CR, Beaudet AL, Sly WS, Valle D (eds) The metabolic and molecular bases of inherited disease, 8th edn. McGraw-Hill, New York, pp 3733–3774

Détaint D, Faivre L, Collod-Beroud G, Child AH, Loeys BL, Binquet C et al (2010) Cardiovascular manifestations in men and women carrying a FBN1 mutation. Eur Heart J 31(18):2223–2229. https://doi.org/10.1093/eurheartj/ehq258

Dietz HC, Cutting CR, Pyeritz RE, Maslen CL, Sakai LY, Corson GM et al (1991) Marfan syndrome caused by a recurrent de novo missense mutation in the fibrillin gene. Nature 352(6333):337–339

Eng CM, Resnick-Silverman LA, Niehaus DJ, Astrin KH, Desnick RJ (1993) Nature and frequency of mutations in the alpha-galactosidase a gene that cause Fabry disease. Am J Hum Genet 53(6):1186

Eng CM, Guffon N, Wilcox WR, Germain DP, Lee P, Waldek S et al (2001) Safety and efficacy of recombinant human α-galactosidase A replacement therapy in Fabry's disease. N Engl J Med 345(1):9–16. https://doi.org/10.1056/NEJM200107053450102

Etter LE, Glover LP (1943) Arachnodactyly complicated by dislocated lens and death from rupture of dissecting aneurysm of aorta. J Am Med Assoc 123(2):88–89. https://doi.org/10.1001/jama.1943.82840370001006

Fabry J (1898) Ein Betrag zur Kenntnis der Ppurpura hammarrhagica nodularis. Arch Dermatol Syphilol 43:187–200

Flugelman MY, Jaklitsch M, Newman KD, Casscells W, Bratthauer GL, Dichek DA (1992) Low level in vivo gene transfer into the arterial wall through a perforated balloon catheter. Circulation 85(3):1110–1117. https://doi.org/10.1161/01.CIR.85.3.1110

Felis A, Whitlow M, Kraus A, Warnock DG, Wallace E (2020) Current and investigational therapeutics for Fabry disease. Kidney Int Rep 5(4):407–413

Forteza A, Evangelista A, Sanchez V, Teixido-Tura G, Sanz P, Gutierrez L et al (2016) Efficacy of losartan vs. atenolol for the prevention of aortic dilation in Marfan syndrome: a randomized clinical trial. Eur Heart J 37(12):978–985. https://doi.org/10.1093/eurheartj/ehv575

Franken R, Radonic T, Den Hartog AW, Groenink M, Pals G, van Eijk M et al (2015) The revised role of TGF-β in aortic aneurysms in Marfan syndrome. Neth Hear J 23:116–121. https://doi.org/10.1007/s12471-014-0622-0

Franken R, Groenink M, de Waard V, Feenstra HM, Scholte AJ, van den Berg MP et al (2016) Genotype impacts survival in Marfan syndrome. Eur Heart J 37(43):3285–3290. https://doi.org/10.1093/eurheartj/ehv739

Garman SC, Garboczi DN (2004) The molecular defect leading to Fabry disease: structure of human α-galactosidase. J Mol Biol 337(2):319–335. https://doi.org/10.1016/j.jmb.2004.01.035

Gibas AL, Klatt R, Johnson J, Clarke JT, Katz J (2008) Disease rarity, carrier status, and gender: a triple disadvantage for women with Fabry disease. J Genet Couns 17:528–537. https://doi.org/10.1007/s10897-008-9179-7

Groth KA, Stochholm K, Hove H, Kyhl K, Gregersen PA, Vejlstrup N et al (2017) Aortic events in a nationwide Marfan syndrome cohort. Clin Res Cardiol 106:105–112. https://doi.org/10.1007/s00392-016-1028-3

Guérard N, Oder D, Nordbeck P, Zwingelstein C, Morand O, Welford RW et al (2018) Lucerastat, an iminosugar for substrate reduction therapy: tolerability, pharmacodynamics, and pharmacokinetics in patients with Fabry disease on enzyme replacement. Clin Pharmacol Ther 103(4):703–711

Habashi JP, Judge DP, Holm TM, Cohn RD, Loeys BL, Cooper TK et al (2006) Losartan, an AT1 antagonist, prevents aortic aneurysm in a mouse model of Marfan syndrome. Science 312(5770):117–121. https://doi.org/10.1126/science.1124287

Hanneman K, Karur GR, Wasim S, Morel CF, Iwanochko RM (2018) Prognostic significance of cardiac magnetic resonance imaging late gadolinium enhancement in Fabry disease. Circulation 138(22):2579–2581. https://doi.org/10.1161/CIRCULATIONAHA.118.037103

Hennermann JB, Arash-Kaps L, Fekete G, Schaaf A, Busch A, Frischmuth T (2019) Pharmacokinetics, pharmacodynamics, and safety of moss-aGalactosidase a in patients with Fabry disease. J Inherit Metab Dis 42(3):527–533. https://doi.org/10.1002/jimd.12052

Hibender S, Franken R, Van Roomen C, Ter Braake A, Van Der Made I, Schermer EE et al (2016) Resveratrol inhibits aortic root dilatation in the Fbn1C1039G/+ Marfan mouse model. Arterioscler Thromb Vasc Biol 36(8):1618–1626. https://doi.org/10.1161/ATVBAHA.116.307841

Huang J, Khan A, Au BC, Barber DL, López-Vásquez L, Prokopishyn NL et al (2017) Lentivector iterations and pre-clinical scale-up/toxicity testing: targeting mobilized CD34+ cells for correction of Fabry disease. Mol Ther Methods Clin Dev 5:241–258. https://doi.org/10.1016/j.omtm.2017.05.003

Hughes DA, Nicholls K, Shankar SP, Sunder-Plassmann G, Koeller D, Nedd K et al (2017) Oral pharmacological chaperone migalastat compared with enzyme replacement therapy in Fabry disease: 18-month results from the randomised phase III ATTRACT study. J Med Genet 54(4):288–296. https://doi.org/10.1136/jmedgenet-2016-104178

Huston MW, Yasuda M, Pagant S, St Martin S, Cao L, Falese L et al (2019) Liver-targeted AAV gene therapy vectors produced by a clinical scale manufacturing process result in high, continuous therapeutic levels of enzyme activity and effective substrate reduction in mouse model of Fabry disease. Mol Genet Metab 126(2):S77. https://doi.org/10.1016/j.ymgme.2018.12.187

Iyer V, Rowbotham S, Biros E, Bingley J, Golledge J (2017) A systematic review investigating the association of microRNAs with human abdominal aortic aneurysms. Atherosclerosis 261:78–89. https://doi.org/10.1016/j.atherosclerosis.2017.03.010

Jiménez-Altayó F, Meirelles T, Crosas-Molist E, Sorolla MA, Del Blanco DG, López-Luque J et al (2018) Redox stress in Marfan syndrome: dissecting the role of the NADPH oxidase NOX4 in aortic aneurysm. Free Radic Biol Med 118:44–58. https://doi.org/10.1016/j.freeradbiomed.2018.02.023

Judge DP, Dietz HC (2005) Marfan's syndrome. Lancet 366(9501):1965–1976. https://doi.org/10.1016/S0140-6736(05)67789-6

Kilpatrick MW, Phylactou LA (1998) Towards an RNA-based therapy for Marfan syndrome. Mol Med Today 4(9):376–381. https://doi.org/10.1016/S1357-4310(98)01326-4

Kumar A, Agarwal S (2014) Marfan syndrome: an eyesight of syndrome. Meta Gene 2:96–105. https://doi.org/10.1016/j.mgene.2013.10.008

Laplante M, Sabatini DM (2009) mTOR signaling at a glance. J Cell Sci 122(20):3589–3594. https://doi.org/10.1242/jcs.051011

Lenders M, Stypmann J, Duning T, Schmitz B, Brand SM, Brand E (2016) Serum-mediated inhibition of enzyme replacement therapy in Fabry disease. J Am Soc Nephrol 27(1):256–264. https://doi.org/10.1681/ASN.2014121226

Lenders M, Schmitz B, Brand SM, Foell D, Brand E (2018) Characterization of drug-neutralizing antibodies in patients with Fabry disease during infusion. J Allergy Clin Immunol 141(6):2289–2292. https://doi.org/10.1016/j.jaci.2017.12.1001

Linthorst GE, Hollak CEM, Korevaar JC, Van Manen JG, Aerts JMFG, Boeschoten EW (2003) Alpha-galactosidase a deficiency in Dutch patients on dialysis: a critical appraisal of screening for Fabry disease. Nephrol Dial Transplant 18(8):1581–1584. https://doi.org/10.1093/ndt/gfg194

MacDermot KD, Holmes A, Miners AH (2001a) Anderson–Fabry disease: clinical manifestations and impact of disease in a cohort of 60 obligate carrier females. J Med Genet 38(11):769–775. https://doi.org/10.1136/jmg.38.11.769

MacDermot KD, Holmes A, Miners A (2001b) Anderson-Fabry disease: clinical manifestations and impact of disease in a cohort of 98 hemizygous males. J Med Genet 38(11):750–760. https://doi.org/10.1136/jmg.38.11.750

Manea A, Manea SA, Gafencu AV, Raicu M, Simionescu M (2008) AP-1–dependent transcriptional regulation of NADPH oxidase in human aortic smooth muscle cells: role of p22phox subunit. Arterioscler Thromb Vasc Biol 28(5):878–885. https://doi.org/10.1161/ATVBAHA.108.163592

McCarthy N (2013) YAP, PTEN and miR-29 size each other up. Nat Rev Cancer 13(1):4–5. https://doi.org/10.1038/nrc3422

McCormick ML, Gavrila D, Weintraub NL (2007) Role of oxidative stress in the pathogenesis of abdominal aortic aneurysms. Arterioscler Thromb Vasc Biol 27(3):461–469. https://doi.org/10.1016/j.mgene.2013.10.008

McKusick VA (1955) The cardiovascular aspects of Marfan's syndrome: a heritable disorder of connective tissue. Circulation 11(3):321–342. https://doi.org/10.1161/01.CIR.11.3.321
Meikle PJ, Hopwood JJ, Clague AE, Carey WF (1999) Prevalence of lysosomal storage disorders. Jama 281(3):249–254. https://doi.org/10.1001/jama.281.3.249
Mizuguchi T, Collod-Beroud G, Akiyama T, Abifadel M, Harada N, Morisaki T et al (2004) Heterozygous TGFBR2 mutations in Marfan syndrome. Nat Genet 36(8):855–860. https://doi.org/10.1038/ng1392
Murdoch JL, Walker BA, Halpern BL, Kuzma JW, McKusick VA (1972a) Life expectancy and causes of death in the Marfan syndrome. N Engl J Med 286(15):804–808. https://doi.org/10.1056/nejm197204132861502
Murdoch JL, Walker BA, McKusick VA (1972b) Parental age effects on the occurrence of new mutations for the Marfan syndrome. Ann Hum Genet 35(3):331–336. https://doi.org/10.1111/j.1469-1809.1957.tb01406.x
Nakao S, Kodama C, Takenaka T, Tanaka A, Yasumoto Y, Yoshida A, Kanzaki T, Enriquez AL, Eng CM, Tanaka H, Tei C, Desnick RJ (2003) Fabry disease: detection of undiagnosed hemodialysis patients and identification of a "renal variant" phenotype. Kidney Int 64(3):801–807. https://doi.org/10.1046/j.1523-1755.2003.00160.x
Okamura H, Emrich F, Trojan J, Chiu P, Dalal AR, Arakawa M et al (2017) Long-term miR-29b suppression reduces aneurysm formation in a Marfan mouse model. Physiol Rep 5(8):e13257. https://doi.org/10.14814/phy2.13257
Opitz JM, Stiles FC, Wise D, von Gemmingen G, Race RR, Sander R et al (1965) The genetics of angiokeratoma corporis diffusum (Fabry's disease), and its linkage with Xg(a) locus. Am J Hum Genet 17:325–342
Parker AL, White KM, Lavery CA, Custers J, Waddington SN, Baker AH (2013) Pseudotyping the adenovirus serotype 5 capsid with both the fibre and penton of serotype 35 enhances vascular smooth muscle cell transduction. Gene Ther 20(12):1158–1164. https://doi.org/10.1038/gt.2013.44
Pompen AW, Ruiter M, Wyers HJ (1947) Angiokeratoma corporis diffusum (universale) Fabry, as a sign of an unknown internal disease; two autopsy reports. Acta Med Scand 128(3):234–255. https://doi.org/10.1111/j.0954-6820.1947.tb06596.x
Poorthuis BJ, Wevers RA, Kleijer WJ, Groener JE, de Jong JG, Van Weely S et al (1999) The frequency of lysosomal storage diseases in The Netherlands. Hum Genet 105:151–156. https://doi.org/10.1007/s004399900075
Robinson PN, Arteaga-Solis E, Baldock C, Collod-Béroud G, Booms P, De Paepe A et al (2006) The molecular genetics of Marfan syndrome and related disorders. J Med Genet 43(10):769–787. https://doi.org/10.1136/jmg.2005.039669
Rodríguez-Marí A, Coll MJ, Chabás A (2003) Molecular analysis in Fabry disease in Spain: fifteen novel GLA mutations and identification of a homozygous female. Hum Mutat 22(3):258–258. https://doi.org/10.1002/humu.9172
Ruiz de Garibay AP, Solinís MÁ, Rodríguez-Gascón A (2013) Gene therapy for fabry disease: a review of the literature. BioDrugs 27:237–246. https://doi.org/10.1007/s40259-013-0032-7
Sakai LY, Keene DR, Engvall E (1986) Fibrillin, a new 350-kD glycoprotein, is a component of extracellular microfibrils. J Cell Biol 103(6):2499–2509. https://doi.org/10.1083/jcb.103.6.2499
Schäfer E, Baron K, Widmer U, Deegan P, Neumann PH et al (2005) Thirty-four novel mutations of the GLA gene in 121 patients with Fabry disease. Hum Mutat 25(4):412–412. https://doi.org/10.1002/humu.9327
Schieppati A, Henter JI, Daina E, Aperia A (2008) Why rare diseases are an important medical and social issue. Lancet 371(9629):2039–2041. https://doi.org/10.1016/S0140-6736(08)60872-7
Schiffmann R, Kopp JB, Austin HA III, Sabnis S, Moore DF, Weibel T et al (2001) Enzyme replacement therapy in Fabry disease: a randomized controlled trial. JAMA 285(21):2743–2749. https://doi.org/10.1001/jama.285.21.2743

Schiffmann R, Goker-Alpan O, Holida M, Giraldo P, Barisoni L, Colvin RB et al (2019) Pegunigalsidase alfa, a novel PEGylated enzyme replacement therapy for Fabry disease, provides sustained plasma concentrations and favorable pharmacodynamics: a 1-year phase 1/2 clinical trial. J Inherit Metab Dis 42(3):534–544. https://doi.org/10.1002/jimd.12080

Schueler U, Kaneski C, Remaley A, Demosky S, Dwyer N, Blanchette-Mackie J et al (2016) A short synthetic peptide mimetic of apolipoprotein A1 mediates cholesterol and globotriaosylceramide efflux from Fabry fibroblasts. JIMD Rep 29:69–75. https://doi.org/10.1007/8904_2015_507

Schuller Y, Linthorst G, Hollak CEM, Van Schaik IN, Biegstraaten M (2016) Pain management strategies for neuropathic pain in Fabry disease—a systematic review. BMC Neurol 16:1–10. https://doi.org/10.1186/s12883-016-0549-8

Seppelt PC, Schwill S, Weymann A, Arif R, Weber A, Zaradzki M et al (2016) Loss of endothelial barrier in Marfan mice (mgR/mgR) results in severe inflammation after adenoviral gene therapy. PLoS One 11(2):e0148012. https://doi.org/10.1371/journal.pone.0148012

Shabbeer J, Robinson M, Desnick RJ (2005) Detection of α-galactosidase a mutations causing Fabry disease by denaturing high performance liquid chromatography. Hum Mutat 25(3):299–305. https://doi.org/10.1002/humu.20144

Silverman DI, Burton KJ, Gray J, Bosner MS, Kouchoukos NT, Roman MJ et al (1995) Life expectancy in the Marfan syndrome. Am J Cardiol 75(2):157–160. https://doi.org/10.1016/S0002-9149(00)80066-1

Slotte JP (1999) Sphingomyelin–cholesterol interactions in biological and model membranes. Chem Phys Lipids 102(1–2):13–27. https://doi.org/10.1016/S0009-3084(99)00071-7

Spada M, Pagliardini S, Yasuda M, Tukel T, Thiagarajan G, Sakuraba H, Ponzone A, Desnick RJ (2006) High incidence of later onset Fabry disease revealed by newborn screening. Am J Hum Genet 79(1):31–40. https://doi.org/10.1086/504601

Sturrock A, Cahill B, Norman K, Huecksteadt TP, Hill K, Sanders K et al (2006) Transforming growth factor-β1 induces Nox4 NAD (P) H oxidase and reactive oxygen species-dependent proliferation in human pulmonary artery smooth muscle cells. Am J Phys Lung Cell Mol Phys 290(4):L661–L673. https://doi.org/10.1152/ajplung.00269.2005

Sweeley CC, Klionsky BL (1963) Fabry's disease: classification as a sphingolipidosis and partial characterization of a novel glycolipid. J Biol Chem 238:PC3148. https://doi.org/10.1016/S0021-9258(18)51888-3

Tahir H, Jackson LL, Warnock DG (2007) Antiproteinuric therapy and Fabry nephropathy: sustained reduction of proteinuria in patients receiving enzyme replacement therapy with agalsidase-β. J Am Soc Nephrol 18(9):2609–2617. https://doi.org/10.1681/ASN.2006121400

Takeda N, Hara H, Fujiwara T, Kanaya T, Maemura S, Komuro I (2018) TGF-β signaling-related genes and thoracic aortic aneurysms and dissections. Int J Mol Sci 19(7):2125. https://doi.org/10.3390/ijms19072125

Teixido-Tura G, Forteza A, Rodríguez-Palomares J, González Mirelis J, Gutiérrez L, Sánchez V et al (2018) Losartan versus atenolol for prevention of aortic dilation in patients with Marfan syndrome. J Am Coll Cardiol 72(14):1613–1618. https://doi.org/10.1016/j.jacc.2018.07.052

Tekin M, Cengiz FB, Ayberkin E, Kendirli T, Fitoz S, Tutar E et al (2007) Familial neonatal Marfan syndrome due to parental mosaicism of a missense mutation in the FBN1 gene. Am J Med Genet A 143(8):875–880. https://doi.org/10.1002/ajmg.a.31660

Wang RY, Bodamer OA, Watson MS, Wilcox WR, ACMG Work Group on Diagnostic Confirmation of Lysosomal Storage Diseases (2011) Lysosomal storage diseases: diagnostic confirmation and management of presymptomatic individuals. Genet Med 13(5):457–484. https://doi.org/10.1097/GIM.0b013e318211a7e1

Welford RWD, Mühlemann A, Garzotti M, Rickert V, Groenen PMA, Morand O et al (2018) Glucosylceramide synthase inhibition with lucerastat lowers globotriaosylceramide and lysosome staining in cultured fibroblasts from Fabry patients with different mutation types. Hum Mol Genet 27(19):3392–3403. https://doi.org/10.1093/hmg/ddy248

Current Insights into the Potential of Gene Therapy to Treat Rare Mitochondrial Diseases

8

Sreyank Tirunagari, Sid Dsa, and Megala Jayaraman

Abstract

Mitochondrial disorders represent a rare category of diseases that arise from abnormalities in essential proteins necessary for the proper functioning and maintenance of mitochondria. These mutations disrupt the proteins and activities associated with mitochondria, resulting in impairments across various systems and organs in the body, leading to debilitating diseases. They are multisystemic in nature and contain mutations in both the nuclear and mitochondrial genomes. Hence, it is highly challenging to detect mitochondrial illnesses, which makes treating them very challenging. In recent times, the growing emphasis on research and clinical trials in the field of mitochondrial diseases has significantly advanced the current understanding of treatment options for these conditions. As a precision medicine approach, gene therapy has become a potential treatment for mitochondrial illnesses. As therapy is still in its early stages, further research is needed to determine its effectiveness and safety. This chapter explores the biology of the mitochondria, mutations, and prospective uses of gene therapy in the future to treat rare mitochondrial diseases.

Keywords

Gene therapy · Mitochondrial diseases · Mitochondrial DNA · Mitochondrial replacement therapy (MRT) · OXPHOS system · Precision medicine

S. Tirunagari · S. Dsa · M. Jayaraman (✉)
Department of Genetic Engineering, Faculty of Engineering and Technology, SRM Institute of Science and Technology, Kattankulathur, Chennai, Tamil Nadu, India
e-mail: megalaj@srmist.edu.in

M. Umair et al. (eds.), *Rare Genetic Disorders*,
https://doi.org/10.1007/978-981-99-9323-9_8

8.1 Introduction

Mitochondria are special organelles that are responsible for energy production and many other cellular processes such as maintaining the oxidative phosphorylation system (OXPHOS), apoptosis, calcium homeostasis, iron-cluster biosynthesis, and other various metabolic pathways (Alston et al. 2017; Duchen 2000). Mitochondria are also governed by their own genome, which is possibly related to their bacterial origins (Ochman and Moran 2001). The sole extranuclear source of DNA in animal cells is found in mitochondria (Nass 1966). It is a double-stranded, circular molecule that is 16 Kb in size and is present in a large copy number, which is generally based on the cell type and its energy demand (Tinker et al. 2021). Primary mitochondrial disorders are those that develop as a result of abnormalities or mutations in the mitochondrial DNA or nuclear DNA, as well as from the dysfunction of mitochondrial proteins or nuclear proteins that are associated with the mitochondria (Tinker et al. 2021). Mitochondrial diseases are primarily associated with irregularities in metabolic pathways, which invariably lead to cell apoptosis.

Mitochondrial DNA is often at risk of mutations or damage due to the genome being very close to the OXPHOS system. The leakage of reactive oxygen species (ROS) during oxidative phosphorylation might cause damage and mutations in the mtDNA (Greaves et al. 2012). The lack of introns also suggests that mutations can easily occur in the coding regions of the mtDNA and can manifest into mitochondrial disease. Therefore, this makes the mtDNA more susceptible to mutation compared to the nuclear DNA (Brown et al. 1979). There are two types of mutations that can occur in mtDNA or nDNA: homoplasmic or heteroplasmic point mutations, and heteroplasmic large-scale rearrangements (Di Donfrancesco et al. 2022). Among hereditary disorders, mitochondrial diseases are a very common group, as their frequency in the population is about 5–20 in 100,000 people (Gorman et al. 2016). It can be challenging to identify mitochondrial illnesses since they impact so many different body systems and tissues and cause patients to have such a wide range of symptoms. The diagnosis of a mitochondrial illness cannot be verified by a single lab or diagnostic test (Anonymous 2023).

8.2 Mitochondrial Genome and Organization

Mitochondrial DNA is maternally inherited extranuclear DNA. Mitochondrial DNA lacks introns, and the only noncoding region is the regulatory region known as the D-loop or displacement loop. Mammalian mtDNA only codes for 37 genes, including 13 mitochondrial OXPHOS complex proteins, 22 transfer RNAs, and 2 ribosomal RNAs (Di Donfrancesco et al. 2022). An additional 1300 nuclear encoded proteins are needed by the Mito proteome to sustain the five multimeric OXPHOS complexes (I–V) and auxiliary mitochondrial activities (Calvo et al. 2016). These genes are translated and transcribed outside of the mitochondria before being imported into them via specific import mechanisms. In contrast, the mitochondrial proteins are synthesized within the mitochondria and are co-translationally inserted into the inner mitochondrial membrane with the help of mitochondrial ribosomes and many other factors (Slone and Huang 2020).

8.2.1 Replicative Segregation

During cellular division, mitochondrial DNA (mtDNA) undergoes a random distribution process to daughter cells. This is due to the higher mutation rate and independent replication of mtDNA from the nuclear genome, resulting in varying mtDNA copy numbers and potential mtDNA variant differences between daughter cells. As a result, we may infer from this system that, in comparison to nuclear DNA, the segregation mechanism in mitochondrial DNA is relatively loosely governed. This process is responsible for mitochondrial bottleneck, heteroplasmy, and homoplasmy, which contribute to the genetic diversity of mtDNA. When the mutations reach a particular threshold, they drift into the daughter cells via this mechanism (Wallace and Chalkia 2013).

8.2.2 Heteroplasmy and Homoplasmy

Mitochondrial DNA is present in large copies in cells. On average, each cell contains between 10^3 and 10^4 copies of the mitochondrial genome, though this number varies between cell type and developmental stage (Rooney et al. 2015). In most of the cells, the mtDNA molecules are similar or the sequence is identical, which is known as homoplasmy. This is a state of the mitochondrial genome where either there are no mutations present in any of the copies of mtDNA that are present in the cell or the mutation is present in all copies of the cell (Chinnery and Hudson 2013). However, in cases of damage to the mtDNA due to irregular repair, reactive oxygen species, or any other environmental agent, a mutation takes place in one or a few copies of the mitochondrial DNA. Because of these mutations, these duplicates can coexist alongside the typical wild-type DNA. This phenomenon is known as heteroplasmy (Stefano and Kream 2016). The level of heteroplasmy might fluctuate across various cells and can influence disease manifestation based on the level of mutated DNA present in each cell. This is known as the threshold value, which signifies the percentage of mtDNA molecules within a cell or tissue that carry a specific mutation. The threshold value determines whether the mutation will cause clinical symptoms. Generally, a higher percentage of mutant mtDNA is associated with a greater risk of developing mtDNA-related disease. The threshold value varies depending on the specific mutation and the affected tissues. Some mutations can cause disease at low levels of heteroplasmy, while others require a higher level. The threshold value is used to monitor disease progression and determine the risk of developing symptoms. However, because of how random these mutations are, they only occur on a very small level and therefore do not manifest on a large scale in the population (Taylor and Turnbull 2005). For the condition to express within the cell, mutations in mtDNA usually need to be present at high levels, usually over 50%. However, certain mtDNA mutations, like those found in mt-tRNA, can cause issues even at very low levels. On the other hand, single, large-scale mtDNA deletions can cause problems when more than 60% of the mtDNA is deleted. This means that the specific mutation and its concentration determine how harmful it is to the cell, and different mutations have different thresholds for causing damage (Gorman et al. 2016).

8.2.3 Maternal Inheritance

Mitochondrial DNA is strictly maternally inherited, and therefore this suggests that the mtDNA is clonal in nature (Giles et al. 1980). Like the oocyte, the sperm likewise contains its own mitochondria. After fertilization, the zygote receives the mitochondria from both the oocyte and the sperm, but the mitochondria from the sperm disappear during early embryogenesis. Although little was known about the process, it was believed to be accomplished by ubiquitination immediately following zygote formation. A study by Sato and Sato (2012) has shown that autophagy is involved in the degradation of the mitochondria in the sperm of *Caenorhabditis elegans* (Sato and Sato 2012). A diluting effect was also thought to contribute to the mitochondria's lack of paternal inheritance since a sperm only contains around 100 copies of mitochondria in comparison to an unfertilized egg that has 100,000 copies (Chinnery and Hudson 2013). During oogenesis, there is a reduction in the number of mitochondrial DNA (mtDNA) molecules in developing oocytes, followed by a subsequent amplification process that results in a high total amount of mtDNA in mature oocytes. This phenomenon is known as the mitochondrial genetic bottleneck. This process plays a crucial role in determining mtDNA variation and the evolution of interactions between mitochondrial and nuclear DNA (Lane 2011). The significant multiplication of the mtDNA is closely linked with the maturity of the oocyte, which induces random accumulation of the mutant mtDNA among the daughter cells, which causes heteroplasmy in the oocytes to vary (Chinnery and Hudson 2013). Uniparental inheritance may result in a steady accumulation of mutations that eventually result in mitochondrial dysfunction. The process was hypothesized to avoid mutational meltdown by causing heteroplasmy to vary by producing either high amounts or a loss of variants. High levels of mutations in the cells would probably lead to the extinction of a certain germ cell lineage and prevent the mutations from being passed on to the following generation (Zhang et al. 2018).

8.3 Mutations in Mitochondrial DNA

Due to the proximity of the mitochondrial DNA to the OXPHOS system and the inability of the repair mechanisms to prevent mutations, reactive oxygen species damage, a lack of histones to protect the DNA, and other environmental factors that harm the mitochondrial DNA can all result in mutations (Greaves et al. 2012; Brown et al. 1979). Mutations in mtDNA can be of two types:

1. **Point Mutations**

   Point mutations in the mitochondria are the root cause of many illnesses in humans. Around 1 in 200 people is thought to have a point mutation, according to population estimates (Alston et al. 2017). About 100 point mutations that cause diseases have been identified in mtDNA. Typically, these mutations are present in tRNAs, rRNAs, and proteins coded by mtDNA and are passed down maternally. Point mutations are seen to be mostly heteroplasmic, often recessive in character, and require a high threshold level of presence to emerge in an individual. A random genetic drift was optimal to explain the dynamics of point mutations and the mosaic nature of them. Early point mutations either lose their effect or spread erratically during mtDNA replication and segregation during cell division. This stochastic process repeatedly cycles through certain cells, allowing altered mtDNA molecules to become the dominant species. This causes a mosaic pattern of cellular OXPHOS abnormalities that were seen in ageing human mitotic tissues (Lawless et al. 2020). m.3243A>G MTTL, which was connected to the condition MELAS (mitochondrial encephalomyopathy, lactic acidosis, and stroke-like episodes), was one of the most prevalent point mutations (Lax et al. 2011).
2. **Mitochondrial DNA Rearrangements/Deletions**

   There are more than 120 significant mtDNA rearrangements known. Large-scale deletions like mtDNA rearrangements typically have a somatic origin (Greaves et al. 2012). As a result, they are not heritable yet frequently cause mitochondrial illness (Chinnery and Hudson 2013). These 1.3–8 kb-sized deletions are uniformly distributed among all the cells of the tissue that has been impacted by them. They are frequently flanked by short direct repeats and mostly occur between the O_H and O_L origins of replication (Bua et al. 2006). Around 50–60% of the threshold value was sufficient to cause mitochondrial disorders in large-scale deletions (Rossignol et al. 2003). There are several suggestions that mtDNA rearrangements result from defective DNA repair mechanisms or are created during the replication process. Additionally, it has been suggested that the accumulation of these mutations was related to neurodegenerative disorders (Lax et al. 2011). Mitochondrial DNA deletions are linked to three primary phenotypes: Kearns-Sayre syndrome, chronic progressive external ophthalmoplegia, and Pearson syndrome (Alston et al. 2017) (Table 8.1).

Table 8.1 Mutations in the OXPHOS system and the phenotypes associated with them

OXPHOS subunit	Mutations	Phenotype	Reference
Complex I	**Core subunits** NDUFS1, NDUFS3, NDUFS4, NDUFS7, NDUFS8, NDUFV1, NDUFA10, NDUFA2, NDUFB3	Leigh or Leigh-like syndrome, mitochondrial cl deficiency	Chinnery and Hudson (2013)
	NDUFS2, NDUFS6, NDUFV2, NDUFA1, NDUFA11, ACAD9	Hypertrophic cardiomyopathy and encephalopathy	Chinnery and Hudson (2013), Alston et al. (2017)
	Assembly factors NDUFAF1, NDUFAF2, NDUFAF3, NDUFAF4, NDUFAF5, NDUFAF6, NDUFAF7, NUBPL, TIMMDC1, FOXRED1, TMEM70	Mitochondrial cI deficiency, mitochondrial encephalopathy, Leigh syndrome, cardiomyopathy, severe neonatal lactic acidosis	Chinnery and Hudson (2013), Fernandez-Vizarra et al. (2021)
Complex II	**Core subunits** SDHA, SDHB, SDHC, SDHD	Mitochondrial cII deficiency, dilated cardiomyopathy, Leigh syndrome, paragangliomas, gastrointestinal stromal tumours	Chinnery and Hudson (2013), Alston et al. (2017), Fernandez-Vizarra et al. (2021)
	Assembly factors SDHAF1, SDHAF2	Leukoencephalopathy, mitochondrial cII deficiency, pheochromocytomas	Fernandez-Vizarra et al. (2021)
Complex III	**Core subunits** MT-CYB, UQCRB, UQCRQ	Mitochondrial cIII deficiency, combined respiratory chain deficiency, LHON, myopathy, hepatopathy, parkinsonism/MELAS overlap syndrome	Alston et al. (2017), Fernandez-Vizarra et al. (2021)
	Assembly factors UQCC2, UQCC3	Lactic acidosis and dysmorphic features, growth retardation, respiratory distress, and seizures	Alston et al. (2017), Fernandez-Vizarra et al. (2021)

(continued)

Table 8.1 (continued)

OXPHOS subunit	Mutations	Phenotype	Reference
Complex IV	**Core subunits, early module, MT-CO1, MT-CO2, MT-CO3 modules** COX4I1, COX4I2, COX5A, MT-CO1, MT-CO2, COX7B, COX8A, MT-CO3, COX6B1, COX6A2	Mitochondrial cIV deficiency, Fanconi anaemia, Leigh-like syndrome, exocrine pancreatic insufficiency, lactic acidaemia, neurological syndromes, cardiomyoencephalopathy, multisystem disorder, LHON	Fernandez-Vizarra et al. (2021)
	Assembly factors TACO1, SURF1, COA3, COX14, COX10, COX15, COX20, SCO1, SCO2, COA6, COXFA4	Mitochondrial cIV deficiency, Leigh syndrome, Charcot-Marie-Tooth syndrome, severe congenital lactic acidosis, encephalopathy, hypertrophic cardiomyopathy, sensorineural hearing loss, metabolic acidosis, growth retardation, hypotonia, cerebellar ataxia	Fernandez-Vizarra et al. (2021)
Complex V	**Core subunits in Fo and F1 domains** MT-ATP6, MT-ATP8, MT-ATP6/8 overlap region, ATP5F1A, ATP5F1D, ATP5F1E	Mitochondrial cV deficiency, neuropathy, ataxia and retinitis pigmentosa (NARP) syndrome, Leigh syndrome, adult-onset ataxia and polyneuropathy, mitochondrial myopathy, lactic acidosis, sideroblastic anaemia, fatal infantile encephalopathy/cardiomyopathy, hypoglycaemia	Fernandez-Vizarra et al. (2021)
	Assembly factors ATPAF2, TMEM70	Mitochondrial cV deficiency, encephalopathy, lactic acidosis, 3-methylglutaconic aciduria, neonatal encephalocardiomyopathy, occasionally facial dysmorphisms, and cI deficiency	Fernandez-Vizarra et al. (2021)

8.4 Diagnosis and Screening of Mitochondrial Diseases

Due to the involvement of several systems and the lack of a single phenotypic manifestation, diagnosing mitochondrial disorders is particularly difficult. It was extremely challenging to identify the underlying source of the dysfunctions since they are accountable for abnormalities in several organs. Leber hereditary optical neuropathy (LHON), a mitochondrial illness that affects the eye, is one of the uncommon mitochondrial disorders to be linked to a particular organ (Gorman et al. 2016).

Few common ways to diagnose a mitochondrial disease are the following:

1. Based on the clinical manifestations: Mitochondrial diseases have a wide range of clinical manifestations. But few symptoms such as vision and hearing loss, encephalopathy and cardiomyopathy, metabolic issues, stroke-like episodes, lactic acidosis, and diabetes overlap with most of the diseases.
2. Based on the affected organ or tissue: Generally, mitochondrial diseases manifest over a range of organs and are multisystemic, but they mainly affect few organs such as the brain, eyes, ears, kidneys, heart, central nervous system, and muscles. Few diseases, such as LHON affect specific organs like the eye, and Friedreich's ataxia affects the spine, CNS, and muscles.
3. Based on the mode of inheritance: Mitochondrial disease-causing mutations can occur in two ways—maternally inherited diseases are caused due to mutations in the mtDNA being passed down to the offspring. Sporadic mitochondrial diseases are caused by mutations that occur during early embryonic development (Chinnery 2015).
4. Based on genetic defect: Mutations in the genes of nuclear DNA (nDNA) or mtDNA, which are important in mitochondrial function, can result in mitochondrial disorders. mtDNA mutations can have an impact on the energy-producing electron transport chain (ETC) complexes in the mitochondria. Numerous aspects of mitochondrial function, including protein import, mtDNA maintenance, fusion and fission, and mitochondrial biogenesis, can be impacted by nDNA mutations (Calvo and Mootha 2010). An intense study of the patient's pedigree was very important to determine the possibility of the disease being liked to a mitochondrial defect. To narrow down the disease's aetiology, several diagnostic procedures must be performed as preliminary testing, including muscle biopsies, neuroimaging, cardiac assessments, and multiple histochemical and biochemical investigations. But they are only preliminary, and molecular diagnostics, which provide more conclusive evidence of a mitochondrial disorder, are preferable (Fig. 8.1).

8.4.1 Molecular Diagnostics

Molecular diagnostics can be most accurately performed by using the DNA extracted from the affected cells. Methods such as southern blotting have been used to detect large-scale deletions, rearrangements and other common mutations (Chinnery and Hudson 2013; Tuppen et al. 2010). Genomics approaches such as candidate gene sequencing, whole-exome sequencing, and next-generation sequencing have also been used to identify the genes responsible for the disorders. A generalized approach was preferred when identifying the genes due to the heterogenic nature of mitochondrial diseases. Because of the mitochondrial genome's short size, sequencing was not extremely challenging, and methods like whole-exome sequencing (WES) have been employed to find mtDNA mutations (Tuppen et al. 2010). Techniques such as RFLP have been utilized to identify the heteroplasmic and

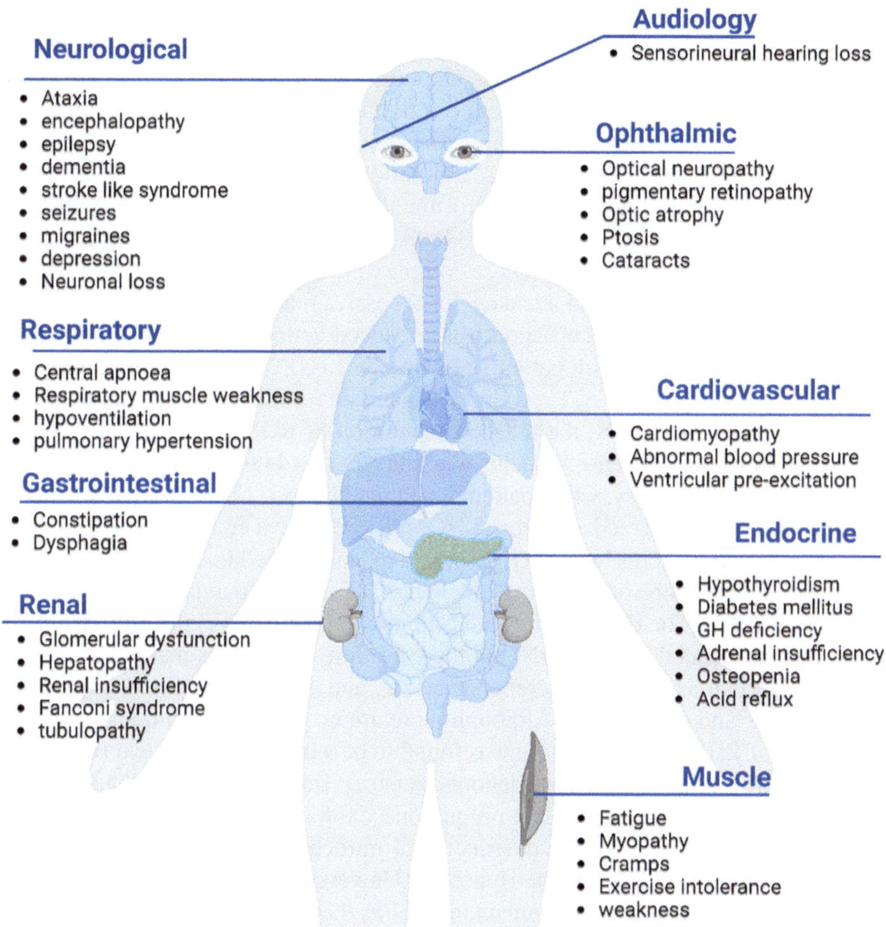

Fig. 8.1 Phenotypic manifestations of mitochondrial diseases

homoplasmic polymorphisms in mtDNA (Thorburn 2000). The understanding of mitochondrial illnesses is still developing, and diagnostic techniques are getting better. To define the complex nature of the disorders, more sophisticated diagnostics are being developed.

8.5 Mitochondrial Diseases

8.5.1 Leber Hereditary Optic Neuropathy (LHON)

LHON is a mitochondrial disease that is associated with the eye and, due to the loss of retinal ganglion cells (RGCs), produces significant vision impairment or even permanent blindness. With different degrees of penetrance in persons with the same mutation, this RGC malfunction affects 50% of the male population and 10–15% of the female population (Kirches 2011). Typically, this begins unilaterally. Within a few months of developing the illness, the other eye invariably becomes impaired as well. In northern Europe, the incidence of LHON has been reported to be between 1 in 30,000 and 1 in 50,000. It was found to be homoplasmic in nature, and therefore all the mtDNA in the tissue comprises mutations (Di Donfrancesco et al. 2022). The majority of the cases are linked to mutations in three genes that synthesize OXPHOS subunit complex I: ND1-G3460A, ND4-G11778A, or ND6-T14484C (Kirches 2011). These mutations cause complex I (NADH-ubiquinone oxidoreductase) to malfunction and cause a leakage of excessive reactive oxygen species, leading to cellular damage and cell apoptosis. Reduced action of antioxidant defences has also been associated with patients affected by LHON (Hayashi and Cortopassi 2015). Although lowered ATP production has been believed to be the cause of the pathogenesis of LHON, the high concentration of ROS plays a greater role in the progression of the disease. Patients with the ND6-T14484C mutation have about a 71% chance of recovery, while patients with the ND4-G11778A have only shown about a 25% chance of recovery from LHON (Riordan-Eva et al. 1995). The substitutions in the amino acids have been theorized to lead the cell to a non-inflammatory apoptotic event, therefore making the death of RGCs painless (Hayashi and Cortopassi 2015). The death of the RGCs in most of the cases is the only definitive phenotype associated with LHON, but there are few other cases that have reported phenotypes such as dystonia, cerebellar ataxia, and other neurodegenerative attributes (McFarland et al. 2007). Although there are no current techniques to reverse the effects of LHON, gene therapy was found to be a potential treatment to help the patients suffering from LHON. Adeno-associated viruses (AAVs) have been made to carry the genes for the NADH-ubiquinone oxidoreductase complex and were infected to the tissues with the dysfunctional mitochondria in attempts to rebuild complex I (Sahel et al. 2021). The Huazhong University of Science and Technology started a trial based on this mechanism in 2011 as they studied nine individuals with LHON whose visual condition was deteriorating. They used a routine intravitreal (IVT) injection to deliver the rAAV2-ND4 to the patients, and they saw that four out of nine had significantly improved conditions. Few other studies have demonstrated that gene therapy is an excellent prospective treatment for LHON (Yang et al. 2016).

8.5.2 Friedreich's Ataxia (FRDA)

FRDA is another mitochondrial disease that is caused due to a mutation in the nuclear DNA coding for a mitochondrial protein called frataxin. Therefore, FRDA is classified as an autosomal recessive disease and is caused due to a homozygous expansion of a guanine-guanine-adenine (GAA) trinucleotide repeat on chromosome 9q13, intron 1 of the frataxin gene (Koeppen 2011). Normal individuals have shown about 20–30 (GAA) repeats, while the affected individuals have shown almost about

100–1500 or more (GAA) repeats. The greater the repeats, the more the deficiency of the frataxin protein. The expansion of these repeats leads to condensation of the gene, which in turn reduces the amount of mRNA coding for frataxin. This condensation was formed due to structures such as triplexes, hairpin loops, R-loops, and heterochromatin (Silva et al. 2015). Due to this, RNA polymerase II activity, transcription, and protein synthesis have all decreased. FRDA has been found to be a rare neurodegenerative disease, and the frequency is about 2 in 100,000 (Cossée et al. 1997). It is an early-onset disease and is most prevalent among children aged 7–15. It affects males and females alike since it is autosomal in nature. The recessive nature of the disease means that both parents either need to be positive for FRDA or should be carriers of the mutated frataxin (FXN) gene. FRDA is characterized by ataxia with respect to the spinal cord and CNS. This leads to the degradation of sensory neurons and spinocerebellar tracts. The frataxin protein is characterized as a small mitochondrial protein that is responsible for the biosynthesis of iron-sulphur clusters, which are very important for metabolic activities. There are other hypothesized activities of frataxin, such as mitochondrial iron transport, iron binding/sequestration, and response to oxidative stress (Puccio and Koenig 2002). Due to the mutation in the FXN gene, a deficiency of frataxin occurs which leads to iron metabolism dysregulation, cellular dysfunction, and death (Campuzano et al. 1996). The (GAA) repeat is not homozygous in all the patients. It has been established that the disease's onset is correlated with the number of repetitions. The lower the age of onset, the higher the number of repetitions. Scoliosis and foot malformations are other early signs, that precede ataxia in individuals affected with FRDA. These symptoms are very common and are found in about 60–80% of the patients. As the condition worsens, the affected children's weak motor skills make it necessary for nearly all of them to be assisted in wheelchairs (Koeppen 2011). It has also been observed that 8–32% of the patients developed diabetes mellitus and these individuals were associated with longer (GAA) repeats on the shorter allele (Nia Monró et al. 1997). The cause of death in most of the patients was cardiomyopathy, and the age of death was also correlated with the number of (GAA) repeats in the individuals. Isnard et al. (1997) showed that patients with fewer trinucleotide repeats were less likely to develop cardiomyopathy because these repeats were linked to the thickening of the left ventricular wall (Isnard et al. 1997). Due to these factors, diagnosis of FRDA is not difficult. Early symptoms, such as spinal cord thinning, have been a defining characteristic that aids in the diagnosis of FRDA. Magnetic resonance imaging (MRI) of the heart has also been used to diagnose cardiomyopathy in patients affected with FRDA.

Various disease models have been developed to understand Friedreich's ataxia and unfold treatments for it. Gene therapy methods have been used to treat these models in attempts to prevent or reduce ataxia, cardiomyopathy, and other symptoms of FRDA. Piguet et al. (2018) have used adeno-associated viruses to inject the mice with functional FXN gene in order to express the frataxin protein. They discovered that during the first few days, the therapy corrected the sensory, neurological, and coordination deficits. However, overexpression of high levels of frataxin proved to be harmful and toxic to mitochondria resulting in severe dysfunctionality of complex I and complex II (Belbellaa et al. 2020). This results in fibrosis, subcellular disarray in cardiomyocytes, and cell death.

8.5.3 Mitochondrial Encephalomyopathy, Lactic Acidosis, and Stroke-Like Episodes (MELAS)

MELAS has been established as a multisystem disease that causes multiple manifestations like lactic acidaemia, stroke-like episodes, dementia, epilepsy, seizures, and cognitive impairment and ataxia (El-Hattab et al. 2015). This multi-organ dysfunction is brought on by damaged mitochondria that are unable to provide the body with enough ATP. Majority of the cases of MELAS have reported a mutation in mitochondrial $tRNA^{leu(uur)}$ gene MT-TL1. A point mutation causing a base substitution from adenine to guanine at position 3243 on the MT-TL1 gene is responsible for MELAS syndrome (Ikeda et al. 2018). Other rare mutations such as m.3271T>C, m.3252A>G, and m.3243A>T have also been associated with MELAS. Patients have been documented to have mutations in different tRNA coding genes and in genes coding for complex I and complex III (El-Hattab et al. 2015; Ikeda et al. 2018). There are multiple clinical manifestations in patients affected by MELAS syndrome. 90% of the patients show stroke-like symptoms, lactic acidosis, seizures, and dementia (El-Hattab et al. 2015). 70–80% of the patients show hemiparesis, muscle weakness, easy fatiguability, hearing impairment, and exercise intolerance. Peripheral neuropathy and other neurological symptoms such as memory loss, ataxia, and cognitive disability are prevalent (up to 30–77%) and generally chronic and progressive (El-Hattab et al. 2015). Other manifestations can include short stature, anxiety, depression, diabetes, optic atrophy, cardiomyopathy, altered consciousness, etc. These symptoms might not correlate with the genetic definition since they differ from patient to patient even though they have the same mutation (Hirano and Pavlakis 1994). MELAS is also associated with excessive production of reactive oxygen species and nitric oxide (NO) deficiency. Therefore, idebenone, a coenzyme Q_{10} analogue, has been used in the treatment of MELAS. CoQ_{10} is a naturally occurring fat-soluble quinone, but idebenone is a water-soluble synthetic analogue. CoQ_{10} works as an antioxidant by enabling electron transportation from complex I to complex III, therefore providing balance to the OXPHOS system (Hirano et al. 2018). Idebenone can also alleviate neurological complexities since it can cross the blood-brain barrier (El-Hattab et al. 2015). Arginine has been found to be a source of nitric oxide and is also used to treat MELAS since NO deficiency plays a major role in the pathogenesis of the disease (Hirano et al. 2018). L-Arginine has been shown to be effective in treating stroke-like episodes. Citrulline administration has shown a greater improvement in NO production compared to arginine administration. Additionally, arginine and citrulline together have also been shown to induce moderation in lactate concentration, therefore suggesting that they may improve lactic acidosis of MELAS (El-Hattab et al. 2015). These therapies are merely short-term solutions, and gene therapy shows the highest potential for treating MELAS in the future.

8.5.4 Alpers Disease

Alpers disease, sometimes known as Alpers–Huttenlocher syndrome, first identified by Bernard Alpers in 1931, is characterized as a hepatopathic or hepatopathic neurodegenerative disease caused by molecular defects in the mitochondria (Alpers 1931). Mainly affecting children, Alpers has been found to be inherited in a mainly autosomal recessive manner (Sandbank and Lerman 1972). Symptoms of the disease are associated with seizures, neuronal dysfunction, ataxia, and degenerative liver disease (Harding et al. 1995; Tzoulis et al. 2006; Wiltshire et al. 2008). Mutations in the mitochondrial DNA associated with POLG or polymerase γ are linked to the expression of Alpers disease (Naviaux and Nguyen 2004; Ferrari et al. 2005; Davidzon et al. 2005). DNA polymerase gamma is responsible for maintaining mitochondrial DNA replication and repair mechanisms (Copeland and Longley 2003). The polymerase gamma gene on chromosome 15q25 encodes for the polymerase gamma protein, which is produced in the cytoplasm and transferred to the inner mitochondrial matrix that binds to mitochondrial proteins. Pol gamma is made up of a 140 kDa alpha subunit that contains DNA polymerase, 30–50 exonuclease, and dRP lyase activities, while the beta subunit comprises a 55 kDa accessory DNA binding factor (Lim et al. 1999). Naviaux et al. introduced the concept of mitochondrial DNA depletion and reduced polymerase gamma activity and later linked it to the phenotypic expression of Alpers-Huttenlocher syndrome (Naviaux et al. 1999; Naviaux and Nguyen 2004). It has been hypothesized that all disorders linked with mutations in the POLG gene share mtDNA depletion as they all originate from defects in the mitochondrial DNA polymerase. In polymerase-g, recessive mutations can exist as homozygous or heterozygous variants. In general, homozygous recessive mutations are linked to milder and later onset disease, but in trans-compound heterozygous mutations, a more severe pathological manifestation is induced. Most of the cases presenting with Alpers disease show one copy of A467T or W748S substitution. In cases where children below the age of 2 express Alpers symptoms, monoallelic expression of a single type of the A467T mutation containing a premature stop codon sequence was found to be the root cause (Chan et al. 2005). Studies have reported that substitutions within the linker region of the POLG gene destabilize the polymerase gamma DNA complex (Chan et al. 2006). These findings underscore the epidemiologic significance of this region and suggest a potential role for it in the disease's pathogenesis (Naviaux and Nguyen 2004; Ferrari et al. 2005; Nguyen et al. 2005). Phenotypic variability can also be attributed to several factors such as epigenetic as well as environmental conditions that can invariably mediate pathogenic manifestations of Alpers disease. One such potential cause of phenotypic variability involves the presence of structural nucleotide variants. Such variants are not expressed unless certain factors that can cause the disease or change the phenotype are present (Saneto and Naviaux 2010). Such mutations have been identified, of which p.E1143G can alter the disease phenotype by increasing enzyme activity and regulating the hepatic sensitivity to valproic acid (Copeland and Longley 2003; Horvath et al. 2006; Stewart et al. 2010). A different study showed the presence of another structural variant p.R964C, which causes mitochondrial toxicity leading to metabolic failure and acidosis when exposed to stavudine (Yamanaka et al. 2007). The main step to diagnose this disease at a molecular level would be to sequence the polymerase gamma gene. Mutational screening and analysis contribute to confirming a diagnosis. Gene sequencing is performed since no physical biomarkers are associated with clinical manifestations of Alpers-Huttenlocher disease. From a therapeutic point of view, treatment for Alpers syndrome was quite limited to supportive and palliative care ranging from gastrotomies (feeding tube for nutrition) to supportive ventilation for collapsed airways. Refractory seizures are very hard to control, even with an established treatment regime. Despite there being no cure, there is emerging therapeutic potential in targeting the regulatory elements that lead to Alpers-Huttenlocher via various genetic therapies discussed further in the chapter.

8.5.5 Leigh's Disease

The mitochondria's main function is to generate energy for cellular metabolism via the TCA cycle, the electron transport chain, and oxidative phosphorylation (DiMauro and Schon 2003). Thus, any damage to the oxidative phosphorylation system will significantly reduce the amount of energy available and thus may cause a harmful phenotypic expression in organ systems that require a lot of energy (LEIGH 1951; Baertling et al. 2014). Generally, starting at the age of 2 years, main symptoms revolve around retardation in neurodevelopment, ataxia, progressive epilepsy, and metabolic failure leading to lactic acidosis. Presentation of Leigh syndrome has been linked to mutations in over 75 nuclear genes (Rahman et al. 2017). Most mitochondrial diseases are associated with deficits in the OXPHOS system that can lead to severe errors in metabolism and the general physiological functioning of the body (Carroll et al. 2014). One such disease that presents itself in early childhood is Leigh syndrome which can cause progressive neuronal dysfunction and leukoencephalopathy. Mutations in the mitochondrial genome leading to complex I dysregulation were said to be the most common cause of disease presentations usually causing oculomotor apraxia, dystonia, and leukoencephalopathy (Baertling et al. 2014; Incecik et al. 2018). Early studies pinpointed mutations such as missense variants of NDUFAF4 and homozygous variants of NDUFV1 that led to complex I deficiency causing neurodevelopmental deficits in cases with Leigh syndrome (Incecik et al. 2018). There are mitochondrial gene variants like MTND (1–6) beyond the nuclear scope of control that causes complex I dysfunction that can cause phenotypic expression very similar to Leigh's symptoms (Spangenberg et al. 2016). Complex V encodes for ATP synthase comprising F_0 and F_1 subunits that generates molecular ATP through the last step of the oxidative phosphorylation pathways. Studies have linked the expression of maternally inherited Leigh syndrome to genotypic variant of MTATP6 m8993 T >G/C that disrupts the F_0 subunit of the ATPase (Ganetzky et al. 2019). mtDNA codes for several components of the OXPHOS system, and the manifestation of a genotypic mutant in any of the complexes leads to DNA depletion causing the establishment of a pathogenic variant

(Nogueira et al. 2011). Depletion syndromes pertaining to not only Leigh syndrome but also other overlapping disorders shed an important light on developing genetic therapies and diagnostic tools to potentially treat it.

There are several methods to investigate and detect the presence of Leigh syndrome using various biochemical and genetic approaches. Any impairment in the mitochondrial OXPHOS system will lead to an increased lactate/pyruvate profile in blood plasma and CSF, leading to metabolic failure and acidosis. Certain causes of Leigh syndrome are curable such as the thiamine transporter deficiencies caused by mutations such as TPK1 and SLC19A3, which can be diagnosed by screening for thiamine pyrophosphate (Ortigoza-Escobar et al. 2016; Bugiardini et al. 2019). Genetic evaluation for the disease includes NGS sequencing as its primary step alongside exome sequencing, gene panels, etc. to find disease-specific mutations (Kremer et al. 2018). Whole-exome sequencing has recently been characterized as the method to genetically evaluate mitochondrial diseases as it would analyse both coding and non-coding sequences. Genetic counselling plays a huge role in diagnosing and carving out a treatment plan for Leigh's disease. Linkage studies and segregation analysis of the mother contribute to prenatal diagnosis of the disease (White et al. 1999; Nesbitt et al. 2014). When an mtDNA mutation is present, the percentage of heteroplasmy in the mother's blood and urine sample, as well as the foetus, can be determined by chorionic villi biopsy (Nesbitt et al. 2014). There are a few treatment options observed in Leigh syndrome associated with thiamine transporter deficiencies caused by mutations in SCL19A3, TPK1, and PDHA1 that include oral dosage forms of thiamine (Brown 2014; Distelmaier et al. 2017). Seizures are a common symptomatic manifestation of Leigh syndrome that can be managed by valproic acid, phenobarbital, and phenytoin despite its cytotoxic effects on the electron transport chain (Finsterer and Zarrouk Mahjoub 2012). Manfredi G. et al. in 2002 induced allotopic gene expression in the mammalian nucleus to preserver MTATP6 functioning, while a study conducted by Tanaka M et al. employed directed restriction endonucleases to target and cut mutated mtDNA (Manfredi et al. 2002; Bacman et al. 2013). Thus, emerging gene therapies will be discussed further in the chapter, which show an incredible potential to manage mitochondrial diseases.

8.6 Gene Therapies to Treat MD

Mitochondrial diseases are caused by a wide array of mutations in both mitochondrial and nuclear DNA that encode primarily for the OXPHOS system, leading to impairment in metabolic function leading to either tissue-specific or systemic failures (Stenton and Prokisch 2020; Angelova 2021). Primary overlapping symptoms are neurological (ataxia, epilepsy) or sensory deficits but can manifest to varying degrees. Gene therapy is considered a relatively newer form of treatment aiming to cure inherited diseases. There are various forms of gene therapy and different modes to deliver it, which will be further explained in the following section. The basic idea of gene therapy was to recombine foreign DNA and express it in mammalian cells to compensate for any genomic deficits (Elizabeth Hunter Szybalska and Szybalski 1908; Borenfreund and Bendich 1961). According to the American Society of Gene and Cell Therapy, the definition of gene therapy is the "introduction, removal, or change in the content of a person's genetic code with the goal of treating or curing a disease." Gene therapy for mitochondrial mutations was particularly difficult due to different copy numbers in different tissue systems. Therefore, effective gene therapy requires the delivery of a specifically designed vector that can attenuate these mutations in all tissues. Advent of gene therapies in the form of genome editing tools such as CRISPR-Cas9 and ZFN/TALEN was supported by their incredible potential to correct pathogenic mutations to restore normalcy (Li et al. 2020). Mitochondrial replacement therapy was theoretically the most direct way of correcting monogenic disorders where the foreign sequence acts as a fresh template to synthesize proteins (Soldatov et al. 2022). Gene therapy for treating mitochondrial diseases can be classified into two approaches: viral vectors that include adeno-associated viruses and non-viral elements such as nano-therapy, particle-based delivery vectors that contain the genetic material, and genome editing tools.

8.6.1 Viral Based Approaches

The most widely used approach for gene replacement therapy comes from the use of non-pathogenic adeno-associated viruses consisting of capsid proteins and single-stranded DNA. The genome is only about 4.7 kb in size, thus giving it the advantage to create AAV vectors for gene delivery by replacing it entirely by the remedial genetic payload (Li and Samulski 2020). However, the transgene with its regulatory elements should be less than 4.7 kb in size limiting its capabilities. Insertional mutagenesis is avoided owing to the lack of integration machine and associated regulatory circuits in AAV (Nakai et al. 2001). AAV transfects a cell in a very elegant manner. The vectors are transported to the nucleus where the ssDNA is unwrapped by helicases, and transformed into dsDNA, and transcribed and translated by the host machinery (Li and Samulski 2020). The presence of these extra-chromosomal DNA in the nucleus causes these vectors to produce therapeutic proteins in vivo. The following section will focus on pre-clinical studies where AAV vector gene therapy is employed to treat Friedrich ataxia, Leigh syndrome, and Leber hereditary optic neuropathy.

8.6.1.1 Friedrich Ataxia

Friedrich ataxia, as discussed earlier, is caused by a trinucleotide amplification in intron 1 of the FXN gene, which codes for the frataxin protein, which results in a reduced amount of FXN transcripts as well as the protein (Cook and Giunti 2017). The frataxin protein is essential for the formation of FeS clusters in the mitochondrial matrix (Delatycki and Bidichandani 2019). A reduced protein content invariably results in deficient ATP production (Delatycki and Bidichandani 2019). Most common manifestations of the disease include cardiomyopathy, dysphagia, and ataxia that eventually leads to death (Tsou et al. 2011; Cook and Giunti 2017;

Delatycki and Bidichandani 2019). Pre-clinical studies using 3-week-old Mck-CKO mice as the animal model showed that treatment with recombined viral vector AAVrh10-CAG-hFXN prevented cardiomyopathy, the most dangerous symptom of Friedrich ataxia. The study showed tissue-specific improvement and fxn protein expression when the hearts of the mice were studied after 32 weeks of treatment (Perdomini et al. 2014). Such evidence points out the potential of treating Friedrich ataxia with AAV vector-based gene therapy to replace the defunct FXN gene. This gene therapy is currently being developed by Voyager Therapeutics (Anonymous 2019). Another conditional knockout mouse model, Pvalb-CKO mice, was used to study the effects of AAV gene therapy to treat the neuromuscular symptoms of FRDA such as ataxia and sensory neuropathy. Treatment of 3.5-week pre-symptomatic Pvalb-CKO mice with AAV9-CAG-hFXN showed less significant results as compared to treatment of 7.5-week-old symptomatic Pvalb-CKO mice with AAVrh10-CAG-hFXN and AAV9-CAG-hFXN. The latter showed increased sensory reflexes and improved co-ordination (Piguet et al. 2018). However, it is important to note that these mice did develop tremors and were euthanized due to seizures (Piguet et al. 2018). This study significantly showed improved management of FRDA symptoms in mouse model.

8.6.1.2 Leigh's Disease

Leigh syndrome is the most prevalent mitochondrial disease observed in infants and young children (Lake et al. 2015). As discussed earlier, 10–20% of LS patients have mitochondrial DNA mutations that lead to progressive neurodegeneration due to metabolic failure in specific tissues (Lake et al. 2015). As discussed earlier, metabolic insufficiency is caused due to dysfunction in components in the regions responsible for the electron transport chain, TCA cycle, and thiamine metabolism (Gerards et al. 2016). Ndusf4 is one of the primary genes responsible for causing Leigh when mutated. A study used Ndufs4 knockout (Ndufs4-KO) mice that showed characteristic Leigh symptoms around 40 days from birth (Di Meo et al. 2017). Upon receiving both intravenous and intracerebral ventricular injections of AAV9-CMV-hNDUFS4 under the regulation of a constitutive CMV promoter, newborn KO mice showed increased survival rates and improved symptoms (Di Meo et al. 2017). The study highlighted the need to treat tissue-specific region like the brain along with the generic course of treatment (Di Meo et al. 2017). There have been some recent developments to target the brain specifically. According to recent studies, genetically engineered vectors have been employed to cross the blood-brain barrier and transduce neuronal and glial cells along with systemic vector delivery (Deverman et al. 2016). Similarly to NDUSF4, NDUSF3 is linked to the regulation of the electron transport chain complex I (Lou et al. 2018). Engineered knockout mice for NDUSF3 showed an increase in myopathy with decreased motor activity (Pereira et al. 2020). Study showed that treatment with AAV9-CMV-Ndufs3 viral vector after 15–18 days of birth significantly prevented myopathy and reduced serum lactate levels, highlighting restored complex I activity (Pereira et al. 2020). They also observed no difference after 3 months of injection between the treated mice and wild-type mice for motor skills, co-ordination, and physical strength

(Pereira et al. 2020). This study comprehensively shows the extent to which AAV vector-based gene delivery can be used to compensate for pathogenic variants.

8.6.1.3 Leber Hereditary Optic Neuropathy (LHON)

LHON, a mitochondrial retinopathy, is being studied as a potential candidate for viral vector-based gene therapy. Progressive vision loss or retinopathy usually sets in teenage or young adulthood and can lead to complete blindness within 2 months of onset of symptoms (Karaarslan 2019). Polymorphisms within specific gene components that codes for mitochondrial NADH dehydrogenase leads to impairment in the complex I of the electron transport chain (Bahr et al. 2020). Recently, GS010 (Lumevoq), a potential gene therapy treatment, was assessed in a number of Phase III studies in the USA but failed to meet its significant endpoint compared to the placebo. It allowed for the delivery of compensatory ND4 protein into the mitochondria (Yu-Wai-Man et al. 2020). Another study reported the use of MTS-AAV-HSP-hMT-ND4 to experimentally create mouse models and study progressive neuropathy in mice models, where they found conclusive and significant prevention of loss of sight and optic nerve atrophy post-injection (Yu et al. 2018). Detectable DNA and RNA levels of the ND4 protein were observed after 13 months of injection (Yu et al. 2018). Such experimental data points to the potential of therapeutically targeting LHON using AAV vector-mediated gene therapy.

8.6.2 Non-viral Approaches

8.6.2.1 Mitochondrial Replacement Therapy (MRT)

Among the most recent developments in gene therapy to treat mitochondrial diseases, MRT has proven to be a very efficient and effective method. We know that mitochondrial diseases are maternally inherited, and therefore, the mutated mtDNA is transferred via mother's oocyte (Tachibana et al. 2013). This approach entails transferring the nuclear genome of the afflicted mother's mtDNA-mutated oocyte into an enucleated oocyte from a healthy donor with a strong mtDNA. The methods involved in this therapy are pronuclear transfer (PNT), spindle chromosome complex transfer (ST), or polar body transfer (PBT) (as seen in Fig. 8.2). The timing of transmission and the nature of the material conveyed differ across each approach (Herbert and Turnbull 2018). This approach results in the offspring inheriting the nuclear DNA from the respective father and mother while they will carry the mtDNA of the healthy donor and therefore are free from the mutations in mitochondrial DNA (Slone and Huang 2020).

- **Spindle Chromosome Complex Transfer (ST):**
- This method of MRT involves the transfer of the nuclear genome before fertilization. Mammalian oocytes are arrested during meiosis II before fertilization in order to prepare for the process. This may be utilized because polarized light birefringence makes it very simple to view the spindle fibres, making it feasible

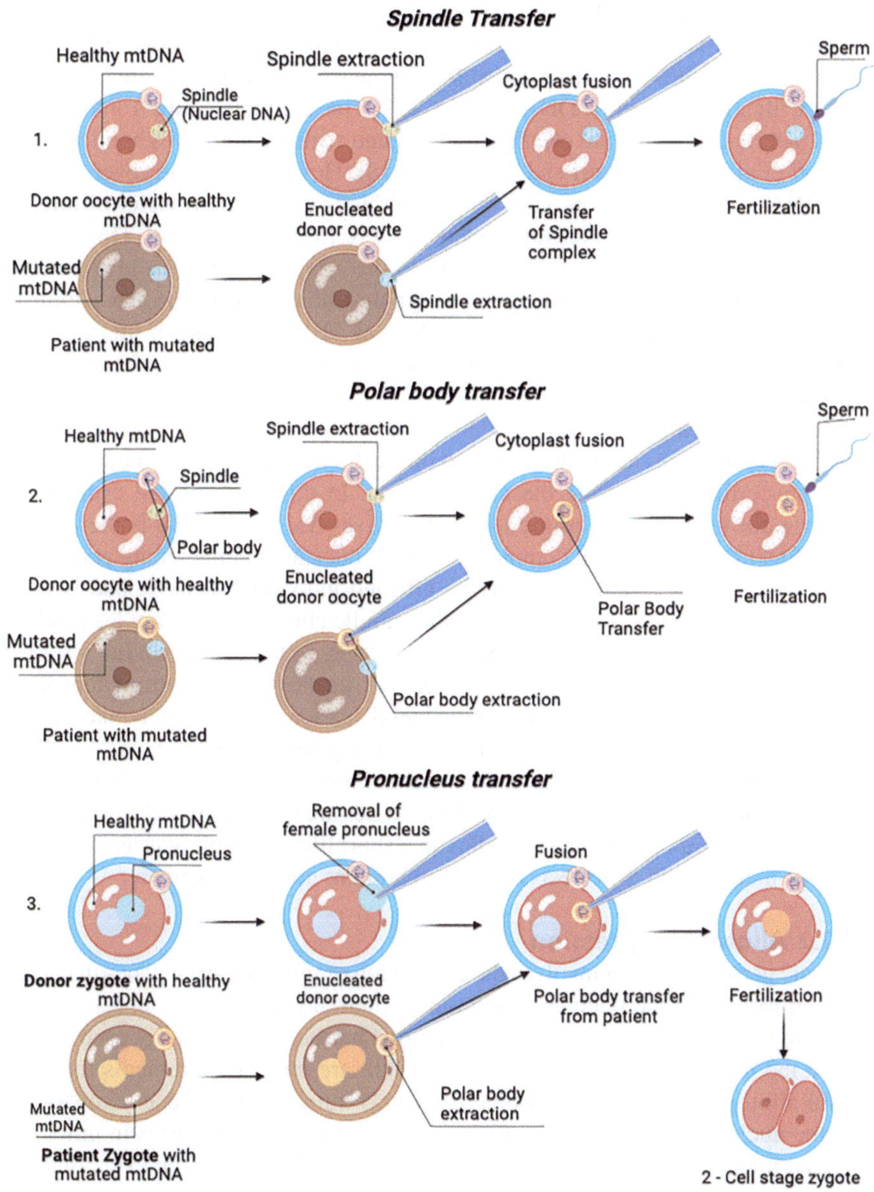

Fig. 8.2 Mitochondrial replacement therapy

to transplant the spindle fibre complex along with the chromosomes (Herbert and Turnbull 2018) (Fig. 8.2.1). However, there is a low carryover percentage of mtDNA that gets transferred along with the spindle chromosome complex. This causes a low amount of heteroplasmy in the offspring (Wolf et al. 2015).

- **Pronucleus Transfer (PNT):**
- This method is employed for the transfer of nuclear genome post-fertilization. The chromosomes are transferred at the zygote stage, where there are two pronuclei with a haploid set of DNAs (Fig. 8.2.3). These pronuclei are from maternal and paternal sides and are clearly differentiated with the help of light microscopy (Wolf et al. 2015; Herbert and Turnbull 2018). However, the carryover levels of mtDNA in this method are very high, and this is due to the transfer of mtDNA from the cytoplasm while transferring the PN from the patients. This level of carryover can prove to be very harmful to the offspring since the magnitude of heteroplasmy can be associated with a high threshold value leading to manifestation of a mitochondrial disease (Samuels et al. 2013). The use of this treatment in people can nonetheless be detrimental, even though a degree of carryover to cause disease is extremely rare.
- **Polar Body Transfer (PBT):**
- This method serves as an alternative to pronuclei transfer. This method is utilized for pre-fertilization during the meiotic cycle. In this stage of cell cycle of the oocyte, the uneven cytoplasmic segregation gives rise to two polar bodies that are non-functional but carry a complement of the nuclear genome (Kang et al. 2016) (Fig. 8.2). A PB is removed from the affected individual's oocyte and is transferred into the donor oocyte in which the spindle chromosome complex has been removed (Wolf et al. 2015) (Fig. 8.2). This results in the oocyte having a copy of nuclear DNA from the affected mother and healthy mtDNA from the donor. Successful trials have been run in animal models such as mice and monkey as they were able to reconstruct the oocyte without any mtDNA mutations (Tachibana et al. 2009). Additionally, very little mtDNA carryover was present in the mouse pups produced using this approach. Polar bodies are non-functional structures and therefore face apoptotic stress, leading to a short lifespan. This acts as a downside to this method, and the effects of PBT on humans are still unknown.

8.6.2.2 Allotopic Expression

Allotopic gene expression involves the expression of mitochondrial genes in nucleus and then localizing the nucleus-coded mitochondrial proteins back into mitochondria from the cytoplasm (Fig. 8.3). This method has been successfully used in trials against the mitochondrial disease LHON. There are many requirements for a successful allotopic expression of proteins, such as mitochondrial gene compatibility with the nuclear genome expression system, codon usage differences in mitochondrial and nuclear translation systems, and proper localization of the expressed protein from cytoplasm to mitochondria membrane (Artika 2020). The expressed proteins must be imported into the mitochondria with the help of incorporating mitochondrial targeting sequence (MTS) into the polypeptide (Di Donfrancesco et al. 2022). This is one of the problems that allotopically expressed proteins encounter most frequently. Most of the proteins encoded in nucleus that are localized into the mitochondria are done via MTS. This localization is done with the help of chaperons such as Hsp7, which are present in both the cytoplasm and mitochondria. Although the proteins follow the MTS pathway for localization, lot of other

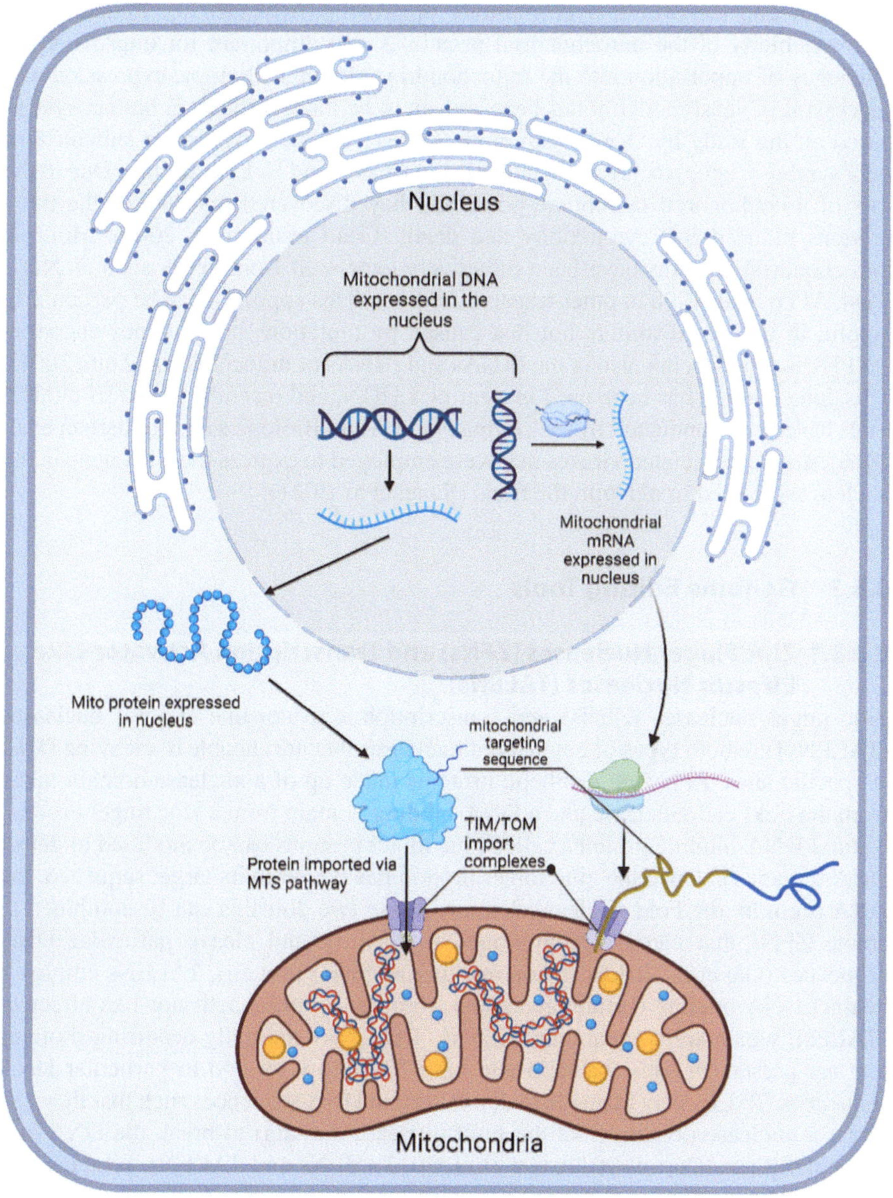

Fig. 8.3 Allotopic expression of mitochondrial proteins in nucleus

factors also play a role in the importation of mitochondrial proteins into mitochondria. These factors include the compatibility and functionality of the protein's MTS pathway, hydrophobicity of the mitochondrial protein, which plays a major role in transportation of the protein in the inner mitochondrial membrane, etc. Other findings have proven that tandem duplication of MTS helps improve the importation efficiency and the overall hydrophilicity of protein (Artika 2020). Reducing the hydrophobicity of the mitochondrial protein is very important for improving the efficiency of importation into the mitochondria. Although allotopic expression was successful in yeast model, it has been proven to be more difficult in human system based on the study by Oca-Cossio et al. (2003) who tried expressing subunit 8 of ATP synthase, apocytochrome b, and ND4 in COS-7 and HeLa cell lines. Due to the loss of mitochondrial membrane potential, they discovered that two of the three proteins induced cell cytotoxicity and death (Oca-Cossio et al. 2003). Multiple mitochondrial proteins have been effectively expressed from genes such as ND1, ND4, ATP6, and ATP8 in other research findings. This approach can be particularly helpful in treating disorders not just caused by mutations in the genes encoding OXPHOS subunits, but also in the tRNAs and rRNAs of mitochondria (Zullo 2001). This gene therapy has been used in treating LHON, and recently, Phase III clinical trials have been conducted by the company GenSight Biologics (Boominathan et al. 2016). Adeno-associated viruses that were employed to express the ND4 gene in the nucleus were used to aid with the trials (Sahel et al. 2021).

8.6.3 Genome Editing Tools

8.6.3.1 Zinc Finger Nucleases (ZFNs) and Transcription Activator-Like Effector Nucleases (TALENs)

Zinc finger nucleases (ZFNs) and transcription activator-like effector nucleases (TALENs) are both types of engineered nucleases that are capable of cleaving DNA at specific sites. ZFNs are synthetic proteins made up of a nuclease domain taken from the FokI endonuclease and a DNA-binding domain from a zinc finger protein. Natural DNA-binding proteins called zinc finger proteins can be modified to detect DNA sequences. Once the zinc finger protein has bound to its target sequence, the DNA is cut by the FokI nuclease domain. These two domains can be combined to create ZFNs that can be programmed to identify and cleave particular DNA sequences (Gaj et al. 2013). In contrast to zinc finger proteins, TALENs employ a distinct DNA-binding domain generated from transcription activator-like effectors (TALEs), which are comparable to ZFNs. TALEs are naturally occurring proteins that are present in some bacteria and have the ability to bind to particular DNA sequences. TALEs may be made to detect certain DNA sequences such that they can direct a nuclease domain (like the FokI nuclease domain) to break the DNA at a particular place (Carroll 2011; Gaj et al. 2013). ZFNs and TALENs, proteins that detect and break DNA in a sequence-specific way, are currently promising technologies. Although ZFNs also require the presence of a nuclear exclusion sequence to ensure that there are no off-target effects in genomic DNA, modification of their N-termini to express the COXVIIIMTS drives efficient localization to mitochondria (Russell et al. 2020). Bacman et al. (2013) initially concentrated on the selective degradation of mtDNA heteroplasmic for either the m.8483 13459del4977 "common deletion" or the m.14459G > A point mutation. These researchers were among the first to use TALENs to eliminate mutant mtDNA genomes. The amount of mutant heteroplasmy was significantly reduced in both cases (Bacman et al. 2013). Adeno-associated viruses have also been used to transport the ZFNs and TALENs that were used to edit the mitochondrial genome. Although several trials have attempted to target the cerebral area, the AAV delivery was unable to cross the blood-brain barrier (Russell et al. 2020).

8.6.3.2 CRISPR-Cas9 (Clustered Regularly Interspaced Short Palindromic Repeats)

The genome editing technique CRISPR-Cas9 employs the RNA-guided endonuclease (Cas9) to cut DNA at specified locations in the genome. It is based on a natural defensive mechanism utilized by bacteria and archaea, which employs the CRISPR system to find and eliminate plasmids or viruses that are invading their territory. Scientists may utilize CRISPR-Cas9 to precisely alter the genome by repairing mutations, introducing new genes, or eliminating undesirable regions by programming the system to target certain DNA sequences.

However, there are many barriers to using gene editing tools for gene therapy of mitochondrial diseases:

- To access the mtDNA, the inner mitochondrial membrane must be penetrated. The mitochondrial DNA's impermeability serves as a barrier to the entry of foreign molecules into the mitochondria. For the foreign molecule to enter the mitochondria, it must be linked to a transporter (Russell et al. 2020).
- Difficulty in specificity of the molecule to target the mutated mtDNA molecules. This helps in the propagation of wild-type DNA over time.
- The transfer of RNA into mitochondria is restricted, and therefore, the guide RNA-based CRISPR system is difficult to utilize for gene therapy of mitochondrial diseases (Fogleman et al. 2016). Fogleman et al. (2016) give insight into the usage of CRISPR-Cas9 system to edit the embryonic mtDNA. This is due to the fact that the number of cells in the embryo is lesser and therefore the mutation load will be lower in case of 8- or 16-cell stage (Fogleman et al. 2016).

Even though the CRISPR-Cas9 system has a lot of potential for treating mitochondrial diseases, the following obstacles must be eliminated before clinical trials can start. The CRISPR-Cas9 system's off-target consequences must be minimized since they might raise moral dilemmas.

8.7 Translational Perspectives and Challenges

Although there are no established protocols to treat mitochondrial disease with gene therapy, it remains the most viable option to treat mitochondrial disorders in terms of ease of delivery and biological specificity. There are not many options to therapeutically cure mitochondrial disorders, while strategies exist to manage their clinical manifestations. One of the key points to remember is that mitochondrial disorders are extremely complex and not linear in their presentations, which makes them harder to diagnose, let alone treat them. A major challenge in targeting mitochondrial disorders is treating them within the established time frame of the disease. These diseases are generally progressive in nature and eventually lead to death; hence, it is imperative to research and find novel therapeutics to retard disease progression and manage symptoms at the same time. One of the important challenges is titrating doses of AAV-based gene therapy (Hanaford et al. 2022). For example, greater amounts of AAV-based gene therapy to produce FXN proteins led to increased symptoms and cardiac toxicity (Belbellaa et al. 2020). Hence, dosage forms and values should be carefully considered in order to reduce the pathogenicity of the disease as well as not to induce cytotoxicity. As mentioned in the chapter, mitochondrial disorders mostly manifest themselves in a systemic manner, and it can be difficult to prioritize which systems to target in case of multisystem presentations. Another important challenge while developing these therapies is the associated costs to discover and develop a therapeutic product. Since mitochondrial diseases are incredibly rare, it is difficult to establish multicentric clinical trials due to small number of patients (Lyseng-Williamson 2016).

Gene therapies can be effectively engineered to target overlapping mitochondrial diseases with similar molecular targets with the intent of increasing cellular ATP production and normalizing metabolic pathways in dysfunctional cells. Along with AAV-based vector therapy and mitochondrial replacement therapy, there is an increased interest in the field of genome editing tools, especially CRISPR, which uses a guide RNA and nuclease to cleave a particular sequence and then induces recombination in a very specific location in the genome. Recent efforts carried out by Hussain et al. (2021) have demonstrated the use of a stem-loop on the guide RNA that facilitates entry into the mitochondria. Such innovations and groundbreaking positive results highlight the potential of investigating gene therapies to target rare mitochondrial diseases.

References

Alpers BJ (1931) Diffuse progressive degeneration of the gray matter of the cerebrum. Arch Neurol Psychiatr 25:469–505. https://doi.org/10.1001/archneurpsyc.1931.02230030027002

Alston CL, Rocha MC, Lax NZ et al (2017) The genetics and pathology of mitochondrial disease. J Pathol 241:236–250

Angelova PR (2021) Sources and triggers of oxidative damage in neurodegeneration. Free Radic Biol Med 173:52–63. https://doi.org/10.1016/j.freeradbiomed.2021.07.003

Anonymous (2019) Voyager's CNS gene therapy. Nat Biotechnol 37:204

Anonymous (2023) Mitochondrial diseases: causes, symptoms, diagnosis & treatment

Artika IM (2020) Allotopic expression of mitochondrial genes: basic strategy and progress. Genes Dis 7:578–584

Bacman SR, Williams SL, Pinto M et al (2013) Specific elimination of mutant mitochondrial genomes in patient-derived cells by mitoTALENs. Nat Med 19:1111–1113. https://doi.org/10.1038/nm.3261

Baertling F, Rodenburg RJ, Schaper J et al (2014) A guide to diagnosis and treatment of Leigh syndrome. J Neurol Neurosurg Psychiatry 85:257–265

Bahr T, Welburn K, Donnelly J, Bai Y (2020) Emerging model systems and treatment approaches for Leber's hereditary optic neuropathy: challenges and opportunities. Biochim Biophys Acta Mol basis Dis 1866:165743. https://doi.org/10.1016/j.bbadis.2020.165743

Belbellaa B, Reutenauer L, Messaddeq N et al (2020) High levels of frataxin overexpression lead to mitochondrial and cardiac toxicity in mouse models. Mol Ther Methods Clin Dev 19:120–138. https://doi.org/10.1016/j.omtm.2020.08.018

Boominathan A, Vanhoozer S, Basisty N et al (2016) Stable nuclear expression of *ATP8* and *ATP6* genes rescues a mtDNA complex V *null* mutant. Nucleic Acids Res gkw756. https://doi.org/10.1093/nar/gkw756

Borenfreund E, Bendich A (1961) A study of the penetration of mammalian cells by deoxyribonucleic acids. J Cell Biol 9:81–91. https://doi.org/10.1083/jcb.9.1.81

Brown G (2014) Defects of thiamine transport and metabolism. J Inherit Metab Dis 37(4):577–585

Brown WM, George M, Wilson AC (1979) Rapid evolution of animal mitochondrial DNA. Proc Natl Acad Sci 76:1967–1971. https://doi.org/10.1073/pnas.76.4.1967

Bua E, Johnson J, Herbst A et al (2006) Mitochondrial DNA–deletion mutations accumulate intracellularly to detrimental levels in aged human skeletal muscle fibers. Am J Hum Genet 79:469–480. https://doi.org/10.1086/507132

Bugiardini E, Pope S, Feichtinger RG et al (2019) Utility of whole blood thiamine pyrophosphate evaluation in TPK1-related diseases. J Clin Med 8:991. https://doi.org/10.3390/jcm8070991

Calvo SE, Mootha VK (2010) The mitochondrial proteome and human disease. Annu Rev Genomics Hum Genet 11:25–44. https://doi.org/10.1146/annurev-genom-082509-141720

Calvo SE, Clauser KR, Mootha VK (2016) MitoCarta2.0: an updated inventory of mammalian mitochondrial proteins. Nucleic Acids Res 44:D1251–D1257. https://doi.org/10.1093/nar/gkv1003

Campuzano V, Montermini L, Molto MD et al (1996) Friedreich's ataxia: autosomal recessive disease caused by an intronic GAA triplet repeat expansion. Science 271(5254):1423–1427

Carroll D (2011) Genome engineering with zinc-finger nucleases. Genetics 188:773–782. https://doi.org/10.1534/genetics.111.131433

Carroll CJ, Brilhante V, Suomalainen A (2014) Next-generation sequencing for mitochondrial disorders. Br J Pharmacol 171:1837–1853

Chan SSL, Longley MJ, Naviaux RK, Copeland WC (2005) Mono-allelic POLG expression resulting from nonsense-mediated decay and alternative splicing in a patient with Alpers syndrome. DNA Repair 4(12):1381–1389

Chan SSL, Longley MJ, Copeland WC (2006) Modulation of the W748S mutation in DNA polymerase γ by the E1143G polymorphism in mitochondrial disorders. Hum Mol Genet 15:3473–3483. https://doi.org/10.1093/hmg/ddl424

Chinnery PF (2015) Mitochondrial disease in adults: what's old and what's new? EMBO Mol Med 7:1503–1512. https://doi.org/10.15252/emmm.201505079

Chinnery PF, Hudson G (2013) Mitochondrial genetics. Br Med Bull 106:135–159

Cook A, Giunti P (2017) Friedreich's ataxia: clinical features, pathogenesis and management. Br Med Bull 124:19–30

Copeland WC, Longley MJ (2003) DNA polymerase gamma in mitochondrial DNA replication and repair. ScientificWorldJournal 3:34–44

Cossée M, Schmitt M, Campuzano V et al (1997) Evolution of the Friedreich's ataxia trinucleotide repeat expansion: founder effect and premutations. Proc Natl Acad Sci 94:7452–7457. https://doi.org/10.1073/pnas.94.14.7452

Davidzon G, Mancuso M, Ferraris S et al (2005) POLG mutations and Alpers syndrome. Ann Neurol 57:921–923. https://doi.org/10.1002/ana.20498

Delatycki MB, Bidichandani SI (2019) Friedreich ataxia—pathogenesis and implications for therapies. Neurobiol Dis 132:104606

Deverman BE, Pravdo PL, Simpson BP et al (2016) Cre-dependent selection yields AAV variants for widespread gene transfer to the adult brain. Nat Biotechnol 34:204–209. https://doi.org/10.1038/nbt.3440

Di Donfrancesco A, Massaro G, Di Meo I et al (2022) Gene therapy for mitochondrial diseases: current status and future perspective. Pharmaceutics 14:1287

Di Meo I, Marchet S, Lamperti C et al (2017) AAV9-based gene therapy partially ameliorates the clinical phenotype of a mouse model of Leigh syndrome. Gene Ther 24:661–667. https://doi.org/10.1038/gt.2017.53

DiMauro S, Schon EA (2003) Mitochondrial respiratory-chain diseases. N Engl J Med 348:2656–2668. https://doi.org/10.1056/nejmra022567

Distelmaier F, Haack TB, Wortmann SB et al (2017) Treatable mitochondrial diseases: cofactor metabolism and beyond. Brain 140:e11

Duchen MR (2000) Mitochondria and calcium: from cell signalling to cell death. J Physiol 529:57–68. https://doi.org/10.1111/j.1469-7793.2000.00057.x

El-Hattab AW, Adesina AM, Jones J, Scaglia F (2015) MELAS syndrome: clinical manifestations, pathogenesis, and treatment options. Mol Genet Metab 116:4–12

Elizabeth Hunter Szybalska B, Szybalski W (1908) GENETICS OF HUMAN CELL LINES, IV. DNA-MEDIATED HERITABLE TRANSFORMATION OF A BIOCHEMICAL TRAIT

Fernandez-Vizarra E, Zeviani M, Zeviani M (2021) Mitochondrial disorders of the OXPHOS system. FEBS Lett 595:1062–1106. https://doi.org/10.1002/1873-3468.13995

Ferrari G, Lamantea E, Donati A et al (2005) Infantile hepatocerebral syndromes associated with mutations in the mitochondrial DNA polymerase-γA. Brain 128:723–731. https://doi.org/10.1093/brain/awh410

Finsterer J, Zarrouk Mahjoub S (2012) Mitochondrial toxicity of antiepileptic drugs and their tolerability in mitochondrial disorders. Expert Opin Drug Metab Toxicol 8:71–79. https://doi.org/10.1517/17425255.2012.644535

Fogleman S, Santana C, Bishop C, et al (2016) CRISPR/Cas9 and mitochondrial gene replacement therapy: promising techniques and ethical considerations

Gaj T, Gersbach CA, Barbas CF (2013) ZFN, TALEN, and CRISPR/Cas-based methods for genome engineering. Trends Biotechnol 31:397–405. https://doi.org/10.1016/j.tibtech.2013.04.004

Ganetzky RD, Stendel C, McCormick EM et al (2019) MT-ATP6 mitochondrial disease variants: phenotypic and biochemical features analysis in 218 published cases and cohort of 14 new cases. Hum Mutat 40:499–515

Gerards M, Sallevelt SCEH, Smeets HJM (2016) Leigh syndrome: resolving the clinical and genetic heterogeneity paves the way for treatment options. Mol Genet Metab 117:300–312

Giles RE, Blanc H, Cann HM, Wallace DC (1980) Maternal inheritance of human mitochondrial DNA. Proc Natl Acad Sci 77:6715–6719. https://doi.org/10.1073/pnas.77.11.6715

Gorman GS, Chinnery PF, DiMauro S et al (2016) Mitochondrial diseases. Nat Rev Dis Primers 2:16080. https://doi.org/10.1038/nrdp.2016.80

Greaves LC, Reeve AK, Taylor RW, Turnbull DM (2012) Mitochondrial DNA and disease. J Pathol 226:274–286

Hanaford AR, Cho YJ, Nakai H (2022) AAV-vector based gene therapy for mitochondrial disease: progress and future perspectives. Orphanet J Rare Dis 17:217

Harding BN, Alsanjari N, Smith SJM et al (1995) Progressive neuronal degeneration of childhood with liver disease (Alpers' disease) presenting in young adults. J Neurol Neurosurg Psychiatry 58:320–325. https://doi.org/10.1136/jnnp.58.3.320

Hayashi G, Cortopassi G (2015) Oxidative stress in inherited mitochondrial diseases. Free Radic Biol Med 88:10–17

Herbert M, Turnbull D (2018) Progress in mitochondrial replacement therapies. Nat Rev Mol Cell Biol 19:71–72

Hirano M, Pavlakis SG (1994) Mitochondrial myopathy, encephalopathy, lactic acidosis, and strokelike episodes (MELAS): current concepts. J Child Neurol 9:4–13

Hirano M, Emmanuele V, Quinzii CM (2018) Emerging therapies for mitochondrial diseases. Essays Biochem 62:467–481

Horvath R, Hudson G, Ferrari G et al (2006) Phenotypic spectrum associated with mutations of the mitochondrial polymerase γ gene. Brain 129:1674–1684. https://doi.org/10.1093/brain/awl088

Hussain SRA, Yalvac ME, Khoo B et al (2021) Adapting CRISPR/Cas9 system for targeting mitochondrial genome. Front Genet 12:627050. https://doi.org/10.3389/fgene.2021.627050
Ikeda T, Osaka H, Shimbo H et al (2018) Mitochondrial DNA 3243A>T mutation in a patient with MELAS syndrome. Hum Genome Var 5:25. https://doi.org/10.1038/s41439-018-0026-6
Incecik F, Herguner O, Besen S et al (2018) Late-onset Leigh syndrome due to NDUFV1 mutation in a 10-year-old boy initially presenting with ataxia. J Pediatr Neurosci 13:205–207. https://doi.org/10.4103/jpn.jpn_138_17
Isnard R, Kalotka H, Dürr A et al (1997) Correlation between left ventricular hypertrophy and GAA trinucleotide repeat length in Friedreich's ataxia. Circulation 95:2247–2249. https://doi.org/10.1161/01.CIR.95.9.2247
Kang E, Wu J, Gutierrez NM et al (2016) Mitochondrial replacement in human oocytes carrying pathogenic mitochondrial DNA mutations. Nature 540:270–275. https://doi.org/10.1038/nature20592
Karaarslan C (2019) Leber's hereditary optic neuropathy as a promising disease for gene therapy development. Adv Ther 36:3299–3307
Kirches E (2011) LHON: mitochondrial mutations and more
Koeppen AH (2011) Friedreich's ataxia: pathology, pathogenesis, and molecular genetics. J Neurol Sci 303:1–12
Kremer LS, Wortmann SB, Prokisch H (2018) "Transcriptomics": molecular diagnosis of inborn errors of metabolism via RNA-sequencing. J Inherit Metab Dis 41:525–532. https://doi.org/10.1007/s10545-017-0133-4
Lake NJ, Bird MJ, Isohanni P, Paetau A (2015) Leigh syndrome. J Neuropathol Exp Neurol 74:482–492. https://doi.org/10.1097/NEN.0000000000000195
Lane N (2011) Mitonuclear match: optimizing fitness and fertility over generations drives ageing within generations. BioEssays 33:860–869. https://doi.org/10.1002/bies.201100051
Lawless C, Greaves L, Reeve AK et al (2020) The rise and rise of mitochondrial DNA mutations. Open Biol 10:200061
Lax NZ, Turnbull DM, Reeve AK (2011) Mitochondrial mutations: newly discovered players in neuronal degeneration. Neuroscientist 17:645–658. https://doi.org/10.1177/1073858411385469
LEIGH D (1951) Subacute necrotizing encephalomyelopathy in an infant. J Neurol Neurosurg Psychiatry 14:216–221. https://doi.org/10.1136/jnnp.14.3.216
Li C, Samulski RJ (2020) Engineering adeno-associated virus vectors for gene therapy. Nat Rev Genet 21:255–272
Li H, Yang Y, Hong W et al (2020) Applications of genome editing technology in the targeted therapy of human diseases: mechanisms, advances and prospects. Signal Transduct Target Ther 5:1
Lim SE, Longley MJ, Copeland WC (1999) The mitochondrial p55 accessory subunit of human DNA polymerase γ enhances DNA binding, promotes Processive DNA synthesis, and confers N-ethylmaleimide resistance. J Biol Chem 274:38197–38203. https://doi.org/10.1074/jbc.274.53.38197
Lou X, Shi H, Wen S et al (2018) A novel NDUFS3 mutation in a Chinese patient with severe Leigh syndrome. J Hum Genet 63:1269–1272. https://doi.org/10.1038/s10038-018-0505-0
Lyseng-Williamson KA (2016) Idebenone: a review in Leber's hereditary optic neuropathy. Drugs 76:805–813. https://doi.org/10.1007/s40265-016-0574-3
Manfredi G, Fu J, Ojaimi J et al (2002) Rescue of a deficiency in ATP synthesis by transfer of MTATP6, a mitochondrial DNA-encoded gene, to the nucleus. Nat Genet 30:394–399. https://doi.org/10.1038/ng851
McFarland R, Chinnery PF, Blakely EL et al (2007) Homoplasmy, heteroplasmy, and mitochondrial dystonia. Neurology 69:911–916. https://doi.org/10.1212/01.wnl.0000267843.10977.4a
Nakai H, Yant SR, Storm TA et al (2001) Extrachromosomal recombinant adeno-associated virus vector genomes are primarily responsible for stable liver transduction in vivo. J Virol 75:6969–6976. https://doi.org/10.1128/jvi.75.15.6969-6976.2001
Nass MM (1966) The circularity of mitochondrial DNA. Proc Natl Acad Sci 56:1215–1222. https://doi.org/10.1073/pnas.56.4.1215

Naviaux RK, Nguyen KV (2004) POLG mutations associated with Alpers' syndrome and mitochondrial DNA depletion. Ann Neurol 55:706–712. https://doi.org/10.1002/ana.20079

Naviaux RK, Nyhan WL, Barshop BA et al (1999) Mitochondrial DNA polymerase γ deficiency and mtDNA depletion in a child with Alpers' syndrome. Ann Neurol 45:54–58

Nesbitt V, Alston CL, Blakely EL et al (2014) A national perspective on prenatal testing for mitochondrial disease. Eur J Hum Genet 22:1255–1259. https://doi.org/10.1038/ejhg.2014.35

Nguyen KV, Østergaard E, Holst Ravn S et al (2005) POLG mutations in Alpers syndrome. Neurology 65:1493–1495. https://doi.org/10.1212/01.wnl.0000182814.55361.70

Nia Monró E, Moltó MD, Martínez F et al (1997) (1997) Phenotype correlation and intergenerational dynamics of the Friedreich ataxia GAA trinucleotide repeat. Am J Hum Genet. 61(1):101–110

Nogueira C, Carrozzo R, Vilarinho L, Santorelli FM (2011) Infantile-onset disorders of mitochondrial replication and protein synthesis. J Child Neurol 26:866–875

Oca-Cossio J, Kenyon L, Hao H, Moraes CT (2003) Limitations of Allotopic expression of mitochondrial genes in mammalian cells. Genetics 165:707–720. https://doi.org/10.1093/genetics/165.2.707

Ochman H, Moran NA (2001) Genes lost and genes found: evolution of bacterial pathogenesis and Symbiosis. Science (1979) 292:1096–1099. https://doi.org/10.1126/science.1058543

Ortigoza-Escobar JD, Molero-Luis M, Arias A et al (2016) Free-thiamine is a potential biomarker of thiamine transporter-2 deficiency: a treatable cause of Leigh syndrome. Brain 139:31–38. https://doi.org/10.1093/brain/awv342

Perdomini M, Belbellaa B, Monassier L et al (2014) Prevention and reversal of severe mitochondrial cardiomyopathy by gene therapy in a mouse model of Friedreich's ataxia. Nat Med 20:542–547. https://doi.org/10.1038/nm.3510

Pereira CV, Peralta S, Arguello T et al (2020) Myopathy reversion in mice after restauration of mitochondrial complex I. EMBO Mol Med 12:e10674. https://doi.org/10.15252/emmm.201910674

Piguet F, de Montigny C, Vaucamps N et al (2018) Rapid and complete reversal of sensory ataxia by gene therapy in a novel model of Friedreich ataxia. Mol Ther 26:1940–1952. https://doi.org/10.1016/j.ymthe.2018.05.006

Puccio H, Koenig M (2002) Friedreich ataxia: a paradigm for mitochondrial diseases. Curr Opin Genet Dev 12(3):272–277

Rahman J, Noronha A, Thiele I, Rahman S (2017) Leigh map: a novel computational diagnostic resource for mitochondrial disease. Ann Neurol 81:9–16. https://doi.org/10.1002/ana.24835

Riordan-Eva P, Sanders MD, Govan GG et al (1995) The clinical features of Leber's hereditary optic neuropathy defined by the presence of a pathogenic mitochondrial DNA mutation. Brain 118:319–337. https://doi.org/10.1093/brain/118.2.319

Rooney JP, Ryde IT, Sanders LH et al (2015) PCR based determination of mitochondrial DNA copy number in multiple species. Methods Mol Biol 1241:23–38

Rossignol R, Faustin B, Rocher C et al (2003) Mitochondrial threshold effects. Biochem J 370:751–762. https://doi.org/10.1042/bj20021594

Russell OM, Gorman GS, Lightowlers RN, Turnbull DM (2020) Mitochondrial diseases: hope for the future. Cell 181:168–188. https://doi.org/10.1016/j.cell.2020.02.051

Sahel J-A, Newman NJ, Yu-Wai-Man P et al (2021) Gene therapies for the treatment of Leber hereditary optic neuropathy. Int Ophthalmol Clin 61(4):195–208

Samuels DC, Wonnapinij P, Chinnery PF (2013) Preventing the transmission of pathogenic mitochondrial DNA mutations: can we achieve long-term benefits from germ-line gene transfer? Hum Reprod 28:554–559. https://doi.org/10.1093/humrep/des439

Sandbank U, Lerman P (1972) Progressive cerebral poliodystrophy—Alpers' disease. Disorganized giant neuronal mitochondria on electron microscopy. J Neurol Neurosurg Psychiatry 35:749–755. https://doi.org/10.1136/jnnp.35.6.749

Saneto RP, Naviaux RK (2010) Polymerase gamma disease through the ages. Dev Disabil Res Rev 16:163–174

Sato M, Sato K (2012) Maternal inheritance of mitochondrial DNA. Autophagy 8:424–425. https://doi.org/10.4161/auto.19243

Silva AM, Brown JM, Buckle VJ et al (2015) Expanded GAA repeats impair FXN gene expression and reposition the FXN locus to the nuclear lamina in single cells. Hum Mol Genet 24:3457–3471. https://doi.org/10.1093/hmg/ddv096
Slone J, Huang T (2020) The special considerations of gene therapy for mitochondrial diseases. NPJ Genom Med 5:7
Soldatov VO, Kubekina MV, Skorkina MY et al (2022) Current advances in gene therapy of mitochondrial diseases. J Transl Med 20:562
Spangenberg L, Graña M, Greif G et al (2016) 3697G>A in MT-ND1 is a causative mutation in mitochondrial disease. Mitochondrion 28:54–59. https://doi.org/10.1016/j.mito.2016.03.006
Stefano GB, Kream RM (2016) Mitochondrial DNA heteroplasmy in human health and disease. Biomed Rep 4:259–262. https://doi.org/10.3892/br.2016.590
Stenton SL, Prokisch H (2020) Genetics of mitochondrial diseases: identifying mutations to help diagnosis. EBioMedicine 56:102784. https://doi.org/10.1016/j.ebiom.2020.102784
Stewart JD, Horvath R, Baruffini E et al (2010) Polymerase γGene POLG determines the risk of sodium valproate-induced liver toxicity. Hepatology 52:1791–1796. https://doi.org/10.1002/hep.23891
Tachibana M, Sparman M, Sritanaudomchai H et al (2009) Mitochondrial gene replacement in primate offspring and embryonic stem cells. Nature 461:367–372. https://doi.org/10.1038/nature08368
Tachibana M, Amato P, Sparman M et al (2013) Towards germline gene therapy of inherited mitochondrial diseases. Nature 493:627–631. https://doi.org/10.1038/nature11647
Taylor RW, Turnbull DM (2005) Mitochondrial DNA mutations in human disease. Nat Rev Genet 6:389–402. https://doi.org/10.1038/nrg1606
Thorburn DR (2000) Practical problems in detecting abnormal mitochondrial function and genomes. Hum Reprod 15:57–67. https://doi.org/10.1093/humrep/15.suppl_2.57
Tinker RJ, Lim AZ, Stefanetti RJ, McFarland R (2021) Current and emerging clinical treatment in mitochondrial disease. Mol Diagn Ther 25:181–206
Tsou AY, Paulsen EK, Lagedrost SJ et al (2011) Mortality in Friedreich ataxia. J Neurol Sci 307:46–49. https://doi.org/10.1016/j.jns.2011.05.023
Tuppen HAL, Blakely EL, Turnbull DM, Taylor RW (2010) Mitochondrial DNA mutations and human disease. Biochim Biophys Acta Bioenerg 1797:113–128
Tzoulis C, Engelsen BA, Telstad W et al (2006) The spectrum of clinical disease caused by the A467T and W748S POLG mutations: a study of 26 cases. Brain 129:1685–1692. https://doi.org/10.1093/brain/awl097
Wallace DC, Chalkia D (2013) Mitochondrial DNA genetics and the Heteroplasmy conundrum in evolution and disease. Cold Spring Harb Perspect Biol 5:a021220–a021220. https://doi.org/10.1101/cshperspect.a021220
White SL, Collins VR, Wolfe R et al (1999) Genetic counseling and prenatal diagnosis for the mitochondrial DNA mutations at nucleotide 8993. Am J Hum Genet 65:474–482. https://doi.org/10.1086/302488
Wiltshire E, Davidzon G, Dimauro S et al (2008) Juvenile Alpers disease. Arch Neurol 65:121–124. https://doi.org/10.1001/archneurol.2007.14
Wolf DP, Mitalipov N, Mitalipov S (2015) Mitochondrial replacement therapy in reproductive medicine. Trends Mol Med 21:68–76
Yamanaka H, Gatanaga H, Kosalaraksa P et al (2007) Novel mutation of human DNA polymerase γ associated with mitochondrial toxicity induced by anti-HIV treatment. J Infect Dis 195:1419–1425. https://doi.org/10.1086/513872
Yang S, Ma S, Wan X et al (2016) Long-term outcomes of gene therapy for the treatment of Leber's hereditary optic neuropathy. EBioMedicine 10:258–268. https://doi.org/10.1016/j.ebiom.2016.07.002
Yu H, Porciatti V, Lewin A et al (2018) Longterm reversal of severe visual loss by mitochondrial gene transfer in a mouse model of Leber hereditary optic neuropathy. Sci Rep 8:5587. https://doi.org/10.1038/s41598-018-23836-y

Yu-Wai-Man P, Newman NJ, Carelli V et al (2020) Bilateral visual improvement with unilateral gene therapy injection for Leber hereditary optic neuropathy. Sci Transl Med 12:eaaz7423. https://doi.org/10.1126/scitranslmed.aaz7423

Zhang H, Burr SP, Chinnery PF (2018) The mitochondrial DNA genetic bottleneck: inheritance and beyond. Essays Biochem 62:225–234

Zullo SJ (2001) Gene therapy of mitochondrial DNA mutations: a brief, biased history of allotopic expression in mammalian cells. Semin Neurol 21(3):327–335

Challenges and Future Opportunities in Rare Genetic Disorders: A Comprehensive Review

9

Nithar Ranjan Madhu, Goutam Biswas, Soumosish Paul, Suman Adhikari, Bhanumati Sarkar, Misbahuddin M. Rafeeq, and Muhammad Umair

Abstract

Rare genetic disorders (RGDs) are a large and diverse group of diseases with low frequency but great clinical and genetic variation. These illnesses collectively affect millions of individuals globally. Genomic and personalised medicine have illuminated RGD aetiology, but many problems and potentials remain. This review discusses RGD challenges, future opportunities and highlights opportunities to improve patient care and research. Patients often face an odyssey to diagnose RGDs. Clinicians sometimes lack experience or cannot obtain diagnostic tools due to the illnesses' rarity and overlapping symptoms. Primary care practitioners' lack of RGD understanding and centralised databases can delay or misdiagnose patients. Due to their limited incidence, RGDs are not economically

N. R. Madhu · G. Biswas · S. Paul
Department of Zoology, Acharya Prafulla Chandra College, New Barrackpore, West Bengal, India

S. Adhikari
Department of Chemistry, Govt. Degree College, Dharmanagar, Tripura, India

B. Sarkar (✉)
Department of Botany, Acharya Prafulla Chandra College, New Barrackpore, West Bengal, India

M. M. Rafeeq
Department of Pharmacology, Faculty of Medicine, Rabigh, King Abdulaziz University, Jeddah, Saudi Arabia
e-mail: marafeeq@kau.edu.sa

M. Umair
Medical Genomics Research Department, King Abdullah International Medical Research Center (KAIMRC), King Saud bin Abdulaziz University for Health Sciences, Ministry of National Guard Health Affairs #(MNGH), Riyadh, Saudi Arabia
e-mail: umairmu@ngha.med.sa

M. Umair et al. (eds.), *Rare Genetic Disorders*,
https://doi.org/10.1007/978-981-99-9323-9_9

viable for therapeutic development. Pharmaceutical companies may avoid treating certain illnesses due to high development costs and little commercial potential. Next-generation sequencing (NGS) has greatly shortened diagnosis time. Genome or exome sequencing can discover causative mutations even in rare or undiagnosed illnesses. This technology and global collaborative networks like the Undiagnosed Diseases Network (UDN) will help many patients end their diagnostic journey. Personalised medicine and tailored therapeutics offer promising RGD treatments. CRISPR/Cas9 gene editing targets genetic abnormalities, allowing for potential treatments. Drug repurposing is a cheap technique to uncover RGD medicines without the lengthy and expensive drug development procedure. Patient advocacy and community research are also rising. Patient organisations fundraise, collaborate on research, and influence policy. More empowered communities fight for patient-centric research and care. In conclusion, RGDs face many hurdles, but technological advances, patient advocacy, and collaborative research give hope. RGDs can be solved using cutting-edge technologies, interdisciplinary cooperation and a patient-centred approach.

Keywords

Rare genetic disorders · RGD · Personalised medicine · Patient therapeutic challenges · Drug repurposing

9.1 Introduction

Rare genetic disorders, sometimes referred to as rare diseases, are a heterogeneous set of problems that collectively afflict a relatively small number of people in the community. Rare genetic disorders are also known as uncommon diseases. Although individual ailments might not be very widespread by themselves, the prevalence of these conditions and the complexity of their symptoms present a considerable obstacle to improving public health (Schieppati et al. 2008). Over the past few years, there has been an explosion of interest in investigating and treating uncommon genetic illnesses, bringing to light many obstacles that must be overcome. This chapter digs into the challenges and future opportunities in this field.

Rare diseases impact a relatively small number of patients, much as their name implies. But when you consider all kinds of rare diseases, the total number of people globally who are afflicted adds up to an astounding 475 million. The National Institutes of Health (NIH) estimates that approximately 25–30 million people in the United States are affected by one of the more than 6800 rare diseases. This represents approximately one in ten individuals. Children make up more than half of those diagnosed with uncommon diseases, meaning that over 200 million youngsters worldwide are enduring pain and suffering right now. It is essential to be aware that more than 80% of rare diseases have a genetic condition as their cause or a predisposing factor. This contributes to the fact that only approximately 5% of rare diseases have treatments that have been clinically approved. Over 36 million people in Europe are afflicted with one of the more than 6000 different uncommon diseases. Remembering that the same disease can be regarded as rare in one location while being considered endemic in another is essential. For instance, thalassemia is brought on by a genetic mutation, which is exceptionally uncommon in Northern Europe but highly common in the Mediterranean region. This is because Northern Europeans tend to have fewer copies of the genes. In the United States, a sickness or condition is considered to be rare if it affects less than 200,000 people (Directorate-General for Research and Innovation (European Commission) 2017; Kaufmann et al. 2018). A disease is considered to be rare in the European Union if it affects no more than 1 person in every 2000 people. Although uncommon when seen as a single entity, the cumulative effects of rare diseases have a substantial impact on the affected population, a profound impact on the quality of life they lead, and a considerable cost to the economy.

The medical community faces a significant challenge in the form of a significant public health issue in the form of rare diseases. Rare diseases, also known as "orphan" diseases, have been ignored for many years. In the end, the Orphan Drug Act of 1983 in the United States was enacted with the assistance of many different organisations. It promulgated many programmes designed to encourage the production of medications for rare diseases. Since the law's introduction, over 600 orphan medicines and biological products have been given the green light for sale in the United States alone. In the current issue of Nature Genetics, Halley et al. (2022) published a comment in which they argue that rare diseases are a serious public health issue that should be given comparable attention for their cumulative effects on individual patients, disease communities, and healthcare systems. As a result, the strategy for researching rare diseases needs to be expanded to provide the greatest possible advantages to the greatest number of people.

Given the high rate of spontaneous mutations and the number of newborns impacted in families with a known risk of a genetic condition, it is likely that India will have a high prevalence of genetic disorders at birth. The principal tool for preventing and controlling genetic disorders is genetic counselling, offered to families with a newborn affected by a genetic disorder or a family history of the disorder. It is difficult to evaluate the prevention of genetic disorders, as represented by the provision of genetic counselling, is challenging from a health systems viewpoint due to the methodological difficulties inherent in investigating rare conditions (Kar et al. 2014).

In recent years, access to chromosomal microarray analysis has allowed for identifying previously unrecognised microdeletion/duplication disorders. In order to diagnose subtelomeric and frequent microdeletion/duplication disorders, several centres rely on multiplex ligation-dependent probe amplification tests (Aggarwal and Phadke 2015). False-positive diagnoses of uncommon diseases may be dramatically reduced because of ongoing, innovative diagnostic and therapeutic initiatives in India. Rare genetic illnesses can now be diagnosed thanks to the development and steadily decreasing cost of DNA sequencing. More research into rare disorders would greatly benefit the general population's health (Whicher et al. 2018). Building a database is crucial for estimating the size of the impacted population. The government must launch a programme to help those living with uncommon diseases get the care they need. Increased efforts in high-throughput genomic sequencing, prenatal diagnosis, disease predisposition, targeted medicines, and gene therapy to understand better and cure uncommon diseases are required (Dolled-Filhart et al.

2012). Because of the rarity of the diseases they treat, clinical trials for orphan drugs require special consideration and creation of novel trial designs. Due to the scarcity of experts in this field, developing innovative treatments for rare diseases requires the cooperation of patient advocacy groups, medical researchers, and government agencies (Choudhury and Chaube 2022).

A sizable number of people in India suffer from rare diseases; estimates range from 72 million to 96 million. India needs to develop its own estimate and definition of uncommon diseases, based mostly on prevalence data, while the one provided by Bhattacharya et al. (2016) is only an approximation. Here, autoimmune disorders, lysosomal storage disorders like Pompe disease, Hirschsprung disease, Gaucher's disease, cystic fibrosis, haemangiomas, and some forms of muscular dystrophies are among the most common rare diseases (Fig. 9.1, Tables 9.1 and 9.2).

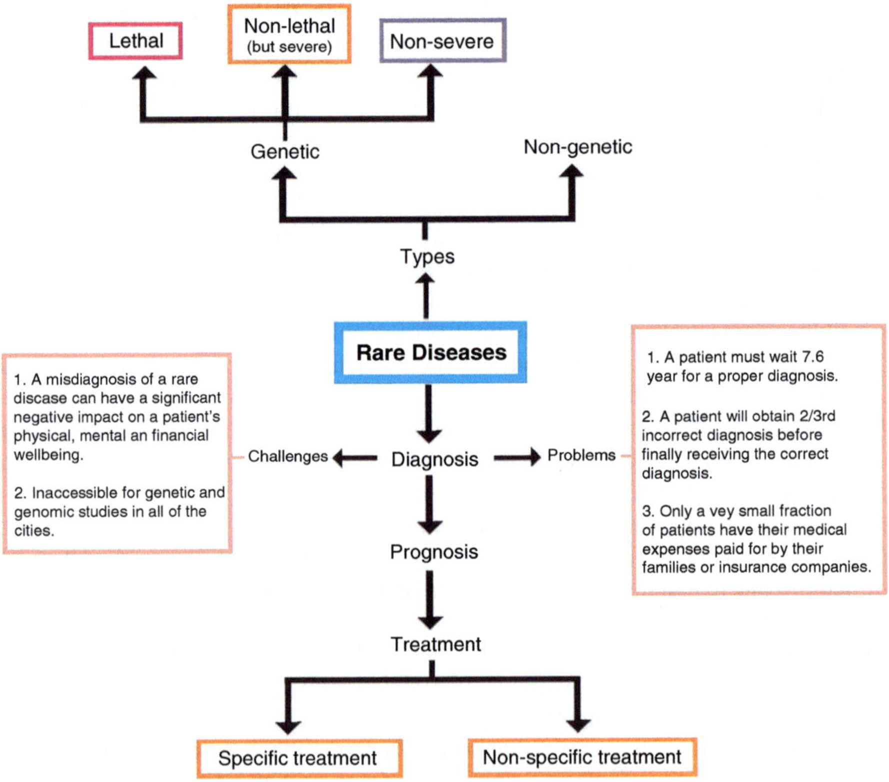

Fig. 9.1 Evaluation steps and disease categories for those with rare conditions. Rare diseases can be broken down into two categories: those that are genetically inherited and those that are not. The symptoms of each type of rare disease, whether hereditary or not, can vary greatly. The processes involved in diagnosing rare diseases are critical because, after a diagnosis, a medical expert can evaluate a patient's prognosis and the treatment options available to them. These options may be disease-specific, in which case they will focus on curing the underlying condition, disease-agnostic and concentrating on relieving the patient's symptoms. Issues in managing rare diseases include various challenges and potential solutions. Diagnosing a rare condition might require significant time, energy, and finances. Here, the primary challenges associated with diagnosing uncommon diseases are illustrated

Table 9.1 Populations affected by rare diseases in various nations

S. No.	Country	Per 10,000 population
1	USA	6.4
2	Europe	5.0
3	Canada	5.0
4	Japan	4.0
5	South Korea	4.0
6	Australia	1.0
7	Taiwan	1.0
8	India (Min and Yurchisin 2015)	1

Source: The I.C. Verma Sub-Committee Report (2017). 'Guidelines for Therapy and Management'

Table 9.2 Types of rare genetic diseases and examples

Type of rare genetic disease	Description	Examples
Single gene mutations	Caused by a mutation in a single gene	Cystic fibrosis, haemophilia (Handyside et al. 1992)
Chromosomal abnormalities	Involves changes in the structure or number of chromosomes	Down syndrome, Turner syndrome (Ogata and Matsuo 1995)
Mitochondrial disorders	Result from mutations in mitochondrial DNA	Mitochondrial myopathy (Holt et al. 1990)
Polygenic/ multifactorial disorders	Caused by a combination of genetic and environmental factors	Autism spectrum disorder, type 2 diabetes (Benayed et al. 2005)
Autosomal dominant	A mutation in one copy of the gene causes the disorder	Huntington's disease (Wexler et al. 1987)
Autosomal recessive	Both copies of the gene must be mutated to have the disorder	Cystic fibrosis (Handyside et al. 1992)
X-linked	Mutations occur on genes located on the X chromosome	Haemophilia (Hoyer 1994)
Deletions/ duplications	Involves missing or extra copies of genetic material	Cri-du-chat syndrome (Niebuhr 1978)
Translocations	Genetic material is transferred between chromosomes	Philadelphia chromosome (CML) (Bartram et al. 1983)
Inversions	Genetic material is inverted in a chromosome segment	Inversion 16 disorder (Morin et al. 2017)
Complex genetic interactions	Multiple genes contribute along with environmental factors	Autism spectrum disorder (Campbell et al. 2008)
Environmental factors	Genetic predisposition combined with external influences	Type 2 diabetes (Horikawa et al. 2000)
Ethnicity related	Prevalence varies among different ethnic groups	Tay-Sachs disease (Weiner 2008)

9.2 Methods

Using search terms such as 'rare genetic disorders', a search was conducted in the published literature contained within the online databases Google Scholar, PubMed, and Medline. It was believed that the chosen papers would lend credibility to the argument being made in the present review. The papers meeting the inclusion criteria were subjected to a comprehensive review.

In addition, the report compiled by the National Commission for the United States Government in 1989 initially brought the challenges faced by the patients with rare genetic diseases to the attention of the general public. The following concerns were brought to light as a result of the Commission's findings:

9.3 Challenges

From the perspectives of clinicians, researchers, and biotech/pharma businesses, rare diseases present significant challenges. Clinicians caring for affected people face difficulties such as a lack of local expertise, expert guidelines, and experience caring for affected patients. Researchers encounter difficulties in securing sufficient patient cohorts and obtaining the necessary financing. Due to the binary nature of the expensive clinical programmes required to treat rare diseases, biotech and pharma frequently view their pursuit of these indications as high-risk bets. Patients and the clinical/scientific communities' dogged determination to address these issues and advance care and knowledge has had a profound effect. There have been many successes, including the establishment of influential patient advocacy bodies that have facilitated partnerships between patients, scientific communities, the government, and pharma/device companies in the interest of early diagnosis, optimal care, and research; the acquisition of funds to support research; the establishment of collaborative consortia of clinicians and scientists; and the general activation of the respective patient communities (Marwaha et al. 2022; Lee et al. 2022).

Researchers confront special difficulties when attempting to study rare diseases because of factors such as the low prevalence of each condition, the complexity of its symptoms, and the difficulty of obtaining relevant data. Patients, their loved ones, support systems, and society all find them overwhelming and expensive (Cutillo et al. 2017). Most people are familiar with diseases like haemophilia, thalassemia, sickle cell anaemia, and primary immunodeficiency, as well as autoimmune disorders and lysosomal storage disorders like Pompe disease, Hirschsprung's disease, Gaucher's disease, cystic fibrosis, haemangioma, and some forms of muscular dystrophy. There is a lack of specialised social insurance policies, plans, and diagnostic facilities, and there is insufficient awareness among the medical community, causing difficulties for patients and carers. The paucity of rare disease databases and counselling services is even more difficult. Perhaps, the biggest problems are with treatment costs and lack of availability (Crowe et al. 2019). The law protecting the rights of people with disabilities does not include extremely rare diseases. Patients with disabilities who are denied access to healthcare or other benefits face no legal recourse. Family members are not permitted to receive any form of comfort care. Drugs for uncommon diseases are currently being produced by only a handful of pharmaceutical companies worldwide (Ghiasvand et al. 2022).

There is an urgent need for more widespread public health initiatives in India to screen newborns for inherited illnesses. Due to a lack of coordination between different organisations, collecting data on a national scale has not been possible so far. In recent years, a non-governmental organisation has also begun maintaining a birth defect registry. Most genetic diagnostic labs nowadays either strictly adhere to international recommendations or make necessary adjustments based on their own quality control standards. Thus, there is a clear need for additional steps to be taken by an independent research centre. Cost is a major barrier to personalised exome sequencing for rare diseases in the Indian context (Choudhury and Chaube 2022). Inadequate bioinformatics infrastructure reduces diagnostic precision. The intricacy of uncommon diseases is exacerbated by the rich cultural diversity of India's population. Fewer than one in ten individuals with a rare disease actually gets the specialised care they need. The administration method itself recognised medicine accessibility as the primary problem. In regions where readily available drugs are prohibitively expensive, putting a severe strain on individual budgets, public healthcare systems, and charitable organisations (Weatherall and Clegg 2001; Kar et al. 2014; Aggarwal and Phadke 2015). The prevalence of knowledge about uncommon diseases in India is low. Patient assistance and support should begin with clinical determination and therapy direction. Despite the importance of reaching an early conclusion, patients and certain doctors often experience delays in diagnosis due to a lack of knowledge. After 7 years of experiencing symptoms, it is safe to classify a condition as extremely uncommon. Patients go through a wide range of symptoms, health problems, anxiety, laboratory testing, and doctor visits throughout this time (Cleary and Chakrapani 2004; Sivasubbu and Scaria 2019). This diagnostic process is time-consuming and expensive since it requires cutting-edge equipment. Many parents suffer from emotional traumas and do not even know it. With a population nearing 1.5 billion and including around 5000 well-characterised subgroups, the Indian subcontinent is one of the most genetically diverse locations on Earth. Boundaries imposed by geography, language, or society, such as prohibitions on intermarriage, increase the likelihood that spouses have a substantial common ancestry. This can cause some recessive disorders to spread and multiply (Boat and Field 2011). Drug companies find it difficult to develop effective treatments for these disorders because they rarely occur (Kaul and Munshi 2012). It is challenging to provide medications due to the scattered nature of patients. Drug manufacturers are deterred by the absence of a centralised database detailing patient conditions and status updates.

9.3.1 Limited Understanding and Research

The lack of a thorough knowledge of the mechanisms at play in rare genetic illnesses is one of the field's biggest obstacles. There are thousands of rare diseases, each with its own specific genetic composition, making it difficult to pinpoint their precise causes and illness pathways. Comprehensive studies are hampered by a lack of patient data and biomaterials as well as the intricacy of genetic variants (Agrawal et al. 2019).

Developing treatments and cures for uncommon diseases is difficult since so little about their pathophysiology and natural history is understood. Rare diseases are challenging to study because of the small patient pool, which sometimes leads to insufficient clinical experience. As a result, the clinical interpretation of rare diseases may be inaccurate or incomplete. Because of the chronic nature of uncommon diseases, long-term follow-up becomes even more of a burden. Therefore, there is a shortage of published data on the long-term treatment outcomes of rare diseases, and these conditions are often only partially defined (Mohanty et al. 2016). As a result, it is important to look at multinational and cross-regional research partnerships and partnerships with the specialists who treat rare diseases and the people whose lives have been impacted by them. The etiology of these diseases and the therapeutic impacts that might significantly improve patients' lives can be better understood in this way. The standards for conducting clinical trials should be evaluated in light of the quality of the medications or diagnostic instruments being tested and revised as necessary.

9.3.2 Diagnostic Delays

Accurate diagnosis is often a protracted process for individuals with rare genetic disorders. Misdiagnoses, lack of awareness among medical professionals, and absence of specific diagnostic tools contribute to delays in identifying the root cause of symptoms. These delays prolong patient suffering and hinder the development of targeted therapeutic approaches.

Delays in achieving an appropriate diagnosis are among the most worrisome issues encountered on the path to diagnosing rare diseases. Orphan diseases, which affect only a small percentage of the population, are classified as rare diseases. Both patients and doctors can be baffled by the varying degrees of complexity displayed by these illnesses. This chapter investigates the serious problem of diagnostic delays in the context of uncommon diseases, discussing their causes and possible remedies (Boulanger et al. 2020).

Each individual affected by one of the thousands of rare diseases in the globe deals with a different set of symptoms and difficulties. Due to their low prevalence, healthcare providers may lack experience dealing with certain illnesses. As a result, people frequently give incorrect diagnoses or brush off symptoms as being caused by more common conditions. Doctors making mistakes in their diagnoses can prolong their patients' pain (Groft et al. 2021).

Several variables cause the diagnostic lag in rare diseases. To begin, delays in diagnosis may occur because of inadequate training and knowledge among medical personnel. Clinicians may not be adequately prepared to recognise and understand the varied presentations of rare disorders since medical school frequently prioritises teaching about more prevalent diseases. This information void highlights the importance of comprehensive, continuing medical education. Second, the extreme difficulty stems from rare diseases being inherently heterogeneous. Diverse patients with the same uncommon disease may experience diverse symptoms, making it difficult to distinguish it from other, more common diseases. Because of this variety, it is challenging to create precise diagnostic criteria. In addition, the problem is exacerbated because there are no universally accepted diagnostic guidelines for rare diseases.

Thirdly, there is typically a lack of diagnostic resources for rare disorders. The high cost of developing reliable diagnostic tests may not be justified for diseases with a low prevalence. This causes patients to endure extensive general testing, some of which may provide equivocal results and extend the diagnostic odyssey (Shen et al. 2019; Martin et al. 2019).

The toll that waiting for a diagnosis takes emotionally and mentally is also significant. Patients with uncommon diseases who have yet to be diagnosed may lack understanding and compassion from their doctors and loved ones. Their medical problems may become worse, and their quality of life may decrease because of this mental anguish.

Rare disease diagnosis delays need to be addressed from multiple angles. First and foremost, it is critical to increase understanding among healthcare providers. Modules on spotting and detecting unusual diseases should be incorporated into doctors' ongoing training to promote a broader comprehension of health issues. Symptoms that span numerous medical disciplines call for a multifaceted approach to diagnosis, which can be achieved by encouraging interdisciplinary collaboration among professionals. Second, it is crucial to fund studies to develop better diagnostic tools. Tests for a wider variety of uncommon diseases need to be developed; thus, governments, pharmaceutical companies, and academic institutions should pool their resources to do so. The diagnostic process can be sped up significantly with the use of cutting-edge technology that takes advantage of recent breakthroughs in genetics and molecular biology. Third, patient advocacy is essential. In order to raise awareness and fund research into rare diseases, patients and their families often take on the role of advocates. These campaigns inform the public and pressure legislators to invest in reducing diagnostic backlogs and enhancing rare illness management (Halley et al. 2022; Bogart et al. 2022).

Diagnostic delays in rare diseases pose a serious problem with far-reaching effects. This issue is exacerbated by several factors, including the difficulty of diagnosing rare diseases and the emotional toll taken on patients due to their diagnosis. However, the landscape of rare disease diagnosis can be modified by improved medical education, more interdisciplinary collaboration, research investments, and patient advocacy. Reducing the time it takes to get a diagnosis can give patients greater control over their care, improve their quality of life, and open the door to more effective treatments and interventions for rare diseases (Han et al. 2022).

9.3.3 Therapeutic Challenges

Due to the various illnesses and underlying genetic abnormalities, developing medicines for uncommon genetic disorders is a complex endeavour. Because of the low potential for profit from a minimal patient group, conventional drug development methods may not be feasible. Another major hurdle is the lack of preclinical models that can be used to evaluate medications that are designed to treat rare disorders. Achieving an appropriate diagnosis is among the most problematic issues when diagnosing uncommon diseases. Orphan diseases, which affect only a small percentage of the population, are rare diseases. Both patients and doctors can be baffled by the varying degrees of complexity displayed by these illnesses (Kaufmann et al. 2018). This chapter investigates the serious problem of therapeutic challenges in the context of uncommon diseases, discussing their causes and possible remedies.

Each individual affected by one of the thousands of rare diseases in the globe deals with a different set of symptoms and difficulties. Due to their low prevalence, healthcare providers may lack experience dealing with certain illnesses. As a result, people frequently give incorrect diagnoses or brush off symptoms as being caused by more common conditions. Doctors making mistakes in their diagnoses can prolong their patients' pain. Several variables cause the diagnostic lag in rare diseases. To begin, delays in diagnosis may occur because of inadequate training and knowledge among medical personnel. Clinicians may not be adequately prepared to recognise and understand the varied presentations of rare disorders since medical school frequently prioritises teaching about more prevalent diseases. This information void highlights the importance of comprehensive, continuing medical education.

Second, the extreme difficulty stems from rare diseases being inherently heterogeneous. Diverse patients with the same uncommon disease may experience diverse symptoms, making it difficult to distinguish it from other, more common diseases. Because of this variety, it is challenging to create precise diagnostic criteria. In addition, the problem is exacerbated because there are no universally accepted diagnostic guidelines for rare diseases (Abozaid et al. 2022).

Thirdly, there is typically a lack of diagnostic resources for rare disorders. The high cost of developing reliable diagnostic tests may not be justified for diseases with a low prevalence. This causes patients to endure extensive general testing, some of which may provide equivocal results and extend the diagnostic odyssey. The toll that waiting for a diagnosis takes emotionally and mentally is also significant. Patients with uncommon diseases who have yet to be diagnosed may lack understanding and compassion from their doctors and loved ones. Because of this mental anguish, their medical problems may worsen and their quality of life may decrease (Bryson et al. 2021).

Rare disease diagnosis delays need to be addressed from multiple angles. First and foremost, it is critical to increase understanding among healthcare providers. Modules on spotting and detecting unusual diseases should be incorporated into doctors' ongoing training to promote a broader comprehension of health issues. Symptoms that span numerous medical disciplines call for a multifaceted approach to diagnosis, which can be achieved by encouraging interdisciplinary collaboration among professionals. Second, it is crucial to fund studies to develop better diagnostic tools. Tests for a wider variety of uncommon diseases need to be developed. Thus governments, pharmaceutical companies, and academic institutions should pool their resources to do so. The diagnostic process can be sped up significantly with the use of cutting-edge technology that takes advantage of recent breakthroughs in genetics and molecular biology. Third, patient advocacy is essential. In order to raise awareness and fund research into rare diseases, patients and their families often take on the role of advocates. These campaigns inform the public and put pressure on legislators to invest in reducing diagnostic backlogs and enhancing rare illness management. Last but not least, diagnostic delays in rare diseases pose a serious problem with far-reaching effects. This issue is exacerbated by several factors, including the difficulty of diagnosing rare diseases and the emotional toll on patients due to their diagnosis. However, the landscape of rare disease diagnosis can be modified by improved medical education, more interdisciplinary collaboration, research investments, and patient advocacy. Reducing the time it takes to get a diagnosis can give patients greater control over their care, improve their quality of life, and open the door to more effective treatments and interventions for rare diseases (Tambuyzer et al. 2020).

9.3.4 Patient Access to Treatment

Providing equal access to therapy for people with uncommon genetic illnesses is difficult even when treatments are available. Health inequalities are exacerbated when patients are unable to obtain potentially life-saving measures due to factors such as high treatment costs, restricted insurance coverage, and regulatory barriers.

9.3.4.1 Unavailability of Treatment

Medicine accessibility and availability are critical in the fight against uncommon disease-related mortality and morbidity. Most uncommon diseases still lack treatment that is either effective or safe, despite recent advances. As a result, even if the right diagnosis is made, no therapy may be available to treat the rare disease. Seven thousand to eight thousand uncommon diseases are estimated to exist, but less than 300 have approved therapeutics (Jackman 2012). Fewer than one in ten patients (19%) receives disease-specific treatment, and over 95% of uncommon diseases have no approved medication. Where medications are an option, they are often unaffordable, putting a major burden on individual budgets, public healthcare systems, and charitable organisations.

9.3.4.2 Treatment Costs That Are Out of Reach

Drug companies are not motivated to discover and commercialise treatments for uncommon diseases because so few people are affected by each condition. This is why medications designed to treat rare disorders are sometimes referred to as 'orphan drugs'. To recuperate the tremendous expense of R&D, companies that produce medications to treat uncommon diseases typically charge exorbitant prices.

Drugs for uncommon diseases are now only being produced by a handful of pharmaceutical companies worldwide, with India lacking any indigenous producers. The government has been unable to supply these medications at no cost because of their high price. As addressed in greater depth in Chap. 3, the annual cost of treating some rare disorders is anticipated to range from Rupees 18 lakhs to 1 crore 70 lakhs for a child weighing 10 kg. In addition, the evaluation approach for uncommon disease treatments is frequently in its infancy, making it difficult to conclude their clinical relevance or cost-effectiveness. Obtaining 'orphan drug' classification in the United States or another country grants the manufacturer of that drug a period of time during which they have sole marketing rights and are shielded from import competition. Because of this, pharmaceutical businesses may confidently set prices for their products. This has altered the process of creating new medicines. Since 2000, the number of applications for orphan drug status has increased by a factor of four. As a direct result, drug sales and earnings have skyrocketed. In fact, over a third of medicines used to treat uncommon diseases now generate more than £1 billion a year. By 2020, the global market for orphan pharmaceuticals will be worth £144 billion, representing 19% of all branded prescription drug sales. The annual cost of treating rare diseases typically surpasses $100,000. While laws like the US Orphan Drug Act have helped speed up the creation of treatments for rare diseases, their inability to rein in skyrocketing pricing has raised serious questions about the long-term financial viability of healthcare systems. The monthly cost of the cystic fibrosis medicine Kalydeco (ivacaftor) is around £14,000. Soliris (eculizumab), the most expensive medicine in the world at £340,000 per patient per year, is one of ten pharmaceuticals all used to treat rare diseases. Although these medications are supplied to fewer patients than blockbuster medications, their extremely high prices can generate earnings on par with the latter. Orphan medication market authorisation holders are more likely to be publicly traded pharmaceutical companies with a higher market value and larger profitability (Hughes and Poletti-Hughes 2016). Many medications that are neither novel nor scientific breakthroughs have been granted 'orphan drug' classification over the years. A low-cost, off-patent medicine that the FDA has only ever approved for one illness can be turned into a lucrative cash cow by being prescribed as an 'off-label' treatment for a rare disease. A drug's price can increase dramatically after it has enjoyed monopoly status for 7–10 years (Côté and Keating 2012; Kesselheim et al. 2012). For example, the annual cost of treating chronic myeloid leukaemia with the ageing drug imatinib can exceed USD 100,000 in industrialised nations. Patent law provisions in India that limit evergreening patents positively affect prices. Concerns regarding the long-term viability of rare illness funding/reimbursement programmes have been raised even in industrialised countries due to the sky-high cost of treatments for uncommon diseases (Bojakowski and Spoors 2013). There have been requests for price transparency and price regulation in the pharmaceutical industry (Simoens 2011), as well as inquiries and investigations at the Congressional level (Health Inc. 2017). Any policy or law implemented in India to encourage medication research for rare diseases should give this issue a serious thought. We urgently need multi-pronged methods to bring down medicine prices, boost generic and local production, and motivate public sector organisations to develop treatments for uncommon diseases.

9.3.5 Collaboration and Information Sharing

Rare illness research is notoriously difficult to coordinate because of its dispersed character, leading to splintered efforts and separate bodies of information. Understanding and treatment can only progress with the help of researchers, clinicians, and patient advocacy groups working together. Data sharing difficulties, privacy concerns, and language barriers all work against efficient teamwork.

9.4 Future Opportunities

In the past, efforts have largely centred on either creating research papers or amassing data on uncommon disorders. The final result of their efforts was a better comprehension of the challenges inherent in clinical studies with limited participant pools. Patient advocacy groups and patient participation in all phases of clinical research have expanded due to current and future modalities. As a result, the scientific community now recognises this as a major public health concern. The next major thrust of rare disease research will be to improve our capacity to understand and interpret data from various sources, such as electronic health records and big data. Personalised genetic therapy combined with improvements in clinical trial design and data analysis will pave the path for future orphan medications. Expanded diagnostics, especially those based on enhanced genetic sequencing technologies, should also facilitate earlier detection of rare disorders. From the clinical research perspective, quicker approvals will be possible due to improved patient recruiting and more flexible approval procedures by regulatory bodies. Children benefit most because early detection and treatment can reduce the likelihood of having any long-term effects from their illnesses. Decentralised clinical trials (DCTs), remote sensors, and the rise of mobile app technology will all contribute to the ongoing tracking of patients using experimental medications. Access to specialist clinicians with a deeper understanding of individual uncommon diseases is already available because of telemedicine. In the end, this means that patients and their loved ones will benefit from a more efficient, varied, and multichannel approach to research.

Communicating effectively in English is vital in today's dynamic global economy, providing entry to a wide range of promising professions and fields. Demand for native speakers of English with a command of the nuances of the language is likely to rise as communication barriers are broken down through technological advancements. Particularly in highly technical disciplines like AI, biotech, and space exploration, where new discoveries need precise articulation, clear communication's importance cannot be overstated (Brasil et al. 2019). Furthermore, as cross-border partnerships become the norm, people who can negotiate international economic deals or build diplomatic ties while fluent in spoken English will find themselves at an advantage. Opportunities also abound in the academic realm, with prestigious institutions looking for researchers and teachers fluent in spoken English to help them communicate with a more culturally diverse student base. In addition, the proliferation of content creation in the digital age requires experts who can harness the power of plain English to convey complicated topics in a way that is accessible to a wide audience. As climate change and public health concerns become increasingly pressing worldwide, it will be increasingly important for professionals to be able to communicate effectively in plain English to develop and disseminate solutions that cut across borders and languages. Though English adds depth and energy to the entertainment business, which always requires fresh perspectives that can write engaging stories, scripts, and dialogue. Though English speakers are in high demand in today's digitally connected world, where their abilities can help brands connect with consumers globally through digital marketing, brand strategy, and influencer engagement. Since legal documents, treaties, and laws are increasingly taking on international dimensions, there is a growing need for people fluent in spoken English to work in the legal and political spheres. Though English is becoming increasingly important as teams work remotely and across borders, depending on effective communication to overcome language and cultural barriers. Those who are up for the task of learning thorough English will be well positioned to take advantage of the many doors that will open in a future where variety and interconnection will be paramount.

9.4.1 Improved Genomic Technologies

High-throughput genomic sequencing tools have dramatically changed how we study and treat uncommon genetic illnesses. Accelerating diagnostic processes and shedding light on novel disease pathways, whole-exome and whole-genome sequencing allow researchers to discover causal mutations quickly. Recent developments in genomic technologies have profoundly impacted the diagnosis and treatment of uncommon diseases. Thanks to these state-of-the-art resources, scientists and medical professionals have been able to understand better the causes and potential treatments for a wide variety of uncommon diseases. Whole-genome sequencing (WGS) and whole-exome sequencing (WES) are types of next-generation sequencing (NGS) that have been important in pinpointing the cryptic genetic abnormalities at the root of many rare disorders. Because of NGS's speed and low cost, it is now possible to analyse huge swaths of a person's genome, opening the door to discovering new disease-associated genes. Single-molecule sequencing technologies have also improved our ability to detect genetic variants to previously unseen levels of precision (Klein et al. 2014; Imperial et al. 2019).

The study of large genomic datasets requires both NGS and cutting-edge bioinformatics methods. Researchers can use these methods to sift through massive amounts of genomic data and identify potentially harmful mutations while discarding benign ones. Identifying genetic drivers for rare diseases has been accelerated by creating databases and platforms that collate genomic and clinical data, promoting collaboration and data sharing among researchers worldwide.

The development of genetic technologies has allowed for better diagnosis and more individualised treatment plans. Thanks to the development of precision medicine, therapeutic approaches can now be more accurately targeted to a patient's specific genetic composition. Correcting the genetic abnormalities causing certain uncommon diseases with gene treatments, such as CRISPR-Cas9-based techniques, may give permanent remedies rather than merely symptom management (Wang et al. 2017; Zhang 2021). Furthermore, pharmacogenomics (the study of how an individual's genetics influence their reaction to pharmaceuticals) allows clinicians to prescribe medications with a higher likelihood of success and fewer adverse effects, reducing the need for ad hoc treatment. Improvements in genetic technology have not only revolutionised medicine and science but also given patients and their loved ones a voice. With easier access to genetic testing, more people suffering from uncommon diseases will be able to receive definitive diagnoses, putting an end to their diagnostic journeys and giving them better insight into their illnesses. With this new information, people will be better able to make health decisions and treatment decisions (van der Lee et al. 2020; Cecchin and Stocco 2020).

But with these developments arise questions of ethics and social responsibility. To ensure that the advantages are spread widely and marginalised groups are not left out, we must properly handle issues like patient consent, data protection, and equitable access to these technologies.

In conclusion, a new era has begun in the study and treatment of uncommon diseases thanks to the rapid development of genomic technologies. Identifying causal genetic alterations, providing personalised therapy options, and empowering patients have all seen significant advances because of the combination of NGS, advanced bioinformatics, and precision medicine approaches. Stakeholders must work together to address ethical, legal, and social consequences as these technologies advance to realise genomics' revolutionary promise without compromising on the ideals of fairness and honesty (Guan et al. 2012; Lauschke and Ingelman-Sundberg 2018).

9.4.2 Precision Medicine

The intricacies of rare diseases provide formidable challenges, but precision medicine, a novel approach to treatment, has great promise. The heterogeneity and lack of knowledge of these illnesses, which each affect a relatively small number of individuals yet have a massive global impact, have long been obstacles. In order to understand the complex molecular and genetic causes of various diseases, precision medicine employs state-of-the-art tools including genomics, proteomics, and sophisticated imaging. Clinicians can select treatments that are specifically targeted to the underlying processes of each patient's ailment by interpreting the unique genetic profiles of affected individuals (Lauschke and Ingelman-Sundberg 2016).

By abandoning the one-size-fits-all mentality of conventional medicine, we can develop more efficient and much less dangerous treatments, reducing or eliminating the side effects that typically accompany broad-spectrum drugs. In addition, patients are given the tools they need to make educated decisions about their healthcare because of precision medicine's emphasis on early and accurate diagnosis. Gene therapies and personalised medication molecules are two examples of cutting-edge treatments being developed thanks to partnerships between academic institutions and pharmaceutical corporations. In spite of challenges, including high costs, restricted access to new diagnostic equipment, and the complex nature of rare diseases, a concerted worldwide commitment to additional research, technology refinement, and fair healthcare access is needed. Precision medicine shines as a guiding light in this effort, showing the way towards improved outcomes in the face of uncommon diseases through more individualised and compassionate treatment (Might and Crouse 2022).

9.4.3 Drug Repurposing

The process of drug repurposing, in which existing pharmaceuticals are identified for new therapeutic objectives, may be useful for treating rare genetic illnesses. Researchers can move more quickly towards developing cures for uncommon diseases if they make use of substances that have already been through safety testing.

Rare diseases affect a sizable fraction of the world's population, and medication repurposing, the process of finding new therapeutic uses for existing pharmaceuticals, shows great potential in combating these conditions. Developing totally novel medications for rare disorders is difficult and expensive. Thus, repurposing provides a faster approach to providing therapy alternatives. This strategy may save considerable time and money by skipping the costly and time-consuming first stages of drug development by building on the success and proven safety profiles of already approved medications. Drug libraries and databases must be combed through carefully to find compounds that have the potential to interact with the molecular pathways underlying the rare disease, which is the basis for repurposing (Roessler et al. 2021). Crucial roles are played by computational approaches, high-throughput screening, and state-of-the-art bioinformatics in this procedure, all of which aid researchers in sifting through massive volumes of data in search of promising candidates. After the drug has been repurposed, it must undergo rigorous preclinical and clinical testing to determine whether or not it is safe and effective for treating the uncommon condition in question. There are also significant caveats to the promising potential of medication repurposing for uncommon diseases. Comprehensive data, in-depth knowledge of the disease's molecular causes, and strong research collaborations are essential for success. Repurposed pharmaceuticals' approval and commercial viability also depend on overcoming regulatory and intellectual property obstacles (Pushpakom et al. 2019). Despite these challenges, drug repurposing provides a glimmer of hope for effectively tackling rare diseases, and it has the potential to revolutionise the landscape of rare disease medicines by making lifesaving treatments more quickly available to those who need them (Zhao et al. 2019; Roessler et al. 2021).

9.4.4 Advocacy and Patient Empowerment

Advocacy groups for people with rare genetic conditions are vital to advancing knowledge, increasing public understanding, and shaping public policy. Patient and family empowerment promotes the growth of patient registries, the collection of useful data, and the development of caring online communities.

Advocating for patients with rare diseases is an important way to help them cope with the difficulties they and their families experience. The physical, emotional, and financial challenges endured by the rare disease population are exacerbated by the limited medical competence, delayed diagnoses, and lack of effective treatments they frequently confront (Aymé et al. 2008). Advocacy that successfully ensures equal access to healthcare, treatment options, and support services for those living with rare diseases entails educating the public about these conditions; encouraging partnerships between patients, healthcare providers, and researchers; and influencing governmental changes. Advocacy campaigns aim to accelerate research and treatment development and encourage information exchange by amplifying the voices of patients and their families affected by rare diseases. In addition, they are crucial in reducing the sense of alienation these people feel, instilling a sense of community, and facilitating the exchange of information and ideas. Efforts to raise awareness and empower patients dealing with rare diseases aim to better the lives of those directly impacted by these conditions while sparking a cultural movement towards greater acceptance, empathy, and medical advancement for all (Koay and Sharp 2013).

9.4.5 Regulatory Incentives

Because of the difficulties inherent in studying rare diseases, regulatory bodies have introduced incentives to spur treatment development for certain ailments. Investment in therapies for uncommon diseases is bolstered by orphan drug designations, streamlined regulatory approval processes, and financial incentives (Song et al. 2012).

There has been a rise in the popularity of regulatory incentives that try to meet the special needs associated with rare diseases. Pharmaceutical companies are the primary target of these incentives, which aim to provide measures to encourage research and development of therapies for rare diseases. Orphan medication classification is a major incentive since it provides additional years of market exclusivity, lower regulatory fees, and tailored direction at every stage of development. Furthermore, rare illness medicines can be reviewed and approved more quickly thanks to fast-track and accelerated approval routes. Access can be made quicker in some situations thanks to conditional approvals based on preliminary evidence (Boycott and Ardigó 2018). In addition, incentive coupons for expedited review of another medicine are of great value. Companies devoted to the study of rare diseases frequently receive financial incentives in the form of tax credits and grants. Collectively, these legal incentives aim to reduce the monetary and conceptual hurdles in the way of new treatments for rare diseases, encouraging creativity and, ultimately, improving the quality of life for individuals who suffer from them (Ollendorf et al. 2018).

9.4.6 Information Sharing and Global Cooperation

Through international partnerships, researchers and medical professionals from all around the world are able to pool their resources and share their knowledge and findings. The general understanding of rare genetic illnesses can be improved by efforts like the Global Alliance for Genomics and Health's standardisation of data sharing.

Information sharing and global collaboration are of utmost importance in the field of rare diseases. Rare diseases afflict a disproportionately small population and are notoriously difficult to treat, necessitating international and interdisciplinary cooperation. Diagnosis, therapy development, and patient care can all benefit from increased communication and sharing of relevant data and information. Gaining a global perspective on rare diseases is possible by connecting researchers, healthcare providers, patients, and advocacy groups. The challenges of scarce materials and dispersed knowledge can be met by simplifying the exchange of clinical observations, genetic information, research results, and best practices. Furthermore, researchers can pool resources, avoid duplicating efforts, and benefit from various viewpoints through international collaboration. Joint research projects, standardised data repositories, and international clinical trials are examples of collaborative activities that can propel progress that would be impossible in siloed settings. Individuals coping with rare diseases can benefit from increased global collaboration because of a greater collective potential to influence policymaking, regulatory frameworks, and financial distribution. The complex problems of rare diseases can be better tackled if we pool our resources and expertise from across countries and fields (Boycott and Ardigó 2018; Dawkins et al. 2018) (Fig. 9.2).

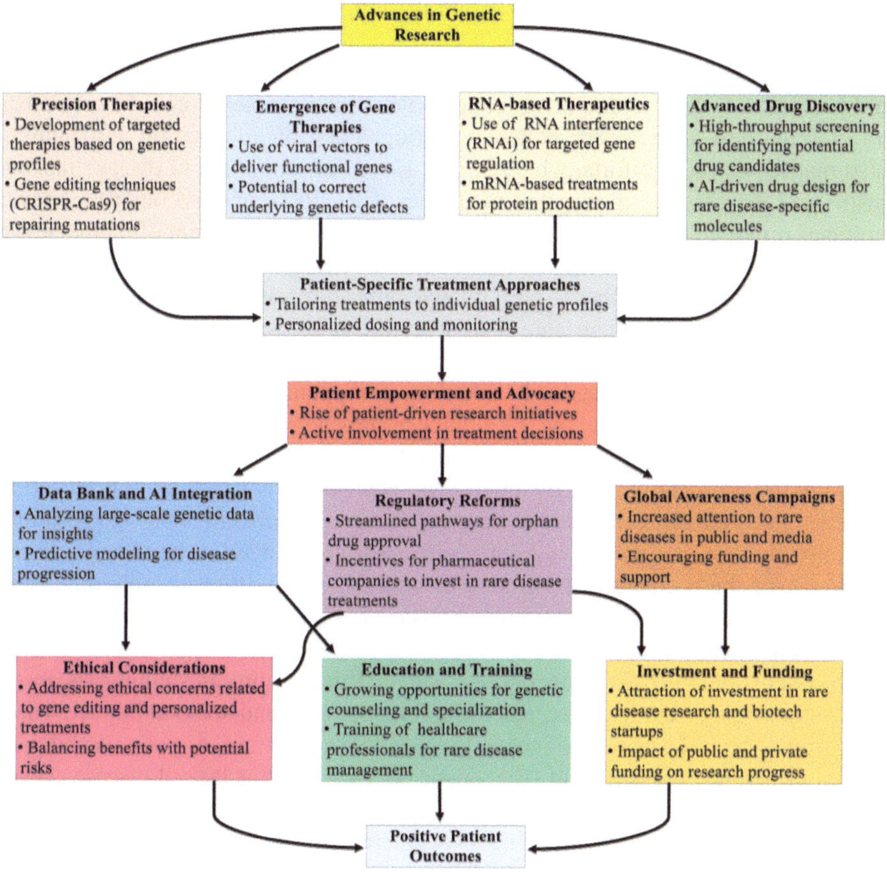

Fig. 9.2 Future opportunities in rare genetic disorders visualising the potential advancements and collaborative efforts in precision therapies, gene editing, drug discovery, and patient-centric approaches, focusing on cutting-edge research, ethical considerations, and patient empowerment. The chart emphasises the collective journey that could bring a brighter future

9.5 Discussion and Conclusion

The rare diseases are complex and diverse, lacking medical and scientific understanding. As new uncommon diseases and conditions are discovered and reported regularly in the medical literature, the landscape of rare diseases is always changing. The discipline is still in its infancy, with the exception of a handful of rare disorders for which significant progress has been made. Prior to very recently, there was no serious investigation into, or public health policy addressing, rare diseases among doctors, scientists, or policymakers. This presents significant obstacles to crafting an all-encompassing policy regarding rare diseases. However, measures should be taken, both immediately and in the long run, to combat uncommon diseases holistically and completely.

The low prevalence and complex nature of rare genetic illnesses present substantial problems. Delays in diagnosis, subpar care, and restricted access to specialists all contribute to unnecessary patient suffering. Lack of information and funding also hinders research efforts, which in turn slows the discovery of disease mechanisms and the creation of targeted medicines. CRISPR-Cas9 and other new technologies provide promise for precise gene editing and development of therapeutics. Researchers, clinicians, pharmaceutical companies, and patient advocacy groups have begun working together to pool knowledge and resources to improve diagnostic turnaround times, patient care, and scientific progress. Personalised medicine and detection of genetic variants will benefit from the growing availability of genomic sequencing databases. In addition, computational biology and artificial intelligence developments allow for a quicker and cheaper path to treatment development by repurposing current medications for rare conditions. More hope can be found in the prospect of gene treatments, in which healthy genes are introduced to patients. In conclusion, despite the fact that rare genetic disorders present formidable obstacles, we are on the cusp of transformative changes in the diagnosis and treatment of these conditions, offering renewed hope to affected individuals and their families as a result of this convergence of technological advancements and collaborative efforts.

In conclusion, the field of uncommon genetic illnesses presents numerous obstacles in the areas of clinical care, scientific investigation, and societal impact. There is an urgent need for better diagnostic tools and infrastructure, as patients and their families often go on a frustrating diagnostic odyssey due to a lack of disease awareness and restricted access to specialised healthcare services. Gene therapies and precision medicine are examples of novel therapeutic techniques necessitated by certain conditions' rarity and genetic heterogeneity, which impede conventional drug development pipelines. To overcome these challenges and expedite the development and approval of targeted medicines, collaboration between researchers, doctors, pharmaceutical companies, and regulatory authorities is essential. Individuals and healthcare systems have a hefty price tag for treating uncommon genetic disorders; advocates should push for more resources, regulatory changes, and insurance coverage to alleviate this problem. Innovations in genomics and data sharing, in particular, provide new opportunities to speed up research and deepen our understanding of the underlying biological pathways. The rise of patient advocacy groups and internet platforms improves patient care and shapes research objectives by fostering a sense of community and allowing the collection of real-world data. A new era of hope and an enhanced quality of life for those afflicted with these mysterious conditions can be ushered in with the concerted application of scientific innovation, collaborative synergy, and societal support as we approach a critical juncture in the study of rare genetic disorders.

9.6 Planning for the Care of Rare Disorders

Rare diseases are typically severe, chronic, disabling, and even fatal; their therapies can be lengthy and specialised. Furthermore, they frequently cause a handicap, occasionally severe. Rare diseases' emotional, financial, and practical toll on their victims is substantial. At least 80% of uncommon diseases may be traced back to a known genetic cause (Global Genes 2017), and as a result, children are disproportionately affected. Half of all new instances occur in children, and among those, 35% occur in those younger than 1 year old, 10% occur in those between 1 and 5 years old, and 12% occur in those more than 15 years of age (National Rare Disease Plan for Ireland 2014–2018). Untreated, rare diseases pose a significant threat to public health and are associated with substantial social and economic costs that extend beyond the affected individual, as family members sometimes have to renounce paid employment to care for sick loved ones. Patients who are receiving treatment, on the other hand, are less likely to necessitate additional expensive procedures like pain management or surgery (Hyry et al. 2013). In addition, without government assistance, it is nearly impossible for most families to pay for the treatment of uncommon diseases. Extremely high treatment costs have a devastating effect on families, draining them emotionally and financially. As a result, parents of children with rare diseases whose treatment costs were not being paid by insurance or otherwise not compensated have gone to the courts to request that the government give the drugs for free so that their children can continue receiving therapy. In India, the Ministry of Health and Family Welfare was ordered to create a 'national policy on the treatment of rare diseases' after the High Court of Delhi issued orders in W.P. (C) No. 4444/2016, W.P. (C) No. 7730/2016, and W.P. (C) No. 7729/2013.

References

Abozaid GM, Kerr K, McKnight A, Al-Omar HA (2022) Criteria to define rare diseases and orphan drugs: a systematic review protocol. BMJ Open 12(7):e062126. https://doi.org/10.1136/bmjopen-2022-062126

Aggarwal S, Phadke SR (2015) Medical genetics and genomic medicine in India: current status and opportunities ahead. Mol Genet Genom Med 3(3):160

Agrawal R, Amaresh Rao M, Brian M, Chowdary G, Gayatri K, Krishnaji Rao M (2019) Baseline knowledge of rare diseases in India—a survey. Int J Rare Dis Disord 2(1):4–11. https://doi.org/10.23937/2643-4571/1710008

Aymé S, Kole A, Groft S (2008) Empowerment of patients: lessons from the rare diseases community. Lancet 371(9629):2048–2051. https://doi.org/10.1016/S0140-6736(08)60875-2

Bartram CR, de Klein A, Hagemeijer A, van Agthoven T, van Kessel AG, Bootsma D et al (1983) Translocation of c-abl oncogene correlates with the presence of a Philadelphia chromosome in chronic myelocytic leukaemia. Nature 306(5940):277–280. https://doi.org/10.1038/306277a0

Benayed R, Gharani N, Rossman I, Mancuso V, Lazar G, Kamdar S et al (2005) Support for the homeobox transcription factor gene ENGRAILED 2 as an autism spectrum disorder susceptibility locus. Am J Hum Genet 77(5):851–868. https://doi.org/10.1086/497705

Bhattacharya S, Katoch VM, Majumder P, Bhattacharya A (2016) Rare diseases in India: current knowledge and new possibilities [Electronic version]. Proc Indian Natl Sci Acad 82(4). https://doi.org/10.16943/ptinsa/2016/48575

Boat TF, Field MJ (2011) Rare diseases and orphan products: accelerating research and development. National Academies Press

Bogart K, Hemmesch A, Barnes E, Blissenbach T, Beisang A, Engel P (2022) Healthcare access, satisfaction, and health-related quality of life among children and adults with rare diseases. Orphanet J Rare Dis 17(1):196. https://doi.org/10.1186/s13023-022-02343-4

Bojakowski S, Spoors J (2013) The funding of orphan medicines in the UK. Br J Healthc Manag 19(7). http://citeseerx.ist.psu.edu/viewdoc/download?doi=10.1.1.694.9509&rep=rep1&type=pdf

Boulanger V, Schlemmer M, Rossov S, Seebald A, Gavin P (2020) Establishing patient registries for rare diseases: rationale and challenges. Pharm Med 34(3):185–190. https://doi.org/10.1007/s40290-020-00332-1

Boycott KM, Ardigó D (2018) Addressing challenges in the diagnosis and treatment of rare genetic diseases. Nat Rev Drug Discov 17(3):151–152. https://doi.org/10.1038/nrd.2017.246

Brasil S, Pascoal C, Francisco R, dos Reis Ferreira V, Videira A, Valadão G (2019) Artificial intelligence (AI) in rare diseases: is the future brighter? Genes 10(12):978. https://doi.org/10.3390/genes10120978

Bryson B, Bogart K, Atwood M, Fraser K, Locke T, Pugh K, Zerrouk M (2021) Navigating the unknown: a content analysis of the unique challenges faced by adults with rare diseases. J Health Psychol 26(5):623–635. https://doi.org/10.1177/1359105319828150

Campbell DB, Li C, Sutcliffe JS, Persico AM, Levitt P (2008) Genetic evidence implicating multiple genes in the MET receptor tyrosine kinase pathway in autism spectrum disorder. Autism Res 1(3):159–168. https://doi.org/10.1002/aur.27

Cecchin E, Stocco G (2020) Pharmacogenomics and personalized medicine. Genes 11(6):679. https://doi.org/10.3390/genes11060679

Choudhury MC, Chaube P (2022) Integrating rare disease management in public health programs in India: exploring the potential of National Health Mission. Orphanet J Rare Dis 17(1):43. https://doi.org/10.1186/s13023-022-02194-z

Cleary M, Chakrapani A (2004) Inherited Metabolic Diseases among South Asians in the United Kingdom. In: Genetic disorders of the Indian subcontinent. Springer, Netherlands, pp 181–205. https://doi.org/10.1007/978-1-4020-2231-9_10

Côté A, Keating B (2012) What is wrong with orphan drugs policies. Value Health 15(8):1185–1191. https://doi.org/10.1016/j.jval.2012.09.004

Crowe AL, McKnight AJ, McAneney H (2019) Communication needs for individuals with rare diseases within and around the healthcare system of Northern Ireland. Front Public Health 7:236. https://doi.org/10.3389/fpubh.2019.00236

Cutillo CM, Austin CP, Groft SC (2017) A global approach to rare diseases research and orphan products development: the international rare diseases research consortium (IRDiRC). In: Rare diseases epidemiology: update and overview. Springer, Cham, pp 349–369

Dawkins HJ, Draghia-Akli R, Lasko P, Lau LP, Jonker AH, Cutillo CM, Rath A, Boycott KM, Baynam G, Lochmüller H, Kaufmann P, Cam YM, Hivert V, Austin CP, International Rare Diseases Research Consortium (2018) Progress in rare diseases research 2010–2016: an IRDiRC perspective. Clin Transl Sci 11(1):11. https://doi.org/10.1111/cts.12501

Directorate-General for Research and Innovation (European Commission) (2017) Rare diseases: a major unmet medical need. Publications Office of the European Union, Brussels

Dolled-Filhart MP, Lordemann A, Dahl W, Haraksingh RR, Ou-Yang CW, Lin JC (2012) Personalizing rare disease research: how genomics is revolutionizing the diagnosis and treatment of rare disease. Pers Med 9(8):805–819

Ghiasvand H, Barnish MS, Moradi T, Nikram E, Naghdi S (2022)·Making orphan drugs and services available and accessible for people who live with rare diseases: what has been done? a systematic scoping review. Expert Opin Orphan Drugs 10(1):55–64. https://doi.org/10.1080/21678707.2022.2153671

Global Genes (2017) Statistics and figures on prevalence of genetic and rare disease. https://globalgenes.org/rare-diseases-facts-statistics/. Accessed 9 Apr 2017

Groft SC, Posada M, Taruscio D (2021) Progress, challenges and global approaches to rare diseases. Acta Paediatr 110(10):2711–2716. https://doi.org/10.1111/apa.15974

Guan YF, Li GR, Wang RJ, Yi YT, Yang L, Jiang D, Zhang X, Peng Y (2012) Application of next-generation sequencing in clinical oncology to advance personalized treatment of cancer. Chin J Cancer 31(10):463. https://doi.org/10.5732/cjc.012.10216

Halley MC, Smith HS, Ashley EA, Goldenberg AJ, Tabor HK (2022) A call for an integrated approach to improve efficiency, equity and sustainability in rare disease research in the United States. Nat Genet 54:219–222. https://doi.org/10.1038/s41588-022-01027-w

Han Q, Fu H, Chu X, Wen R, Zhang M, You T et al (2022) Research advances in treatment methods and drug development for rare diseases. Front Pharmacol 13:971541. https://doi.org/10.3389/fphar.2022.971541

Handyside AH, Lesko JG, Tarín JJ, Winston RM, Hughes MR (1992) Birth of a normal girl after in vitro fertilization and preimplantation diagnostic testing for cystic fibrosis. N Engl J Med 327(13):905–909. https://doi.org/10.1056/NEJM199209243271301

Health Inc. (2017) Sen. Grassley launches inquiry into orphan drug law's effect on prices. http://www.npr.org/sections/health-shots/2017/02/10/514373480/sen-grassley-launches-inquiry-into-orphan-drug-laws-effect-on-prices

Holt IJ, Harding AE, Petty RK, Morgan-Hughes JA (1990) A new mitochondrial disease associated with mitochondrial DNA heteroplasmy. Am J Hum Genet 46(3):428

Horikawa Y, Oda N, Cox NJ, Li X, Orho-Melander M, Hara M et al (2000) Genetic variation in the gene encoding calpain-10 is associated with type 2 diabetes mellitus. Nat Genet 26(2):163–175. https://doi.org/10.1038/79876

Hoyer LW (1994) Hemophilia a. N Engl J Med 330(1):38–47. https://doi.org/10.1056/NEJM199401063300108

Hughes DA, Poletti-Hughes J (2016) Profitability and market value of orphan drug companies: a retrospective, propensity-matched case-control study. PLoS ONE 11(10):e0164681. https://doi.org/10.1371/journal.pone.0164681

Hyry HI, Roos JCP, Manuel J, Cox TM (2013) The legal imperative for treating rare disorders. Orphanet J Rare Dis 8:135. https://doi.org/10.1186/1750-1172-8-135

Imperial R, Nazer M, Ahmed Z, Kam AE, Pluard TJ, Bahaj W et al (2019) Matched whole-genome sequencing (WGS) and whole-exome sequencing (WES) of tumor tissue with circulating tumor DNA (ctDNA) analysis: complementary modalities in clinical practice. Cancers 11(9):1399. https://doi.org/10.3390/cancers11091399

Jackman D (2012) People with rare diseases face a host of unique challenges. BIOtechNOW. http://www.biotech-now.org/health/2012/05/people-with-rare-diseases-face-a-host-of-unique-challenges. Accessed Apr 9 2017

Kar A, Phadnis S, Dharmarajan S, Nakade J (2014) Epidemiology and social costs of haemophilia in India. Indian J Med Res 140(1):19

Kaufmann P, Pariser AR, Austin C (2018) From scientific discovery to treatments for rare diseases—the view from the National Center for Advancing Translational Sciences–Office of Rare Diseases Research. Orphanet J Rare Dis 13:1–8. https://doi.org/10.1186/s13023-018-0936-x

Kaul S, Munshi A (2012) Genetics of ischemic stroke: Indian perspective. Neurol India 60(5):498

Kesselheim AS, Myers JA, Solomon DH, Winkelmayer WC, Levin R, Avorn J (2012) The Prevalence and Cost of Unapproved Uses of Top-Selling Orphan Drugs. PLoS One 7(2):e31894. https://doi.org/10.1371/journal.pone.0031894

Klein HG, Bauer P, Hambuch T (2014) Whole genome sequencing (WGS), whole exome sequencing (WES) and clinical exome sequencing (CES) in patient care. J Lab Med 38(4):221–230. https://doi.org/10.1515/labmed-2014-0025

Koay PP, Sharp RR (2013) The role of patient advocacy organizations in shaping genomic science. Annu Rev Genomics Hum Genet 14:579–595. https://doi.org/10.1146/annurev-genom-091212-153525

Lauschke VM, Ingelman-Sundberg M (2016) Precision medicine and rare genetic variants. Trends Pharmacol Sci 37(2):85–86. https://doi.org/10.1016/j.tips.2015.10.006

Lauschke VM, Ingelman-Sundberg M (2018) How to consider rare genetic variants in personalized drug therapy. Clin Pharmacol Ther 103(5):745–748. https://doi.org/10.1002/cpt.976

Lee J, Liu C, Kim J, Chen Z, Sun Y, Rogers JR, Chung WK, Weng C (2022) Deep learning for rare disease: a scoping review. J Biomed Inform 135:104227. https://doi.org/10.1016/j.jbi.2022.104227

Martin AR, Williams E, Foulger RE, Leigh S, Daugherty LC, Niblock O et al (2019) PanelApp crowdsources expert knowledge to establish consensus diagnostic gene panels. Nat Genet 51(11):1560–1565. https://doi.org/10.1038/s41588-019-0528-2

Marwaha S, Knowles JW, Ashley EA (2022) A guide for the diagnosis of rare and undiagnosed disease: beyond the exome. Genome Med 14(1):1–22. https://doi.org/10.1186/s13073-022-01026-w

Might M, Crouse AB (2022) Why rare disease needs precision medicine—and precision medicine needs rare disease. Cell Rep Med 3(2):100530. https://doi.org/10.1016/j.xcrm.2022.100530

Min S, Yurchisin J (2015) Problem based learning: jeanswear design for rare disease day. Iowa State University, Digital Repository. https://doi.org/10.31274/itaa_proceedings-180814-84

Mohanty R, Barick U, Gowda A, Nair A, Mittal S, Patil A (2016) Scope for patient registries in India. Int J Med Res Health Sci 5(4):58–61

Morin SJ, Eccles J, Iturriaga A, Zimmerman RS (2017) Translocations, inversions and other chromosome rearrangements. Fertil Steril 107(1):19–26. https://doi.org/10.1016/j.fertnstert.2016.10.013

National Rare Disease Plan for Ireland (2014–2018). http://health.gov.ie/wp-content/uploads/2014/07/EditedFile.pdf

Niebuhr E (1978) The Cri du Chat syndrome: epidemiology, cytogenetics, and clinical features. Hum Genet 44:227–275. https://doi.org/10.1007/BF00394291

Ogata T, Matsuo N (1995) Turner syndrome and female sex chromosome aberrations: deduction of the principal factors involved in the development of clinical features. Hum Genet 95:607–629. https://doi.org/10.1007/BF00209476

Ollendorf DA, Chapman RH, Pearson SD (2018) Evaluating and valuing drugs for rare conditions: no easy answers. Value Health 21(5):547–552. https://doi.org/10.1016/j.jval.2018.01.008

Pushpakom S, Iorio F, Eyers PA, Escott KJ, Hopper S, Wells A, Doig A, Guilliams T, Latimer T, McNamee C, Norris A, Sanseau P, Cavalla D, Pirmohamed M (2019) Drug repurposing: progress, challenges and recommendations. Nat Rev Drug Discov 18(1):41–58. https://doi.org/10.1038/nrd.2018.168

Report of I.C. Verma Sub-Committee (2017) 'Guidelines for Therapy and Management'

Roessler HI, Knoers NV, van Haelst MM, van Haaften G (2021) Drug repurposing for rare diseases. Trends Pharmacol Sci 42(4):255–267. https://doi.org/10.1016/j.tips.2021.01.003

Schieppati A, Henter JI, Daina E, Aperia A (2008) Why rare diseases are an important medical and social issue. Lancet 371(9629):2039–2041. https://doi.org/10.1016/S0140-6736(08)60872-7

Shen F, Zhao Y, Wang L, Mojarad MR, Wang Y, Liu S, Liu H (2019) Rare disease knowledge enrichment through a data-driven approach. BMC Med Inform Decis Mak 19:1–11. https://doi.org/10.1186/s12911-019-0752-9

Simoens S (2011) Pricing and reimbursement of orphan drugs: the need for more transparency. Orphanet J Rare Dis 6:42. https://doi.org/10.1186/1750-1172-6-42

Sivasubbu S, Scaria V (2019) Genomics of rare genetic diseases—experiences from India. Hum Genomics 13(1):1–8

Song P, Gao J, Inagaki Y, Kokudo N, Tang W (2012) Rare diseases, orphan drugs, and their regulation in Asia: current status and future perspectives. Intract Rare Dis Res 1(1):3–9. https://doi.org/10.5582/irdr.2012.v1.1.3

Tambuyzer E, Vandendriessche B, Austin CP, Brooks PJ, Larsson K, Miller Needleman KI et al (2020) Therapies for rare diseases: therapeutic modalities, progress and challenges ahead. Nat Rev Drug Discov 19(2):93–111. https://doi.org/10.1038/s41573-019-0049-9

van der Lee M, Kriek M, Guchelaar HJ, Swen JJ (2020) Technologies for pharmacogenomics: a review. Genes 11(12):1456. https://doi.org/10.3390/genes11121456

Wang HX, Li M, Lee CM, Chakraborty S, Kim HW, Bao G, Leong KW (2017) CRISPR/Cas9-based genome editing for disease modeling and therapy: challenges and opportunities for non-viral delivery. Chem Rev 117(15):9874–9906. https://doi.org/10.1021/acs.chemrev.6b00799

Weatherall DJ, Clegg JB (2001) Inherited haemoglobin disorders: an increasing global health problem. Bull World Health Organ 79:704–712

Weiner MF (2008) Perspective on race and ethnicity in Alzheimer's disease research. Alzheimers Dement 4(4):233–238. https://doi.org/10.1016/j.jalz.2007.10.016

Wexler NS, Young AB, Tanzi RE, Travers H, Starosta-Rubinstein S, Penney JB et al (1987) Homozygotes for Huntington's disease. Nature 326(6109):194–197. https://doi.org/10.1038/326194a0

Whicher D, Philbin S, Aronson N (2018) An overview of the impact of rare disease characteristics on research methodology. Orphanet J Rare Dis 13(1):14

Zhang B (2021) CRISPR/Cas gene therapy. J Cell Physiol 236(4):2459–2481. https://doi.org/10.1002/jcp.30064

Zhao W, Hou X, Vick OG, Dong Y (2019) RNA delivery biomaterials for the treatment of genetic and rare diseases. Biomaterials 217:119291. https://doi.org/10.1016/j.biomaterials.2019.119291

MIX
Papier aus verantwortungsvollen Quellen
Paper from responsible sources
FSC® C105338

If you have any concerns about our products,
you can contact us on
ProductSafety@springernature.com

In case Publisher is established outside the EU,
the EU authorized representative is:
Springer Nature Customer Service Center GmbH
Europaplatz 3, 69115 Heidelberg, Germany

Printed by Libri Plureos GmbH
in Hamburg, Germany